Recent Developments in Analytical Techniques for Corrosion Research

Ihsan ulhaq Toor
Editor

Recent Developments in Analytical Techniques for Corrosion Research

Springer

Editor
Ihsan ulhaq Toor
Mechanical Engineering Department
King Fahd University of Petroleum & Minerals
Dhahran, Saudi Arabia

ISBN 978-3-030-89100-8 ISBN 978-3-030-89101-5 (eBook)
https://doi.org/10.1007/978-3-030-89101-5

This Springer imprint is published by the registered company Springer Nature Switzerland AG
The registered company address is: Gewerbestrasse 11, 6330 Cham, Switzerland

Contents

Advancement in Electrochemical Techniques

Introduction

Ihsanulhaq Toor

Abstract Asset integrity is a key for smooth industrial operations, however material degradation, i.e. corrosion is major threat to infrastructure reliability and performance. In order to fight against corrosion, a deeper understanding of the underlying mechanisms and material performance is required which can be best achieved by employing advance analytical and modeling techniques in corrosion science and engineering. An in-depth understanding and utilization of such analytical tools, from electrochemical to in-situ/ex-situ surface characterization techniques as well as modeling of corrosion systems can help in problem identification as well as damage prediction. These areas are of considerable importance for the researchers in the industry as well academia, not only to develop new corrosion resistant materials but also to devise a suitable asset integrity strategy. Therefore, the primary aim of this book is to provide an in-depth information regarding the emerging developments in electrochemical, analytical and modeling tools available for corrosion research, so that it can enable its readers to use these techniques to overcome material degradation challenges. Here, in this chapter, a preview of 12 worthy contributions of this book will be presented.

Keywords Localized corrosion · Passivity · Atomic force microscopy · Transmission electron microscopy · Electrochemical impedance · Cyclic voltammetry · Hydrogen embrittlement · Erosion corrosion · Inhibitor evaluation · Scanning electrochemical microscopy (SCEM) · Stress corrosion cracking

I. Toor (✉)
Mechanical Engineering Department, King Fahd University of Petroleum & Minerals (KFUPM), P. O. Box. 1308, Dhahran 31261, Saudi Arabia
e-mail: ihsan@kfupm.edu.sa

Interdisciplinary Research Center (IRC) for Advanced Materials, King Fahd University of Petroleum & Minerals (KFUPM), P. O. Box. 1308, Dhahran 31261, Saudi Arabia

I. Toor (ed.), *Recent Developments in Analytical Techniques for Corrosion Research*,
https://doi.org/10.1007/978-3-030-89101-5_1

1 Introduction

Infrastructure reliability considering its economic and safety impact is a major concern for most of the industrial applications. Material degradation, under various environmental conditions is a challenge and forcing huge investments by the companies to keep the operations running without any major setbacks. The researchers are working to develop new materials, innovative analytical tools to monitor material degradation and modeling tools for the corrosion systems for better corrosion monitoring and prediction. There are many challenges to address in terms of materials and environment interactions. The type of the materials (metal alloys) and how their microstructure/composition interacts with different environments, requires a considerable back and forth research and optimization. This requires utilizing modern lab facilities for alloy development, metallurgical characterizations and advance analytical tools to study their degradation performance both in-situ/ex-situ in variety of environments. This development and optimization also requires modern simulation and modeling tools for speedy optimizations. The experimental design/design of experiments is a challenging task and has to be carefully planned for the desired outcomes. With the advancements in industrial growth, materials are being exposed and required for extreme environments both at land, sea and in space etc. Material degradation can be of different types depending on the environmental considerations such aqueous vs non-aqueous, ambient vs high temperature and acidic vs alkaline type of corrosion. The infrastructure under atmospheric conditions (open air, sea) and in industries such as oil/gas, aerospace, nuclear, petrochemical, etc., has its own definite challeneges. The environmental paparmeters such as temperature pressure, pH, fluid flow etc., along with the material chemistry can have a direct effect on the material degradation. So the tools used in this complex cycle of material development, analysis/interpretation, simulation/modeling and ultimately corrosion prediction, have a huge significance and must be well understood. As scientist/researchers and engineers, we need to underatnd this complex cycle of material/environment interaction to fight against these degradation challenges at all stages from material development till successful application for desired life time of the infrastructure. This material environmental interaction suggests that research in corrosion science and engineering is quite complicate and requires interdescilpinary approach to come up with solutions of the engineering problems. The material/environment interactions vary over the period of time, and so again it is challenging to determine these variations both in-situ and ex-situ. It is not easy to simulate complex environments experimentally, so simulation tools need to be employed to examine material degaration modes at high tmepeature, pressure, pH etc. Considering these challenges, the contribtions for this book are carefully selected in the areas of electrochemistry, in-situ/ex-situ surface analytical tools along with modeling techniques to model material environment iteractions.

1.1 *Themes and Methodology for Selected Contributions*

This book was planned to be organized in three sections and the first theme was about the "advancement in electrochemical techniques and the target topics fall in this category were, "in-situ Photo-electrochemical/Mott-Schottky anlysis Techniques, Voltammetry Techniques, Scanning electrochemical microscopy (SCEM), Electrochemical Impedance Spectroscopy (EIS) and related techniques. The second theme was about the "advancements in spectroscopic and microscopic techniques" and the targetted topics were; "X-Ray Spectroscopy (XPS), Electrochemical Sanning Tunneling Mircoscopy (EC-STM), Atomic force microscopy (AFM), Liquid-Cell/Scanning Transmission Electron Microscopy (LC-STEM) and related ideas. Finally the third theme was about the "advancements in corrosion simulation & modeling" and the focussed topics were "hydrogen induced corrosion/hydrogen embrittlement modeling, stress corrosion cracking modeling, erosion-corrosion models and modeling for inhibitor design along with related ideas. Based on these themes, we received eighteen proposals and based on initial review, twelve proposals were selected for detailed manuscript preparation to be included in the first edition of this book. This was followed by book chapter draft preparation and drafts were peer reviewed and enriched further based on reviewer feedback.

2 Advancement in Electrochemical Techniques for Coroisson Research

2.1 *In-Situ Photoelectrochemcao Spectrocopy and Mott-Schottky Analysis*

In this chpater, IH Toor discussed in detail the applciations of PES and Mott-Schottkya techniques in evaluating the performance of metal alloys. It was discussed in detail how the alloying additions can change the defect concentration and chemistry of the oxide films on metals and alloys. As the metal/metal oxide/solution interface is a very complex interface, so instead of using ex-situ characterizations tools to examine the oxide films, it is quite demanding to use in-situ characterization techniques such as photoelectrochemical spectroscopy (PES) and Mott-Schottky analysis. The basis to use these techniques is the fact that oxide films on metals and alloys behave like a semiconductor. So therefore, if we can measure the defect concentration and semiconducting properties of the films, then those can be correlated with the physicochemical structure of the oxides and that will ultimately help us to understand the corrosion performance of that particular material.

2.2 *Voltammetry Techniques in Corrosion Research*

K. S. Raja, author of Chap. "Voltammetry and In-Situ UV-Vis Absorbance Studies for Understanding Corrosion Mechanisms" (in press) discussed in detail the applications of voltammetric technques in corroison research. The author mainly focussed on two different techniques; (a) voltammetry, and (b) in-situ optical absorbance spectrometry for studying corrosion phenomena of metals and non-metals. In this chapter, the fundamentals of different voltammetric techniques such as linear potential sweep voltammetry, cyclic voltammetry, and pulse voltammetry are discussed. The chpater presents the voltammetry results with reference to various reaction paths such as electrochemical, chemical, and coupled reactions and different processes such reversible, irreversible, and quasi reversible etc. along with the the basics of UV-Vis optical absorbance spectroscopy. The last part of the chapter focuses on in-situ spectroscopy techniques and interpretation of results with reference to corosion studies on FTO in different pH conditions.

2.3 *The Utilization of Scanning Electrochemical Microscopic (SECM) Techniques in Corrosion Monitoring*

AM Kumar, author of Chap. "The Utilization of Scanning Electrochemical Microscopic (SECM) Technique in Corrosion Monitoring" (in press) discussed here several electrochemical methods such as; scanning reference electrode (SRET), scanning vibrating electrode (SVET), localized electrochemical impedance spectroscopy (LEIS), and scanning electrochemical microscopy (SECM). Amongst the methods, SECM has been presented to assess the corrosion phenomenon during early stages at the micrometer scale, as it delivers the electrochemical behavior and surface topographic evidence about the electrochemical reactions at the interface in the solution. Owing to their higher spatial resolution and electrochemical sensitivity, the application of SECM has helped studying metal/electrolyte interface. This method has previously exhibited exact exciting evidence regarding the beginning of localized corrosion of numerous metallic materials, the formation of anodized layers on metallic samples, and the illustration of thin inhibitor films on various metallic materials and galvanic corrosion of Zn-Fe cell. Moreover, the usage of SECM to investigate the degradation of protective coatings on various metallic substrates is very promising in investigating the multifaceted reactions happening in the aggressive environment.

2.4 *Electrochemcial Impedence Spectroscopy for Monitoring the Performance of Corrosion Inhibitors*

S. Zehra, R. Aslam and M. Mobin, authors of Chap. "Electrochemical Impedance Spectroscopy: A Useful Tool for Monitoring the Performance of Corrosion Inhibitors" (in press) discussed the basics of EIS and its applications in evaluating the performance of corrosion inhibitors. EIS technique employs a small amplitude AC signal over a very wide range of frequencies and is considered to be a very practical approach to evaluate the persistency of the films formed by corrosion inhibitors on the metal substrate. The authors discussed the use of this technique to study the underlying mechanisms of corrosion protection by different types of inhibitors.

3 Advancements in Spectroscopic and Microscopic Techniques for Corrosion Research

3.1 *Applications of Liquid Cell-TEM in Corrosion Research*

Khalid Hattar and Raymond R. Unocic, the authors of Chap. "Applications of Liquid Cell-TEM in Corrosion Research" (in press) disucussed here in detail liquid cell-TEM and its applications in corroison research. The authors provided a brief history of this technique and its development over the past couple of decades. Ove rthe past few years in-situ LC-TEM is being employed to explore corrosion initiation in differnet environments such as deionized water, brine, or acetic acid etc. There are recent studies in which experiments were performed on complex environments like site-specific regions of pipeline steel in a range of tailored corrosive environments. The authors at the end also discussed in detail the possible applications of this technique to study more complex phenomenon such as stress corrosion cracking and biofouling etc.

3.2 *Electrochemical Scanning Tunneling Microscopy to Study Corrosion*

M. B. Haider, S. Kunwar, and K. Alam, authors of Chap. "Applications of Electrochemical Scanning Tunneling Microscopy in Corrosion Research" (in press) discussed in detail the applications of EC-STM in corrosion research. They have presenetd the basis of this technqiue, as it is used to image atomic resolution on conducting surfaces, find the density of states of the sample surfaces and atomic manipulation of species on the sample surfaces. They focus on EC-STM, which is a specialized scanning tunneling microscopy, where the tip and the sample, both are in in an electrolyte environment. It is highlighted that EC-STM not only has the tunneling current but also Faraday's current and capacitative current. Special care is given to keep the Faradic current and capacitative current at the lowest, so mostly tunneling current is achieved to image the surface. This is achieved by coating the scanning wire with insulating material and keep the tip of the wire exposed so the tunneling current is through the tip of the wire only. Electrochemical scanning tunneling microscopy is based on the continuous oxidation and reduction of the electrolyte between the tip and the sample which creates the tunneling current. The authors have discussed different applications of EC-STM to study the surfaces like nickel, copper and gold in different chemical environments to understand the importance of this technique to study corrosion of metal alloys.

3.3 *Recent Trends in Applications of X-ray Photoelectron Spectroscopy (XPS)*

The authors of Chap. 8 "Recent Trends in Applications of X-ray Photoelectron Spectroscopy (XPS) Technique in Coatings for Corrosion Protection" (in press) discussed in detail the applciations of XPS in corroison studies, especially its applciation to study the performance of the coatings. The technique can be used to study the compositional analysis of various types of coatings (metallic and polymeric) that are actively investigated for corrosion protection. The technique can also be employed to explore smart polymeric coatings confirming the self-release of various inhibitors and self-healing agents loaded in nanocarriers. The authors have also discussed the role of XPS in post corrosion analysis as it provides ample information to understand the corrosion protection mechanism.

3.4 *Applications of Atomic Force Microscopy in Corrosion Research*

S. Akhtar, author of Chap. "Applications of Atomic Force Microscopy in Corrosion Research" (in press) discussed the use of atomic force microscopy (AFM) to analyze the surface features and topography of materials at a very high lateral resolution. It is discussed that the resolution of the AFM (fraction of nanometer) is better than the optical limit of diffraction and AFM machine produces two dimensional and three dimensional images of the samples. These images are then used to study topographical structures, such as surface roughness, surface pits, cracks, and other features of the corroded specimen. The authors then discussed diferent AFM techniques such as scanning Kelvin probe force microscopy (SK-PFM), which is used to produce Volta potential distribution maps, in-situ AFM, which is used to study the electrochemical processes and in-situ AFM scratching technique, which was developed to simulate the de-passivation of protective coatings.

4 Advancements in Corrosion Modeling & Prediction

4.1 A Review of Erosion-Corrosion Models for the Oil and Gas Industry Applications

A. Nassef, M. Keller, S. Hassani, S. Shirazi and K. Roberts, authors of Chap. "A Review of Erosion-Corrosion Models for the Oil and Gas Industry Applications" (in press) discussed in detail the current understanding of the mechanisms and modeling related to erosion, CO_2 corrosion, and erosion-corrosion. In oil and gas production, erosion-corrosion is caused predominately by the impact of entrained solid particles e.g. sand; $CaCO_3$ particles etc., on carbon steel piping and equipment materials. CO_2 and H_2S are two common types of corrosion encountered in oil and gas production. However, CO_2 is the most commonly associated with erosion-corrosion. The combined effect of erosion and corrosion can increase metal degradation rate while negatively affecting corrosion mitigation films, such as removing a protective scales e.g. iron carbonate ($FeCO_3$); and iron sulfide (FeS), or stripping away an inhibitor film through solid particle erosion. These combined effects can lead to higher corrosion rates, pitting and material failure. So there is a serious need to predict erosion-corrosion and its effects on corrosion enhancement. The use of chemical inhibitors is particularly important when the wells are deep, off-shore or difficult to reach regions, because coupon testing is impractical and replacement costs are high. The authors have reviewd in detail the procedure for predicting erosion-corrosion penetration rate known as SPPS: E-C by integrating the three E/CRC models including SPPS, CO_2, SPPS and Threshold velocity model.

4.2 Effect of Hydrogen and Defects on Deformation and Failure of Austenitic Stainless Steel

E. Ogosi, A. Siddiq, U. B. Asim and M. E. Kartal, the authors of Chap. "Effect of Hydrogen and Defects on Deformation and Failure of Austenitic Stainless Steel " (in press) discussed the effects of hydrogen in stainless steels, with a focus to understand the effect of crystal orientation and subsequently hydrogen influence on plastic deformation and void growth. They have performed simulations for a variety of stress triaxilaities, load parameters, hydrogen concentrations and observed that the initial crystal orientation has a varied effect regarding the influence of hydrogen on plastic deformation and void growth. They have highlghted that hydrogen in trap distribution at various stages of the deformation process, was influenced by intial crystal orientation. Hydrogen affects the evolution of crystal rotation during deformation but was not found to significantly affect the general pattern of crystal orientation evolution.

4.3 Molecular Modeling for Corrosion Inhibitor Design

I. B. Obot and A. Q. Alsayoud, authors of Chap. "Molecular Modeling for Corrosion Inhibitor Design" (in press) discussed in detail the applications of molecular modeling in corrosion inhibitor design. With the advancement in computer hardware and software development, the interactions between metal surface and corrosion inhibitor can be computed and quantify easily using varity of available tools. These tools will be helpful in the design and development of new and improved corrosion inhibitors. Authors discussed molecular dynamics (MD) simulation, a powerful tool for the design and mechanistic study of corrosion inhibitors. So in this chpater, the introduction and theory of MD simulation and their applciations in the design of corrosion inhibitor for the most important industrial metals such as steel, copper and aluminum is discussed in detail.

4.4 Hydrogen Embrittlement in Nickel-Base Superalloy

H. Khalid and B. Mansoor, authors of Chap. "Hydrogen Embrittlement in Nickel-Base Superalloy 718" (in press) discussed the hydrogen embrittlement (HE) problem in Ni base alloy 718, especially in hydrogenating environments. Thye have reviewed and presented differnet theories, which have been put forward to explain the precise mechanism by which hydrogen deteriorates mechanical behavior of structural materials, such as Hydrogen enhanced localized plasticity (HELP) and hydrogen enhanced decohesion (HEDE). These two are popular theories as there is considerable evidence that suggests that the presence of hydrogen in the metallic lattice enhances mobility of dislocations, leading to regions of enhanced plasticity, which in turn leads to regions of highly localized strains resulting in early crack initiation. On the other hand, HEDE mechanism, which relies on the ability of hydrogen to reduce bond strength, especially at weaker interfaces, resulting in decohesion and early fracture. Authors discussed that the various suggested mechanisms may be active simultaneously, with one or more of them dominating based on the weak sites in the material and the specific environmental conditions. So, in this chapter, the authors introduced the microstructure of alloy 718 and then studied the interplay between microstructural features and mechanical properties of alloy 718 in the presence of hydrogen.

Applications of Photoelectrochemical Spectroscopy (PES) and Mott-Schottky Techniques in Corrosion Research

Ihsanulhaq Toor

Abstract It is the passive film (thin oxide), which significantly improves the corrosion resistance of passive metals and alloys such as Ni, Al, and stainless steels being used in many industrial applications. The science behind the development of corrosion resistance alloy (CRAs) depends a lot on the careful selection of alloying elements, which can improve the qualities of this thin protective oxide. These additions can alter the film structure and composition to either improve or degrade the overall performance of the newly developed alloys. The metal/metal oxide/solution interface is a very complex interface, as many electrochemical reactions are occurring at this interface. Therefore, it would be great to use some in-situ characterization techniques to examine the metal/metal oxide/solution interface for ongoing changes that are taking place at this interface. Therefore, both the photoelectrochemical spectroscopy (PES) and Mott-Schottky analysis techniques can be utilized for in-situ investigation of these oxide films. In this chapter, the applications of PES and MS analysis will be discussed along with brief introduction of the both techniques.

Keywords Corrosion · Photoelectrochemistry · Mott-Schottky · Passive film · Semiconductor

1 Introduction

Many of the passive metals and their alloys such as stainless steels, Al alloys and Ni based corrosion resistant alloys (CRAs) are being used in many industrial applications due to their excellent corrosion resistance and mechanical properties. During the alloy design, development and optimization process, different alloying additions such as Si, Cu, Mn, Mo, W, N etc. are made to achieve the required corrosion resistance and

I. Toor (✉)
Mechanical Engineering Department, King Fahd University of Petroleum & Minerals (KFUPM), P. O. Box. 1308, Dhahran 31261, Saudi Arabia
e-mail: ihsan@kfupm.edu.sa

Interdisciplinary Research Center (IRC) for Advanced Materials, King Fahd University of Petroleum & Minerals (KFUPM), P. O. Box. 1308, Dhahran 31261, Saudi Arabia

I. Toor (ed.), *Recent Developments in Analytical Techniques for Corrosion Research*,
https://doi.org/10.1007/978-3-030-89101-5_2

mechanical properties. It is important to study the impact of these alloying additions on the passive film structure and composition of these newly developed alloys in order to understand their corrosion resistance performance. The addition of alloying elements can modify (increase or decrease) the defect chemistry of the passive film and thus can change the properties of the film [1]. There can be different types of defects such as oxygen vacancies, metal interstitials and metal vacancies up to very high concentrations such as 10^{19}–10^{21} cm^{-3} and these defects one way or the other control the passive film properties. These defects or impurities can be divided in to donors and acceptors. The role of donors is to give up extra electrons to the conduction band of a semiconductor material and by doing so they create n-type behavior. On the other hand, the acceptors accept the valence electrons and create p-type semiconductivity [2].

So studying the defect concentration of the oxide films can help us understand the performance of these newly developed materials, as how the addition of alloying elements might have affected the behavior of these alloys. For instance, if the addition of some element/s decrease the defect concentration, then it is expected that the material will have a better film in terms of its corrosion resistance properties and vice versa. The compactness and porosity of the films can be disturbed due to defect chemistry of the film [3–5]. Over the years, passive films on FeCr alloys were investigated using different ex-situ characterization tools and it is believed that they are Cr enriched with oxides/hydroxides of iron and chromium [6–11]. It is quite obvious that analysis of passive films is very challenging not only because of its thinness (few nanometer (nm) thickness) but also due to its complex nature (metal/oxide film/environment). The traditional ex-situ characterizations tools may not provide the true nature of the film, as there can be some compositional changes when the oxide film is removed out for such examination [12]. Therefore, it is important to use some in-situ tools to characterize these oxide films to have the most accurate information on passive film structure and composition. Over the past few years, two such techniques, i.e. photoelectrochemistry and Mott-Schottky are considerably employed for in-situ passive film structure and composition investigations with considerable success [12–14]. The basis to use these techniques is the fact that oxide films on metals and alloys behave like a semiconductor. So therefore, if we can measure the semiconducting properties of the films, then those can be correlated with the physicochemical structure of the oxides and that will ultimately help us to understand the corrosion performance of that particular material.

Photoelectrochemical spectroscopy (PES) is considered to be an effective tool for in-situ characterizations of the passive film and can be used to evaluate the electrical and/or optical behavior of the oxide films on metallic alloys [12]. Its applications in studying the passive behavior of metals are based on the fact that oxide films exhibit semi-conducting behavior and so therefore, when light energy with a specified band gap (E_g) irradiate the metal/metal oxide/solution interface, then photons are absorbed and electron-hole pairs are generated [2, 15]. The electron-hole pairs, which are created in the bulk of the semiconductor are combined together by giving off the heat. On the other hand, by using space charge field, the electron-hole pairs created in the space charge region, can be separated in to electrons and holes. These holes are then forced towards the oxide/solution interface and cause the oxidation of R to O, however the electrons travel using the external circuit through the semiconductor electrode lead. So when photocurrent of a specified band gap is used to illuminate an n-type semiconductor, anodic photocurrent is generated, while cathodic photocurrent for a p-type semiconductor. However, in a p-type semiconductor, the space charge moves the electrons towards oxide/solution interface and holes towards the bulk material, which is vice versa for n-type semiconductor. The photocurrent (i_{ph}) and absorption coefficient ($\underline{\alpha}$) of a crystalline material are related to each other [16]. It is clear from Eq. 1, that i_{ph} is directly proportional to $\underline{\alpha}$, which depends on the photon energy as below;

$$\underline{\alpha} = A\left(h\nu - E_g\right)^n / h\nu \tag{1}$$

where A is a constant, h is the Plank constant, ν is photon frequency ($h\nu$ = photon energy), and band gap energy is represented by E_g. On the other hand, for direct and indirect transition, the value of n will be 1/2 and 2 respectively. For the amorphous materials/semiconductors, it is common to use n = 2, however the transitions in amorphous materials are called non-direct transitions to avoid any confusions with crystalline materials [17]. The band gap energy of the semiconductor under investigation can be calculated by plotting $(i_{ph}\ h\nu)^{1/n}$ versus hν using Eq. 1.

On the other hand, as per ideal Schottky barrier model approximation, both charges (charge in the space charge region and Helmholtz layer) are equal and opposite. Based on this, space charge capacitance can be calculated and this capacitance is dependent on applied voltage, therefore it is called the differential capacitance. The depletion layer capacitance per unit of surface is shown in Eq. 2 [15, 18]:

$$C = \frac{dQ_{SC}}{dV} \tag{2}$$

where dQ_{SC} is the increase in charge per unit area. The derivation of C is based on the Poisson's equation expressed in the form of for one dimensional application with (x), the charge density at a position x, and $\Phi(x)$, the potential at the same position. Integration of this equation, taking account of boundary conditions, leads to an important relationship between the differential capacitance and the applied potential known as the Mott-Schottky relation (Eq. 4a, 4b);

$$\frac{d^2\Phi(x)}{dx^2} = -\frac{\rho(x)}{\varepsilon\varepsilon_o} \tag{3}$$

$$\frac{1}{C_{SC}^2} = \left(\frac{2}{\varepsilon\varepsilon_o e N_D}\right)\left(E - E_{FB} - \frac{kT}{e}\right) \quad \text{for an n-type semiconductor} \tag{4a}$$

$$\frac{1}{C_{SC}^2} = -\left(\frac{2}{\varepsilon\varepsilon_o e N_A}\right)\left(E - E_{FB} + \frac{kT}{e}\right) \quad \text{for a p-type semiconductor} \tag{4b}$$

Here N_D and N_A represent the donors and acceptors density respectively, whereas E_{FB} represents the flat band potential. Here ε represents dielectric constant of the oxide (15.6 for Cr-substituted γ-Fe_2O_3) [19]; while ε_0 = vacuum permittivity ($8.854*10^{-14}$ F cm^{-1}); e = electron charge; N_D = donor density; N_A = *Acceptor density*; E = applied potential, E_{FB} = flat band potential; k = Boltzmann constant. So based on these equations it is clear that for n-type and p-type semiconductors, C^{-2} versus E will be linear with positive and negative slopes that are inversely proportional to the N_D and N_A, respectively. In-order to carry out the Mott-Schottky analysis, the interfacial capacitance (C) *can be obtained using* $C = 1/\omega Z''$, where ω = angular frequency, Z'' = imaginary part of the specific impedance. The measured capacitance (C) can be considered to be the same as of space charge capacitance (C_{SC}), if we assume that double layer capacitance is very high and so can be ignored in a series combination with the space-charge capacitance.

2 Applications of Photoelectrochemical Spectroscopy and Mott-Schottky Analysis

2.1 *Applications of Photoelectrochemical Spectroscopy*

In order to conduct the photocurrent experiments, a special instrumentation is required, and that contains a three-electrode cell with one-liter multi neck flask specifically designed to provide an additional quartz window for photon inlet. Before performing the PES experiments, passive film is formed on the materials under investigation by selecting a suitable potential in the passive range. Figure 1 shows the application of PES to analyze the passive film structure and composition on two

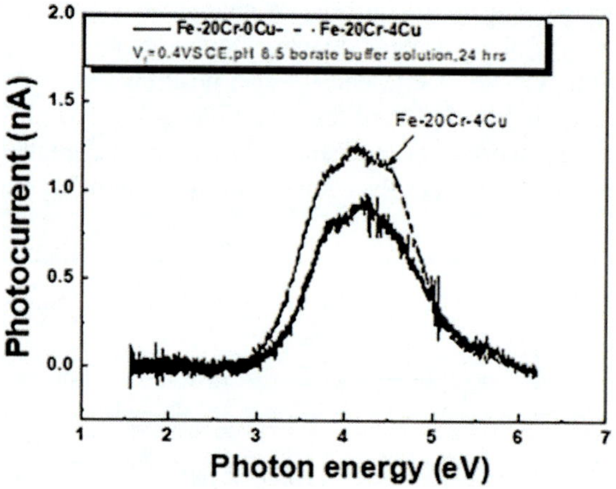

Fig. 1 The photocurrent spectra of oxide films on Fe20Cr alloys with and without copper

different Fe–Cr alloys with and without any copper additions. The photocurrent set-up used in the measured is discussed in detail elsewhere [1, 20]. For light source, a 300 W Xenon arc lamp was used and digital monochromator was employed to provide a monochromatic light with a wavelength of 200–800 nm and during this experimentation, continuous illumination technique was used rather than using a lock-in technique. The working electrode with an exposed area of 0.2 cm^2 was used and passive film was formed on the materials by stepping the applied potential from −1.0 VSCE to a film formation potential (V_f) in the passive region. To obtain the photocurrent spectra shown in Fig. 1 (photocurrent vs. photon energy), the passive film was formed on both alloys in the passive range (0.4 V_{SCE}) for one day in pH 8.5 borate buffer solution. This passive range was selected based on potentiodynamic polarization experiments conducted in pH 8.5 borate buffer solution at room temperature. Subsequently the photocurrent spectra for the passive films was measured at potentials in the passive region.

Based on the results obtained in Fig. 1, it is clear that both the alloys (with and without Cu) exhibited somewhat similar photocurrent spectra, except a small difference in photocurrent intensity of both materials. The passive film on both alloys were found to be n-type based on anodic photocurrent data of Fig. 1. It was observed that the anodic photocurrent was decreased after the peak current value and this decrease is believed to be due to an increase in the surface recombination rate of photo-excited carriers at higher photon energies [12, 21, 22]. Both the alloys exhibited an increase in photocurrent starting from 2.5 eV and a shoulder appeared at 3.7 eV. The value of the peak current was found to be 4.3 eV. It can be deduced from these results that the oxide film on both of these alloys has similar base structure and generation of photocurrent on both alloys is from same electron transition sources. It was reported by other researchers regarding the oxide film on FeCr alloys as a function of film formation potential, that oxide film had a base structure of crystalline spinel γ-Fe_2O_3. This base structure involves two different electronic excitation processes. The first excitation process is the "d-d transition", which represent the electronic excitation from Fe^{3+} band to Fe^{2+} band; $2Fe^{3+} \rightarrow Fe^{2+} + Fe^{4+}$ and the second process is "p-d transition" which specifies the excitation from the valence band (O-2p) to the conduction band (Fe-3d); $O^{2-} + Fe^{3+} \rightarrow O^- + Fe^{2+}$ [1, 12]. Figure 1 shows that there was an increase in the photocurrent intensity of copper containing alloys owing to its thick film as compared to copper free alloys. These results also suggest that oxide film on copper containing is (Cu + Cr)-substituted γ-Fe_2O_3 as compared to (Cr)-substituted γ-Fe_2O_3 on copper free alloys.

The Eq. (5) shown below shows that photocurrent is proportional to the optical absorption coefficient;

$$i_{ph} = A\frac{(h\nu - E_g)}{h\nu}, \tag{5}$$

where A is a constant; hv = incident photon energy; band gap energy is represented by E_g, and n describes the type of transition between the valence and the conduction band in a crystalline material. This equation can be used to calculate band gap energy

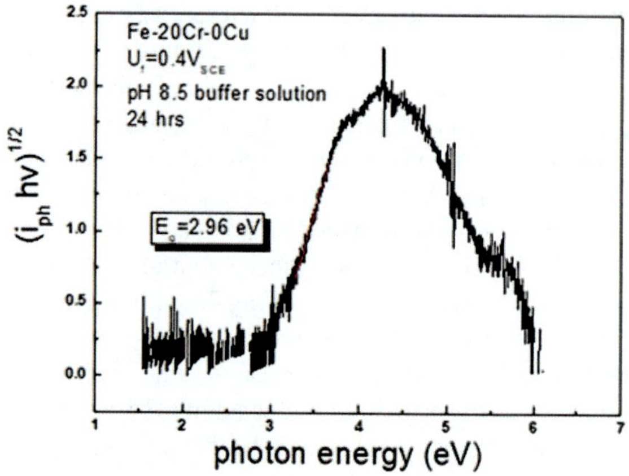

Fig. 2 The band gap calculation using Eq. 2 on copper containing alloys

of the oxide films from $(i_{ph}hv)^{1/2}$ versus hv plot and no considerable change was observed in the band gap energy of the two alloys (Eq. 5).

These results suggest that Cu increased the photocurrent spectra probably because of less Cr^{+3} ion in the (Cu + Cr) substituted γ-Fe_2O_3 passive film (which serves as a recombination site). Secondly the photocurrent of the passive film on copper containing alloy was generated by the electron transition sources of Cr-substituted $\gamma\text{-}Fe_2O_3$ involving d-d and p-d transitions.

Many other researchers used photoelectrochemical analysis to determine the properties of the oxide film on different alloys. Jang et al. [23] analyzed the passive films on Ni using photoelectrochemistry and Mott-Schottky analysis and proposed that the passive film on Ni is of duplex nature with inner NiO and outer $Ni(OH)_2$ and both showed a p-type semiconductivity with different values of flat band potential. In other studies PES and Mott-Schottky techniques were successfully employed to examine the passive film structure and composition of FeCr alloys with additions of Mo and Ni [24]. It was suggested that passive films on such alloys are composed of (Cr, Ni, Mo)-substituted γ-Fe_2O_3. While conducting PES experiments on the passive films, it is important to employ proper instrumentation as otherwise data will not be complete and that will lead to conflicting results. For example, researchers have reported differently about the structure, composition and semiconducting properties of passive films on Ni [8–10]. Some reported that it is consists of only NiO, while others reported that it is duplex layered with inner NiO and outer $Ni(OH)_2$ and vice versa with different band gap energies. It is reported by Jang et al. [23] that such dissimilarities can arise when researchers use chopped light/lock-in-amplifier [13] and by doing so the data is not enough for interpretation. It was suggested, instead to use an experimental set-up where the light wavelength can be continuously changed using a monochromator controlled by a stepping motor. This kind of experimental set-up will enable the researchers to generate more experimental data, i.e., photocurrent data over a broader range of photon energy and so will help in better data collection and interpretation. Another advantage of using such a technique is that it will be useful to get rid of the effects of impurity states on the band structure as under continuous illumination; small changes in the charge carrier density, which are due to localized energy levels, cannot be detected. On the other hand, such a technique will also be very useful because under such conditions, only base structure of the passive film is examined and photocurrent data will be a true representative of the passive film base structure and will provide more accurate band gap energy.

2.2 *Applications of Mott-Schottky Analysis*

On the other hand, in-situ Mott-Schottky analysis can provide accurate information about the defect chemistry/semiconducting properties of the oxide films. In order to carry out the Mott-Schottky analysis, passive film is formed on the alloys under investigation at a specified film formation potential, which was determined in a separate potentiodynamic polarization experiment. Toor et al. [25] investigated the passive film properties of FeCr alloys with and without copper using Mott-Schottky analysis as shown in Fig. 3a, b at two different film formation potentials. It is clear from the figures that there was no significant difference between the two alloys and addition of copper did not affect the passive film base structure. The figures exhibited two linear regions with positive and negative slopes and they are related to n-type and p-type semiconducting behavior respectively. It is worth noting that n-type behavior if related to (Fe, Cr)-oxide, while p-type is related to inner Cr enriched oxide for both alloys [1, 26]. It is well established that donor species in n-type oxide film are oxygen vacancy, V_O^{2+}, whereas metal vacancy V_{Cr}^{-3} is dominant in p-type passive film.

The results of acceptor and shallow/deep donor densities are summarized in Table 1, which are calculated based on Eqs. (4a, 4b) as stated earlier.

Based on Mott-Schottky analysis, it was found in Fig. 4a, b that copper containing alloys may not perform well considering their corrosion performance, the reason being an increase in their donor and acceptor densities. An increase in acceptor

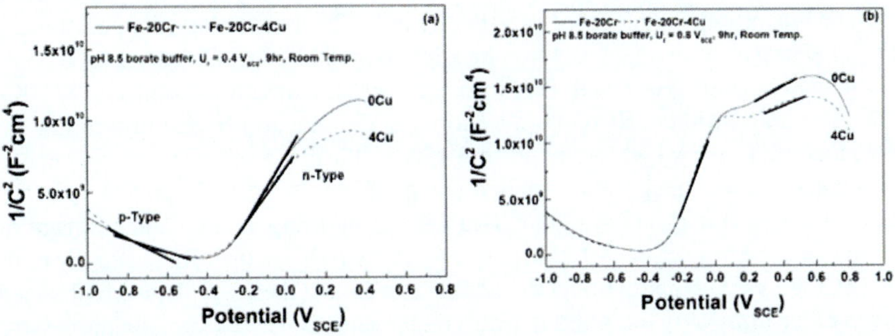

Fig. 3 Mott–Schottky plots for passive films formed on Fe–20Cr and Fe–20Cr–4Cu alloys different potentials [25-reused with permission]

Table 1 Donor and acceptor density of Fe–Cr alloys in deaerated borate buffer solution at room temperatures, based on data of Fig. 3 [25-reused with permission]

Samples	$U_f = 0.4$ V(SCE)		$U_f = 0.8$ V(SCE)		
	N_D/cm^{-3}	N_A/cm^{-3}	N_{D1}/cm^{-3}	N_{D2}/cm^{-3}	N_A/cm^{-3}
Fe20Cr	4.5×10^{20}	2.28×10^{20}	2.78×10^{20}	1.27×10^{21}	1.99×10^{21}
Fee20Cr4Cu	5.8×10^{20}	3.01×10^{21}	2.75×10^{20}	1.74×10^{21}	2.41×10^{21}

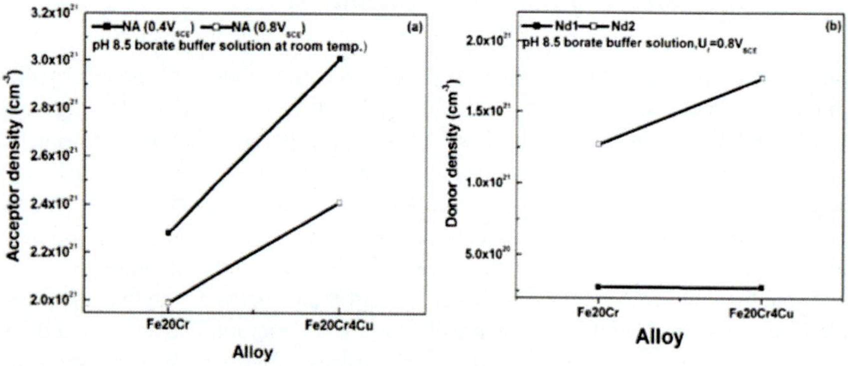

Fig. 4 Acceptor density and shallow/deep donor densities based on Mott–Schottky of passive films formed on Fe20CrxCu alloys [21-reused with permission]

density means a decrease Cr^{3+} concentration the passive film. Similar results were obtained when analysis was performed at 0.8 V_{SCE}, however there was an observed increase in deep donor density (Cr^{+6}, N_{D2}), which attributes to charge neutrality in the film surface. These results of Mott-Schottky analysis clearly represent that copper was dissolved in the Cr-substituted γ-Fe_2O_3. As per charge neutrality concepts, it is expected that Cu^{2+} ions replaced the Fe^{3+} in the passive film and that results din decreased positive charge in the film and this decrease was compensated by an increase of Cr^{6+} in the film. Therefore, it is concluded that some copper is enriched/incorporated/dissolved in γ-Fe_2O_3 in Fe-20Cr-xCu alloys by substituting Fe^{3+} and passive film on these alloys becomes (Cu + Cr) substituted γ-Fe_2O_3.

In another study, Toor et al. [27] investigated the oxide films formed on Fe20Cr alloys with and without Si addition by forming the film at different film formation potentials (0.4 ~ 0.8 V_{SCE}) for a period of 9 h. After the film was made, the capacitance was measured by changing the applied potential from the film formation potential to −1.0 V_{SCE} at a specified rate of 1 mV/s and by using an excitation voltage of 10 mV and the frequency of 1 kHz [28]. Figure 5a, b shows the Mott-Schottky plots of FeCr alloys with and without the addition of Si and these graphs were obtained when oxide films were formed at two different potentials, i.e. 0.4 V_{SCE} (in the passive region) and 0.8 V_{SCE} (in the trans-passive region). These potentials were selected based on polarization experiments conducted in same solution at room temperature. Again, like in the previous study [25], it is clear from Fig. 5 that there is no significant

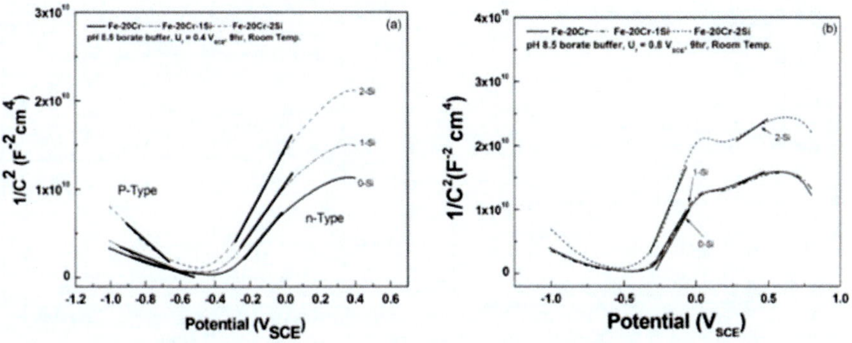

Fig. 5 Mott–Schottky plots for passive films formed on Fe–20CrxSi alloys different potentials [27-reused with permission]

change in the behavior of both alloys (with and without Si addition) and oxide film on both of these alloys seems to have the same base structure. Two linear regions are clear in Fig. 5a with a positive slope at potentials greater than $-0.4\ V_{SCE}$ and it represents n-type semiconducting behavior. This n-type behavior is correlated to (Fe, Cr)-oxide. On the other hand, a negative slope at potentials $<-0.4\ V_{SCE}$ was observed (Fig. 5a) and it represents p-type semiconducting behavior. This p-type behavior is related to inner Cr enriched oxide for both alloys [1, 26]. So the oxygen vacancy, V_O^{2+} and metal vacancy, V_{Cr}^{-3} were the main donor and acceptor species in n-type and p-type semiconductor respectively [1, 26]. Figure 5b shows results at $0.8\ V_{SCE}$ film formation potential and here it is clear that there are two linear regions with two positive slopes above $-0.4V_{SCE}$ and $0V_{SCE}$ respectively. It is believed that the second liner region represents the presence of Cr^{6+} ions, which are acting as deep donor in the oxide film [26]. These results are summarized in Table 2.

As per the results summarized in Table 1, it is obvious that Mott-Schottky analysis is very helpful in identifying the semiconducting properties of the materials. It is clear that the addition of Si in the alloys decreased the donor density of the alloys at 0.4 V and that suggest that the addition of Si decreased the concentration of V_O^{2+} of the passive films and somewhat similar behavior was observed at higher potentials (0.8 V_{SCE}) as both the shallow donor density (N_{D1}, V_O^{2+}) and deep donor density (Cr^{+6}, N_{D2}) was decreased. These results can be further elaborated based on principles of

Table 2 Donor and acceptor density of FeCrxSi alloys in deaerated borate buffer solution at room temperatures, based on data of Fig. 5. [27-reused with permission]

	$U_f = 0.4V_{SCE}$		$U_f = 0.8V_{SCE}$		
	N_D	N_A	N_{D1}	N_{D2}	N_A
Fe20Cr	4.82E20	2.10E21	2.78E20	1.27E21	2.10E21
Fe20Cr1Si	3.90E20	1.26E21	2.80E20	1.24E21	1.25E21
Fe20Cr2Si	3.05E20	1.20E21	1.77E20	9.50E20	1.20E21

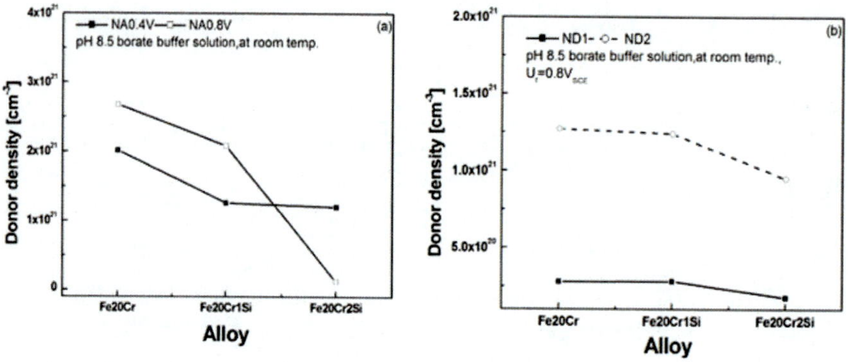

Fig. 6 Acceptor density and shallow/deep donor densities based on Mott–Schottky of passive films formed on Fe20CrxSi alloys [27-reused with permission]

charge neutrality. In case of these FeCrxSi alloys, the Si^{+4} was found enriched in the Cr-substituted γ-Fe_2O_3 structure by replacing Fe^{+3} and so its presence increased the positive charge and decreased the donor density. Thus, the oxide film of Si containing alloys will be now Si enriched/incorporated along with Cr and Fe and that ultimately will be helpful to improve the corrosion performance of Si containing alloys. Figure 6a, b shows that concentration of V_O^{2+} in the passive films on Si containing alloy was decreased as donor density is decreased, while acceptor density was also decreased and this decrease in acceptor density is linked to an increase in Cr^{3+} concentration of the passive film, which is good considering the corrosion resistance of these alloys.

3 Conclusions

Based on above discussion on the applications of photoelectrochemical spectroscopy and Mott-Schottky analysis, it is quite clear that these two techniques can be successfully employed to examine the passive films on meals and alloys. They provide accurate information by in-situ electrochemical analysis in a variety of environments and for different alloys. The defect chemistry and semiconducting properties measured using these techniques can be successfully, correlated with the corrosion performance of these alloys. There can be challenges in the photoelectrochemical experimental setup, so proper instrumentation is required for continuous illuminations over a wide range of wavelength, so that a detailed photoelectrochemical spectrum can be generated for analysis and interpretation.

References

1. E.A. Cho, H.S. Kwon, D.D. Macdonald, Electrochim. Acta **47**, 1661 (2002)
2. A.J. Bard, L.R. Faulkner, *In Electrochemical Methods* (John Wiley & Sons, New York, 1980), p. 630N
3. U. Stimming, J.W. Schultz, Electrochim. Acta **24**, 859 (1979)
4. J.D. Dunitz, L.E. Orgel, in *Solid State Chemistry and its Applications*, ed by A.R. West (Wiley (SEA), Singapore, 1989), p. 313
5. M.D.C. Belo et al., Thin Solid Films **515**, 2167m (2006)
6. P. Schmuki, H. Bohni, J. Electrochem. Soc. **139**, 1908 (1992)
7. M.D.C. Belo, B. Rondot, C. Compere, M.F. Montemor, A.M.P. Simoes, M.G.S. Ferreira, Corros. Sci. **40**, 481 (1998)
8. N. Sato, Corros. Sci. **42**, 1957 (2000)
9. S. M. Wilhelm, N. Hackerman, J. Electrochem. Soc. **128**, 1668 (1981)
10. C. Sunseri, S. Piazza, F. Di Quarto, Mater. Sci. Forum **185–188**, 435 (1995)
11. S. Fujimoto, H. Tsuchiya, M. Sakamoto, T. Shibata, K. Asami, 201st Meeting of the electrochemical society, Philadelphia, Pennsylvania, 12–17 May 2002, p. 278
12. J. S. Kim, E. A. Cho, H. S. Kwon, Corros. Sci. **43**, 1403 (2001)
13. R.H. Bube, *Photoelectronic Properties of Semiconductors* (Cambridge University Press, Cambridge, UK, 1992), p. 121
14. J.S. Kim, E.A. Cho, H.S. Kwon, Corros. Sci. **43**, 1403 (2001)
15. M.D.C. Belo, Photoelectrochemistry: theoretical basis", in *Electrochemical and Optical Techniques for the Study and Monitoring of Metallic Corrosion*, ed. by M.G.S. Ferreira, C.A. Melendres (Kluwer Academic Publishers, Dordrecht, The Netherlands, 1991), p. 191
16. W.W. Gatner, Phys. Rev. **116**, 84 (1959)
17. C. Sunseri, S. Piazza, A. Di Paola, F. Di Quarto, J. Electrochem. Soc. **134**, 2410 (1987)
18. N.E. Hakiki, M.D.C. Belo, A.M.P. Simoes, M.G.S. Ferreira, J. Electrochem. Soc. **145**, 3821 (1998)
19. Di Paola, Electrochim. Acta **34**, 203 (1989)
20. H. J. Jang, C. J. Park, H. S. Kwon, Electrochim. Acta **50**(16–17), 3503–3508
21. J.F. McAleer, L.M. Peter, Faraday Discuss. **70**, 67 (1980)
22. B. Danzfuss, U. Stimming, J. Electroanal. Chem. **164**, 89 (1984)
23. H.J. Jang, C.J. Park, H.S. Kwon, Photoelectrochemical analysis on the passive film formed on Ni in pH 8.5 buffer solution. Electrochim. Acta **50**(16–17), 3503–3508 (2005)
24. H.J. Jang, H.S. Kwon, In situ study on the effects of Ni and Mo on the passive film formed on Fe–20Cr alloys by photoelectrochemical and Mott-Schottky techniques. J. Electroanal. Chem. **590**(2), 120–125 (2006)
25. I.H. Toor, M. Ejaz, H.S. Kwon, Mott-Schottky analysis of passive films on Cu containing Fe–20Cr–xCu (x = 0, 4) alloys. Corros. Eng. Sci. Technol. **49**(5), 390–395 (2014). https://doi.org/10.1179/1743278214Y.0000000154
26. N.E. Hakiki, S. Boudin, B. Rondot, M.D.C. Belo, Corros. Sci. **37**, 1809 (1995)
27. I.H. Toor, J. Electrochem. Soc. **158**(11), C391–C395 (2011)
28. S.J. Ahn, H.S. Kwon, Electrochim. Acta **49**, 3347–3353 (2004)

Voltammetry and In-Situ UV–Vis Absorbance Studies for Understanding Corrosion Mechanisms

K. S. Raja

Abstract This chapter will discuss two different techniques viz., (a) voltammetry, and (b) in-situ optical absorbance spectrometry for studying corrosion phenomena of metals and non-metals. The basic principles of different voltammetric techniques such as linear potential sweep voltammetry, cyclic voltammetry, and pulse voltammetry are described. Interpretation of the voltammetry results with reference to various reaction paths such as electrochemical, chemical, and coupled reactions; and different processes such reversible, irreversible, and quasi reversible are discussed. Examples of voltammetric techniques include corrosion of steels and fluorine doped tin oxide (FTO). Basic principles of UV–Vis optical absorbance spectroscopy is discussed. The techniques of carrying out in-situ spectroscopy, and interpretation of results is described with an example of corrosion studies on FTO in different pH conditions.

Keywords Cyclic voltammetry · Spectro-electrochemistry · UV–Vis optical absorbance · Oxide film · Passivation · Corrosion · Localized corrosion

1 Introduction

Voltammetry is a generic name of different types of potential sweep techniques that are applied to investigate electrochemical reaction mechanisms of an electrochemical system. In these techniques, potential of the working electrode is scanned in both cathodic and anodic directions at different scan rates covering a wide range of potential window and the resultant current is recorded. The potential and current are plotted in a linear scale which is called as voltammogram. This potential sweep method is useful for evaluating both thermodynamic (in terms of reduction or oxidation potentials) and kinetic parameters (such as diffusion coefficients, rate constants, number of electrons involved, etc.) of the electrochemical reactions being investigated. Voltammetry is typically applied to study the electrochemistry of electroactive

K. S. Raja (✉)
Department of Chemical and Materials Engineering, University of Idaho, Moscow, ID 83843, United States
e-mail: ksraja@uidaho.edu

I. Toor (ed.), *Recent Developments in Analytical Techniques for Corrosion Research*,
https://doi.org/10.1007/978-3-030-89101-5_3

species present in an electrolyte. Electrodes employed in voltammetric techniques are generally inert (such as: glassy carbon, or platinum) and do not participate in the electrochemical reaction. The electrodes conduct electrons and act as a support for the electrochemical reactions to take place. On the other hand, when potential sweep techniques are applied to study corrosion behavior of a solid specimen, the working electrode participates in the electrochemical reaction. In this case, the specimen itself is a working electrode and it is not inert because the working electrode undergoes an electrochemical reaction. Therefore, the potential sweep techniques involving electrochemically active electrodes for understanding corrosion behavior are given different names such as potentiodynamic polarization, and cyclic polarization. The current density recorded during these tests is typically plotted in log scale, while the potential is plotted in a linear scale. The log scale of current density helps present both the active and passive states of the specimen as a function of potential in the same plot. Conventional voltammograms plotted in a linear scale may not be very useful to understand the corrosion behavior of a material. However, if one wants to study the half-cell reactions of the corrosion process individually and understand the reaction mechanisms, then linear voltammetric techniques are required. The voltammetric techniques help understand the steps involved in an electrochemical reaction and help derive the reaction kinetics parameters. Therefore, it is important to refresh the basics of electrode kinetics.

2 Electrode Kinetics

Voltammetry is used for studying the electrode kinetics. The reaction occurring at an electrode is considered heterogeneous electron transfer to or from the electrode surface. The reaction can be written in general form as:

$$\mathrm{Ox} + \mathrm{e}^- \leftrightarrow \mathrm{Red} \tag{1}$$

In particular, the following reactions can be considered as examples:

$$\mathrm{Fe}^{3+} + \mathrm{e}^- \rightarrow \mathrm{Fe}^{2+} \tag{2}$$

$$2\mathrm{H_2O} + 2\mathrm{e}^- \rightarrow \mathrm{H_2} + 2\mathrm{OH}^- \tag{3}$$

Here, we consider reduction of ions that are present in the electrolyte. In order for Fe^{3+} to undergo reduction at the electrode, it must diffuse from bulk solution to the electrode surface or within a critical distance to the electrode surface for facilitating electron tunneling and electron transfer [1]. The electron transfers result in flow of current which could be related to the flux of reactants participating in the electrochemical reaction by the relation:

$$i = nF_j \tag{4}$$

where, i = current density in A/cm^2, n = number of electrons, F = Faraday's constant in coulomb/mol., j = flux of ions in mol/cm^2.s

The flux is given by the reaction [1]:

$$j = k_n[\text{concentration of reactant}]_o^n \tag{5}$$

k_n = nth order rate constant cm/s or s^{-1}, C_o = concentration near electrode mol/cm^3 or surface concentration of reactants mol/cm^3 or mol/cm^2.

The rate constants vary with potential and given by the relation:

$$k_c = k_o exp(-\alpha_c nFE/RT) \text{ for cathodic reaction} \tag{6}$$

$$k_a = k_o\, exp(\alpha_a nFE/RT) \text{ for anodic reaction} \tag{7}$$

where α_a and α_c are transfer coefficients that vary from 0 and 1. The sum of transfer coefficients is given as 1 for a simple electron transfer reaction ($\alpha_a + \alpha_c = 1$). These relations lead to the Butler Volmer equation for electrode kinetics:

$$I = I_o[exp(\alpha_a nf\eta/RT) - exp(-\alpha_c nF\eta/RT)] \tag{8}$$

where, I_o = exchange current density, η = over potential to drive the anodic or cathodic reaction ($E - E^O{}_{eq}$).

2.1 Reaction Rate Constants

The slow and fast reaction kinetic could be associated with energy barrier of which the electron should surmount in order for the transfer to occur in either direction as schematically shown in Fig. 1 [2].

The energy barrier can be understood based in the Franck–Condon principle [3, 4]. If electron transfer occurs by tunneling from the ground state of reactant (R), then the product (P) must be formed in an excited state as shown in Fig. 1. This excitation is achieved by thermal vibration of the reactant that provides the required activation energy (E_{act}) to the cross over point of the two potential energy curves of R and P as seen in Fig. 1. The vibrationally excited energy of the product is dissociated by thermal deactivation to its ground state through collision with solvent molecules. In effect, the electron transfer of reaction (1) takes place by a vertical transition of the potential energy from ground state to cross over potential energy by thermal activation. This occurs by change in the bond lengths, bond angles within reactant species. If the change in the reaction coordinate between reactant and product is small, the activation barrier (E_{ac}t) is also small. Here, the reaction coordinate implies

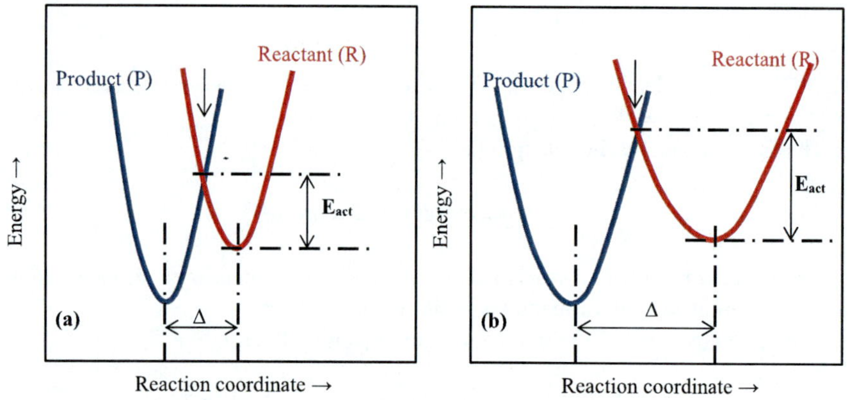

Fig. 1 Schematic illustration of increase in activation energy (E_{act}) of a reaction with increase in the reaction coordinates: **a** the change in reaction coordinates (Δ) is small, **b** Δ is large [2]

bond length and/or angle of the reactant or product species. If both the reactant and product have similar structures and bond lengths, then the rate constant k^0 will be large corresponding to a low activation barrier for reaction. On the other hand, when the R and P are structurally dissimilar or have significantly different bond lengths then the rate constant will be small and the activation barrier will be large. The standard heterogeneous rate constant varies from 10^{-8} to 10 cm/s.

2.2 *Mass Transport*

In an electrochemical system, mass transport occurs by three anodes modes such as (1) Diffusion, (2) Migration and (3) Convection. This chapter will focus on mass transport by diffusion only.

Steady state diffusion: Mass transport by diffusion occurs from high concentration to low concentration. The diffusive flux is given by Fick's first law:

$$j = -D\frac{\partial C}{\partial x} \tag{9}$$

where D = Diffusion coefficient in cm^2/s, $\partial c/\partial x$ = local concentration gradient. The diffusion coefficient is sensitive to temperature and given by the Arrhenius type relationship:

$$D = D_o \exp(-E_D/RT) \tag{10}$$

where, D_0 = Pre-exponent diffusivity in cm^2/s, E_D = activation energy for diffusion in $J.mol^{-1}$, R = universal gas constant in $J.\ mol^{-1}.K^{-1}$, and T = temperature in K.

Diffusion controlled current under steady state condition is given by the relation:

$$I = \text{n FD}\left(\frac{dC_o}{dx}\right)_0 = \frac{\text{nFD}C_0^{\infty}}{\delta} \quad (11)$$

δ = thickness of diffusion layer.

Non-steady state condition: The concentration at a point may vary with time under non-steady state condition. Therefore, the temporal variation of the concentration should be considered which is given by Fick's Second Law:

$$\frac{\partial C}{\partial t} = D\frac{\partial^2 C}{\partial x^2} \quad (12)$$

To solve the above equation a new variable, τ, is introduced as:

$$\tau = \frac{x}{2\sqrt{DT}} \quad \text{and} \quad \tau^2 = \frac{x^2}{4DT} \quad (13)$$

Now the Eq. (11) can be written in the form of:

$$\frac{\partial^2 C}{\partial \tau^2} + 2\tau\frac{\partial C}{\partial \tau} = 0 \quad (14)$$

Integration of (14) results in:

$$\frac{\partial C}{\partial \tau} = aExp\left[-\tau^2\right] \quad (15)$$

$$\int_c^{C*} dc = a\int_\tau^{\infty} \exp(-\tau^2)d\tau \quad (16)$$

After applying appropriate boundary conditions, the concetration profile can be given by the relation:

$$C = C^*\frac{2}{\sqrt{\pi}}\int_0^{\tau} \exp(-\tau^2)d\tau = C^*erf\left(\frac{x}{2\sqrt{Dt}}\right) \quad (17)$$

Since the diffusion-controlled current is given by the relation: $I = nFAj$.

Where, $j = \left(\text{D}\frac{\partial C}{\partial x}\right)_{x=0} = \frac{D}{2\sqrt{Dt}}\frac{dC}{d\tau}$, $\tau = \frac{x}{2\sqrt{Dt}}$; and C* = bulk concentration of electroactive species,

The current, I, as a function of time can be given by the relation known as Cottrell equation:

$$I = \frac{nFA\sqrt{D}C^*}{\sqrt{\pi t}} \quad (18)$$

3 Voltammetry and Electrode Reactions

3.1 Types of Voltammetry

The different types of voltammetry techniques can be listed as.

(a) Linear sweep voltammetry
(b) Cyclic voltammetry
(c) Potential step voltammetric techniques
(d) Differential pulse voltammetry
(e) Square wave voltammetry
(f) Stripping voltammetry.

Typically, linear sweep voltammetry is carried out by scanning the potential between two set points E_1 and E_2 at a desired scan rate and the resultant current is recorded. In case of the cyclic voltammetry, as the name indicates, the potential scan is cycled between two set points. The potential window can range from anodic and cathodic potentials with reference to the open circuit potential of the working electrode in the electrolyte. The extreme potential window will generally be within the stability potentials of the electrolyte. The potential can be scanned only in the anodic or cathodic direction if one wants to study only the oxidation or reduction of the species, respectively.

A three-electrode configuration is used for the voltammetry techniques. Inert type solid working and counter electrodes are used. The reference electrode could be silver/ silver chloride or saturated calomel electrode for aqueous electrolytes. At alkaline pH conditions Hg/HgO reference electrode is recommended. For non-aqueous, silver/ silver ion reference electrode is typically used which contains the same test solution in the reference compartment with silver salt ($AgNO_3$, $AgClO_4$ etc.) dissolved in the non-aqueous solvent. Inert metal wires such as Pt or Au could be used a s quasi reference electrode. For high temperature molten salt experiments Ag/Ag^+ electrode (AgCl coated Ag wire immersed in 1 wt% AgCl containing the test molten salt), Ni/Ni^{2+} ionic electrode (Ni wire immersed in nickel salt containing test molten salt), Pt/Pt^{2+}, Li-Al/ Li^+ and $Ni/Ni_3S_2/S^{2-}$ electrodes are used [5]. Voltammetry is carried out at different scan rate spanning three decades such as 10 mV/s to several volts/s. The recorded voltammograms will be a function of:

(A) The standard electrochemical rate constant (k^o)
(B) Formal potential of the reactant/ product couple
(C) The diffusion coefficients of reactant and product
(D) The potential scan rate and the potential window.

In addition to the above, the voltammograms depend on the type of electrochemical reactions such as reversible, irreversible and quasi-reversible and reaction mechanisms such as single step electrochemical reaction, multi-step electrochemical reaction, electrochemical-chemical (EC) reaction, electrochemical-chemical-electrochemical (ECE) etc.

3.2 *Electrode Reactions*

3.2.1 Reversible Reactions

The redox reaction, $O + ne^{-} \rightleftharpoons R$ is considered reversible when the forward and the reverse reaction rate constants are very high and are equal [6]. The rate constant should be higher than the rate of mass transport [7], $k^{o} > > m_{T}$, where $m_{T} \approx D/\delta \approx \sqrt{D/\left(\frac{RT}{Fv}\right)}$ where v = scan rate. Cyclic voltammograms at different potential scan rates for a reversible reaction are shown in Fig. 2. If a very slow linear scan rate is employed, the recorded current will reach a steady value as a function of potential because beyond a certain distance from the electrode the concentration will be uniform due to natural convection. Also, a very slow scan rate leads to a steady state condition where the concentration gradient in the Nernst diffusion layer becomes linear. At very slow scan rates a plateau current profile is observed as function of increased over potentials rather than a peak current profile. This observation is attributed to a decrease in the surface concentration of reactant with increase in the over potential for a reaction which eventually becomes zero. Further increase in the over potential does not change the surface concentration and hence the concentration gradient $(\partial c/\partial x)$. Therefore the current reaches a plateau or limiting the value given by the relation:

$$I_{lim} = nFAC_{0}^{\infty}/\delta_{eq} \quad (19)$$

where, δ_{eq} = equilibrium diffusion layer thickness.

When the scan rate is increased, the diffusion layer thickness is altered significantly due to diffusion of reactants from bulk to the electrode surface. The diffusion gradient is not linear. The current varies with potential. Both the change in diffusion layer and electrode potential with time leads to a peak shaped current potential

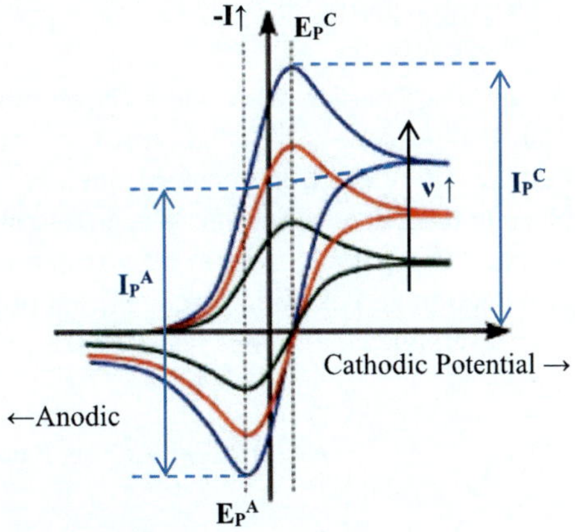

Fig. 2 Cyclic voltammograms at different potential scan rates (ν) of a reversible process $Ox + e^{-} \rightarrow R$

profile. The peak current increases with the scan rate, and given by Randles–Sevcik equation [8, 9]:

$$i_P = 0.4463nF\left(\frac{nF}{RT}\right)^{1/2} C_o^{\infty} D^{1/2} \vartheta^{1/2} \tag{20}$$

where i_p = peak current density (A/cm^2), D is in cm^2/s, scan rate in v = V/s, bulk concentration C_o^{∞} is in mol/cm^3.

When the potential scan is reversed, the current is expected to trace back the forward current profile when the scan rate is slow. On the other hand, when the scan rate is fast the current potential profiles will be as seen in Fig. 2. During reverse scan, a reduction current is observed until the potential reached the equilibrium redox potential (E_e^o). The oxidation current starts to flow when the 'R' present near the electrode starts to be oxidized to 'O'. The potentials of the peak currents of reverse scan ($E^A{}_P$) will be different from the potentials of forward peak currents ($E^C{}_P$. The difference in the peak potentials ($\Delta E_p = E_P^C - E_P^A$) helps determine the reversibility. The diagnostic tests for reversible processes are listed below [6]:

1. $\Delta E_p = E_p{}^{anodic} - E_p{}^{cathodic} = 59/n$ mV at 25 °C
2. $|E_p - E_{p/2}| = 59/n$ mV at 25 °C or 2.218 RT/F
3. $|I_p^a / I_p^c| = 1$
4. $I_p \propto v^{1/2}$
5. E_p is independent of scan rate
6. At potentials beyond E_p, $I^{-2} \propto t$.

Not following one or more of the above criteria indicates that the electrochemical reaction is not reversible within the scan rates employed. The formal potential, E_o^f(O/R) can be calculated as the potential between the reduction and oxidation peaks when $D_o = D_R$.

3.2.2 Irreversible Reactions

As discussed earlier, when the electron transfer rate is much slower than the mass transfer rate, irreversible condition exists. At very slow scan rates there is a possibility that the rate of electron transfer is greater than the rate of mass transfer which may blead to reversible condition. When the scan rate is increased, the mass transfer rate becomes comparable to that of electron transfer rate, that leads to wider separation of reduction and oxidation peak current potentials. The peak current density under the irreversible condition is given by the following relations:

$$i_P = 0.227 n_\alpha F C_0^{\infty} k_s Exp\left[\frac{-\alpha n_\alpha F(E - E_e^0)}{RT}\right] \tag{21}$$

Fig. 3 Peak current as a function of square root of the scan rate for $\alpha_c = 0.5$

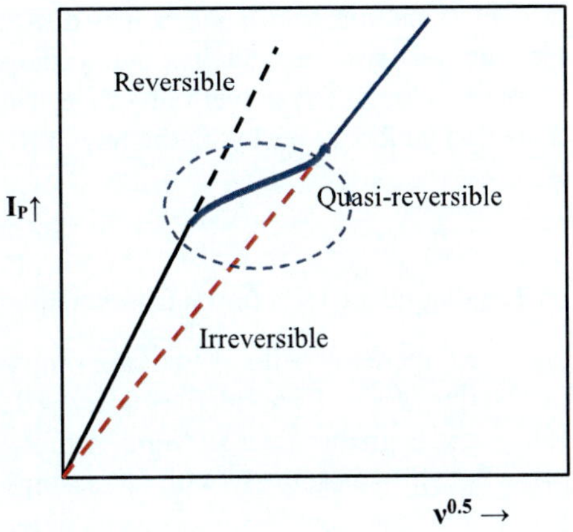

$$i_P = 0.496 n_\alpha F \left(\frac{\alpha n_\alpha F}{RT}\right)^{1/2} C_o^\infty D^{1/2} \vartheta^{1/2} \tag{22}$$

where n_α = number of electrons transferred in the rate determining step. The relation (22) indicates that the irreversible peak current density will be about 22% less than that of reversible condition. A linear relation will be observed when the peak current density is plotted as a function of $v^{1/2}$. The diffusivity can be calculated from the slope of the i_p versus $\upsilon^{0.5}$ plot. The plot will run through the origin for both the reversible and irreversible cases as seen in Fig. 3.

The cyclic voltammogram of a totally irreversible system will not have a reverse peak current. The diagnostic tests for a totally irreversible system at 25 °C are listed below [6]:

1. No reverse peak
2. $I_p^c \propto v^{1/2}$
3. For each decade increase in the scan rate the E_p^c shifts by $\frac{-30}{\alpha_c n_\alpha} mV$
4. $\left|E_p - E_{p/2}\right| = \frac{48}{\alpha_c n_\alpha} mV$.

A plot of ln (i_p) Vs $E_p - E^o$ at different scan rates will be straight line with a slope proportional to αn_o and the intercept proportional to the rate constant following the relation (21). Therefore, various kinetic parameters such as diffusivity, rate constant, transfer coefficient and number of electrons can be determined from the plots of current density versus scan rate and natural logarithmic of current density versus over potential.

Figure 3 schematically depicts i_p versus $v^{1/2}$ pots of both reversible and irreversible conditions. As noted, the reversible current is higher than that of irreversible case. These plots follow the relation of Randles–Sevcik. It is noted that a reversible process at low scan rate changes to irreversible condition at higher scan rates by transitioning through a region called 'quasi-reversible' as depicted in Fig. 3.

This transition occurs when it is difficult to maintain Nernstian equilibrium at the electrode surface, which is due to the slower rate of electron transfer than that of mass is due to the slower rate of electron transfer than that of mass transfer. The transition generally occurs in the region [6]:

$$0.3v^{1/2} \geq k^o \geq 2 \times 10^{-5} v^{1/2} \text{ cm/s}$$

The diagnostic tests for quasi-reversible condition are [6]:

(1) $|I_p|$ increases with $v^{1/2}$ but not proportional to it.
(2) $|I_p^a - I_p^c| = 1$ provided $\alpha_c = \alpha_a = 0.5$
(3) ΔE_p is greater than $59/n$ mV which increases with increasing v
(4) E_p^c shifts negatively with increasing v.

A parameter '∅' is introduced to follow the transition between the reversible and irreversible limits: The parameter ∅ is defined as [1, 7]:

$$\emptyset = \frac{k^0}{\left(\frac{FD\vartheta}{RT}\right)^{1/2}} \quad (23)$$

The following ranges of ∅ are assigned to classify the reaction types at stationary macroelectrodes [1, 7] at 298 K and for $\alpha = 0.5$.

Reversible: $\emptyset \geq 15$, $\emptyset \geq 0.3\ v^{1/2}$ cm.s^{-1}
Quasi-reversible: $15 > \emptyset > 10^{-3}$ $0.3\ v^{1/2} > k^o > 2$ X $10^{-5}\ v^{1/2}$ cm.s^{-1}
Irreversible: $\emptyset \leq 10^{-3}$ $k^o \leq 2$ X $10^{-5}\ v^{1/2}$.

3.2.3 Coupled Reactions

Cyclic voltammetry is a useful technique for investigating coupled chemical reactions with reaction intermediates. The coupled reactions are combinations of chemical and electrochemical reactions occurring at different sequences. Kinetic parameters can be derived out of the cyclic voltammetry data taken from the first cycle. The subsequent cycle provides mechanistic insight.

The CE mechanism

The initial reaction is of chemical type followed by an electrochemical reaction. The chemical reaction occurs as denoted below:

$$\mathrm{Y} \underset{K_{-I}}{\overset{K_I}{\rightleftharpoons}} \mathrm{O}$$

Which is followed by an electrochemical reaction

$$\mathrm{O} + \mathrm{ne}^- \rightleftharpoons \mathrm{R}$$

The diagnostic tests for CE mechanism are [6]:

1. $\frac{I_p^C}{\nu^{1/2}}$ decreases as the scan rate increases
2. $\left|\frac{I_p^A}{I_p^C}\right|$ increases with the scan rate
3. $\left|\frac{I_p^A}{I_p^C}\right| \geq 1$.

When the electron transfer is fast (reversible condition) and the chemical reaction step is very slow, the current will not be controlled by diffusion of reactants. Therefore, no peak current will be recorded in the voltammogram but a steady state limiting current wave will be observed. The other extreme condition is noted when the chemical step is very fast, and the electron transfer is very slow. This case is diffusion controlled because of availability of 'O" species. Due to irreversible condition of electron transfer, the reverse oxidation peak due to $\mathrm{R} \rightarrow \mathrm{O} + \mathrm{e}^-$ will be absent. Therefore, diagnostic tests mainly focus on the forward peak current $I_p^{C.}$ As the scan rate increases to higher values the chemical reaction rate (and therefore concentration of "O" (C_o^∞)) may not be sufficient to show an increase in the current peak. Therefore, $\frac{I_p^c}{\nu^{1/2}}$ decreases with the scan rate.

EC Reaction

The initial reaction is electrochemical in nature which is followed by a chemical reaction where R is converted to Y as given by:

$$\mathrm{O} + \mathrm{ne}^- \rightleftharpoons \mathrm{R}$$

$$\mathrm{R} \underset{K_{-1}}{\overset{K_1}{\rightleftharpoons}} \mathrm{Y}$$

If the electron transfer is irreversible, the chemical reaction will not affect the shape of the voltammogram and no data on chemical reaction can be obtained. This situation is similar to a single step irreversible electrochemical reaction. If the electron transfer is reversible and the chemical reaction is irreversible, then the shape of the voltammograms will depend on the scan rates. The coupled chemical reaction decreases the concentration of R due to its irreversible nature. The electrode reaction has to be faster in order to maintain Nernstian equilibrium at the surface. The diagnostic tests for EC reactions are [6]:

1. $\left|\frac{I_p^A}{I_p^c}\right| < 1$, but increases to unity with increase in the scan rate
2. $\frac{I_p^c}{\nu^{1/2}}$ decreases with increase in scan rate
3. E_p^C shifts negatively with increasing scan rate.

The ECE Mechanism

Three types of reactions occur in sequence as given below:

Electrochemical step: $O + n_1e^- \rightleftharpoons R_1$
Chemical step: $R_1 \xrightarrow{k} O^1$
Electrochemical step: $O + n_2e^- \rightleftharpoons R_2$.

The diagnostic tests for ECE reaction are [6]:

1. $\frac{I_p^C}{v^{1/2}}$ varies with scan rate
2. $\left|\frac{I_p^C}{v^{1/2}}\right|$ (low v) > $\left|\frac{I_p^C}{v^{1/2}}\right|$ (high v)
3. $\left|\frac{I_p^A}{I_p^C}\right|$ increases with scan rate and reaches unity at a high scan rate.

The peak current density associated with an ECE mechanism can be given as [10]:

$$I_p{}^C = (v\pi)^{1/2}\, FC_o^\infty (nFD/RT)^{1/2}\{n_{1.X.}C_o^\infty + n_2\phi C_o^\infty\} \tag{24}$$

The values of X and ϕ are tabulated in by Nicholson and Shain [10] a function of (kRT/nFv).

3.3 *Surface Processes*

Cyclic voltammetry is a useful technique for analyzing surface processes such as adsorption, deposition and passivation. Let us assume that an electrochemical reaction is adsorption based:

$$A_{(ads)} + e^- \rightleftharpoons B_{(ads)} \tag{25}$$

The amount of adsrobates on the electrode is quantified by the surface coverage θ, given as mol/cm^2. Let us assume that:

$$\theta_A(t) + \theta_B(t) = \theta_{total} \tag{26}$$

The current is related by the equation

$$I/FA = \partial\theta_B(t)/\partial t = -\partial\theta_A(t)/\partial t \tag{27}$$

The peak current is given by the relation [11]

$$I_p = \frac{F^2}{4RT} vA \cdot \theta_{total} \tag{28}$$

In the adsorption process, the peak current is directly proportional to the scan rate in contrast to square root of the scan rate observed for diffusion-controlled reactions. Figure 4 shows a typical voltammogram for a reversible on electron A/B couple where both A and B are adsorbed on the electrode surface. The peak width at half height of full width at half maximum (FWHM) is given by the relation [11]:

$$FWHM = \frac{3.54RT}{F} \tag{29}$$

The cyclic voltammogram of adsorption-controlled process with reversible electron transfer is symmetric about zero current axis and the potentials of the forward and reverse peaks are identical due to absence of hysteresis associated with diffusion process.

The diagnostic tests for a reversible one electron transfer electrochemical reaction involving adsorption of reactant and product can be given as:

1. Peak current is linearly proportional to the scan rate ($I_P \alpha \nu$).
2. Potentials associated with the peak currents are identical for both reduction and oxidation $\left(\Delta E_P = E_P{}^C - E_P{}^A = 0\right)$.

The peak potential is given by the relation [11]:

$$E_P = E_e^o - \left(\frac{RT}{nF}\right)\ln\frac{b_O}{b_R} \tag{30}$$

where $b_O = \beta_O\theta_O$, $b_R = \beta_R\theta_R$, $\beta_i = e^{(-\Delta G_i/RT)}$, θ_i = coverage of adsorbate.

When $b_O = b_R$, the peak potential is equal to the equilibrium redox potential as seen in Fig. 4. When there is a stronger absorption of one type of species than the other, the peak potentials will be shifted from the equilibrium potential. When $b_O > b_R$ (strong absorption of O), the peak potential of reduction wave will be more negative than E^0. Similarly, when the R is adsorbed strongly, the peak potential of oxidation wave will be more positive than the E^0.

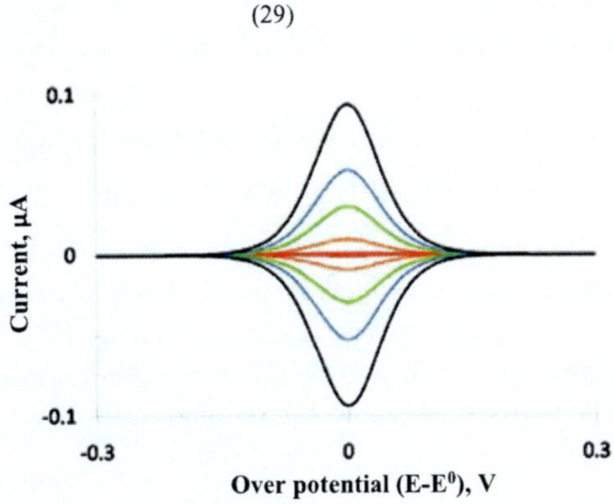

Fig. 4 Voltammogram for a reversible one electron reaction $\left(O + e^- \rightarrow R\right)$ where the O and R adsorbed on the electrode surface

3.3.1 Laviron Formalism

The effect of surface adsorption of the electroactive species is extended to thin layer electrolytes where the mass transport is not diffusion controlled. The heterogeneous electron transfer rate constant (k_s) and the charge-transfer coefficient (α) are determined from the peak potentials based on Laviron's formalism [12]:

$$E_p = E_o + \frac{RT}{\alpha nF}\left\{\text{In}\left(\frac{RTk_s}{\alpha nF} - \text{In}v\right)\right\} \tag{31}$$

For a reversible surface reaction the peak current is given by:

$$I_p = \frac{n^2F^2vA\theta}{4RT} = \frac{nFQv}{4RT} \tag{32}$$

where, surface coverage, $\theta = \frac{Q}{nFA}$, Q = charge obtained by integrating the cathodic or anodic current wave, and A is the area of working electrode. The Eq. (32) is similar to the Eq. (28). The cathodic and anodic peak potentials are given by the relation:

$$E_{pc} = E^o + \frac{RT}{\alpha F}\text{In}\frac{RTk_s}{\alpha Fv} \tag{33}$$

$$E_{pa} = E^o\frac{RT}{(1-\alpha)F}\text{In}\frac{RTk_s}{(1-\alpha)Fv} \tag{34}$$

$$\text{Log } k_s = \alpha\log(1-\alpha) + (1-\alpha)\log\alpha - \log\frac{RT}{nFv} - \frac{\alpha(1-\alpha)nF\Delta Ep}{2.3\ RT} \tag{35}$$

where, ΔE_p is peak potential difference. The E_{pa} and E_{pc} values can be experimentally determined at different scan rates (ν). Plotting E_{pa} or E_{pc} as a function of the logarithm of the scan rates will result in a straight line whose slope will be equal to $\frac{2.3RT}{(1-\alpha)nF}$ or $\frac{-2.3RT}{\alpha nF}$ for the anodic or cathodic peaks respectively. From these slopes, the values of α (transfer coefficient) and k_s (heterogeneous rate constant) can be calculated provided if the number of electrons is known. Otherwise, α can be assumed to be 0.5 and the values of n and k_s can be estimated.

4 Differential Pulse Voltammetry

This technique is used as an electro analytical technique to detect species present in an electrolyte. The detection limit is 10^{-7} M. In this technique the potential is scanned in the form a staircase by super imposing a wave form as depicted in Fig. 5. The wave form can be square or sinusoidal.

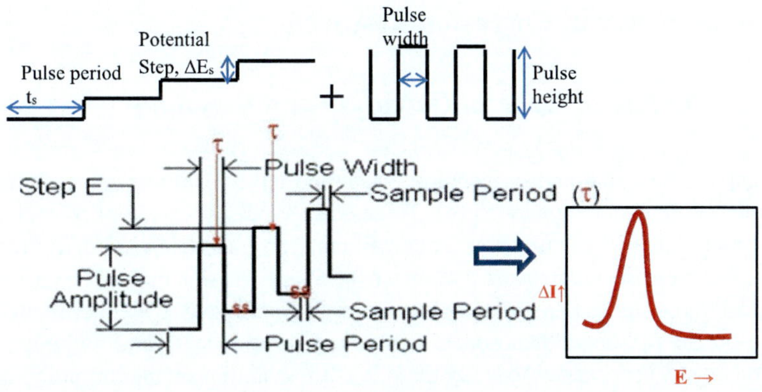

Fig. 5 Potential wave forms and the recorded differential current versus potential plot of differential pulse voltammetry

The experiment records the current just before end of each pulse as shown in point 2 and the current just before applying the pulse at point1. The differential current ΔI is given as I_2-I_1 and plotted as a function of stair case potential. A bell shaped differential current profile is recorded as seen in Fig. 5. The pulse duration, t_p is at least ten times shorter than the period of the staircase wave form, t_s. The differential pulsing reduces the error introduced by double layer changing and other non-faradaic processes. One of the important variations in the differential pulse voltammetry is square wave pulse voltammetry. As the name indicates this technique employs a square wave pulse that is superimposed on a staircase potential ramp. The square wave voltammetry parameters are pulse height (ΔE_p), the stair-case height or potential step (ΔE_s), pulse time (t_p) and cycle period or pulse period (t_s). The ΔE_s is much smaller than $\Delta E_{p.}$ The potential shifts by ΔE_s at start of each cycle which is related to scan rate by the relation: $v = \Delta E_s/2tp = f.\ \Delta E_s$, Where f = frequency = 1/2 t_p. The half wave potential is given by the relation [6, 11]:

$$E_{1/2} = E^o + \frac{RT}{nF}\text{In}\left(\frac{D_R}{D_o}\right)^{1/2} \tag{36}$$

where, D_R and D_O are diffusivities of R and O species, respectively.

The differential peak current Δi_p is given by the relation [13]:

$$\Delta I_P = \frac{nFAC_0^{\infty}D_O^{1/2}}{\sqrt{\pi t_p}}\Delta\varphi_P \tag{37}$$

where $\Delta\Psi_p$ is a dimensionless parameter that depends on $n\Delta E_p$ and $n\Delta E_s$ and tabulated in reference [13].

5 Voltammetry in Corrosion Research

5.1 Evaluation of General Corrosion and Dissoltion

A potential step voltammetry technique was reported to estimate corrosion rate of steel reinforcements in concrete [14]. Three voltage pulses were applied each in the anodic and cathodic directions and the resultant currents were recorded. From the potential-current data, corrosion parameters such as corrosion current density (i_{corr}), polarization resistance (R_P), and tafel slopes were determined. CV experiments were carried out in carbonate/bicarbonate containing solution with and without sulfide addition on steel specimens to study the role of these anions on the formation of iron oxide and iron sulfide films [15]. The CV results clearly indicated three regions at different potentials in the HCO_3^-/CO_3^{2-} solutions at pH 8–10, as depicted in Fig. 6. Region 1 (between −0.92 and −0.75 V_{SCE}) showed formation of Fe(I) intermediare in the form of $FeOH^-{}_{ads}$ leading to formation of $Fe(OH)_2$, region 2 (between −0.8 and −0.6 V_{SCE}) showed an increase in current due to dissolution of Fe^{2+} by complexation with HCO_3-:

$$Fe(OH)_2 + HCO_{3^-} \leftrightarrow FeHCO_{3^+} + 2OH^- \tag{38}$$

Region 3 (at potentials more positive than -0.3 VSCE) indicated film breakdown as seen from the CV. Addition of sulfide (10^{-5} M) to carbonate/bicarbonate increased the anodic current and resulted in CE or EC type reactions such as:

$$SH^- + FeOH^-{}_{ads} \rightarrow FeSH_{ads} + OH^- \tag{39}$$

$$FeSH^-{}_{ads} \rightarrow FeSH_{ads} + e^- \tag{40}$$

$$FeSH_{ads} + HCO_3{}^- \rightarrow FeHCO_3{}^+ + SH^- + e^- \tag{41}$$

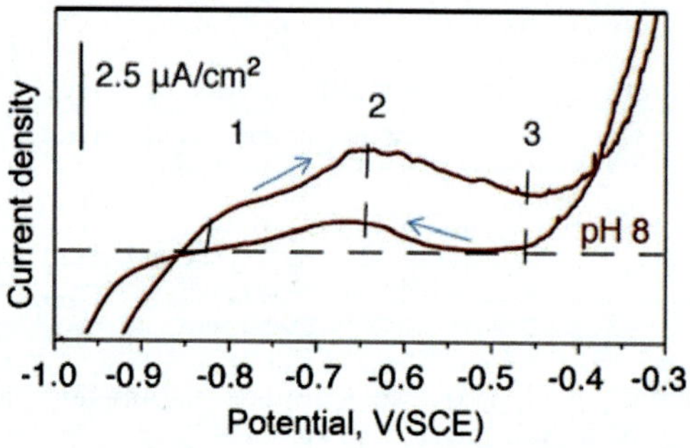

Fig. 6 Cyclic voltammogram of a carbon steel specimen in 0.2 M carbonate/bicarbonate solution of pH 8 Ref. [15]

In another study, CV was used for investigating corrosion of Al alloy 6061 in 3.5% NaCl solution containing 100–500 ppm of $CeCl_3$ [16]. The potential window was between −0.4 and 1.8 V_{SCE}, and the scan rate was 5 mV/s. The CV results indicated the role of cerium salts on the inhibition of pitting corrosion. Sometimes, the term cyclic voltammetry is mentioned as synonymous to cyclic polarization which is misleading [17]. Typically, cyclic voltammetry is carried out at faster scan rates (> 10 mV/s), whereas cyclic polarization is performed at a slower scan rate (between 0.167 and 1 mV/s). The cyclic polarization can be interrupted if the reverse scan current profile intercepts the forward scan indicating repassivation of pits that occur below a pitting protection potential. Cyclic voltammetry was carried out to investigate the role of Fe3+ ions on the reduction of concentrated nitric acid [18]. Nitrate ions are oxidizing in nature when present in an environment that shifts the corrosion potential to more positive values. Generally, reduction of nitrate ions occurs by the following reactions:

$$NO_3- + 2H^+ + 2e^- \leftrightarrow NO_2- + H_2O \tag{42}$$

$$NO_2- + 7H^+ + 6e^- \leftrightarrow NH_4OH + H_2O \tag{43}$$

When Fe^{3+} is present in the solution, the following reduction reactions (electrochemical and chemical) were proposed with the help of CV results:

$$Fe^{3+} + e^- \leftrightarrow Fe^{2+} \tag{44}$$

$$2Fe^{2+} + 3H^+_{(aq)} + NO_3- \leftrightarrow 2Fe^{3+}_{(aq)} + HNO_{2(aq)} + H_2O \tag{45}$$

$$HNO_{2(aq)} + H^+{}_{(aq)} + e^- \leftrightarrow NO_{(ads)} + H_2O \tag{46}$$

$$2NO_{(ads)} + H^+{}_{(aq)} + NO_3-{}_{(aq)} + H_2O \leftrightarrow 3HNO_2 \tag{47}$$

$$2NO_{(ads)} + 2H^+{}_{(aq)} + 2e^- \leftrightarrow N_2O_{(ads)} + H_2O \tag{48}$$

In another study cyclic voltammetry was used for understanding the dissolution behavior of fluoride doped tin oxide coating specimens [19]. Fluorine doped tin oxide (FTO) is a transparent conducting oxide which is widely used in different applications ranging from sensors, digital displays, smart window coatings, catalysts supports, and electrode substrates for various electrochemical applications. Since FTO is used as an electrode as well as a catalyst in various electrochemical applications, stability against corrosion is an important issue. FTO is considered to have a chemical stability greater than tin oxide due to fluorine doping which results in ionic type bonding between Sn-F [20]. Cyclic stability of the FTO was evaluated by carrying out cyclic voltammetry from -2 to 2 V_{RHE} in 0.1 M HCl, 0.1 M NaCl and 0.1 M NaOH electrolytes at

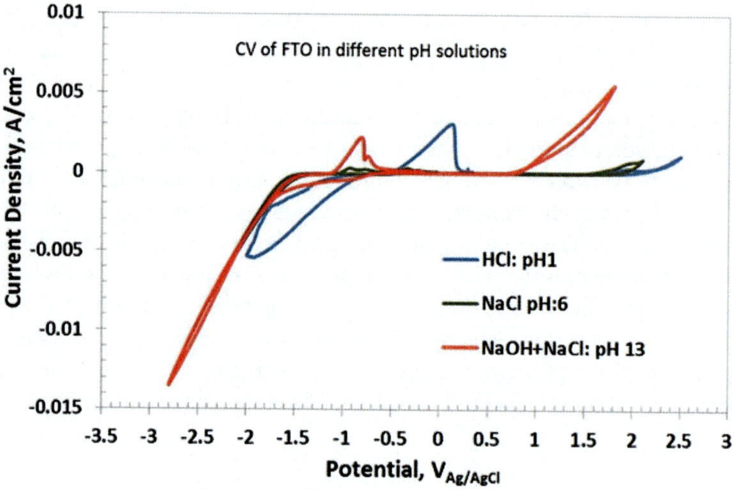

Fig. 7 Cyclic voltammograms of FTO in chloride solutions at different pH conditions

different scan rates between 0.01 and 1 V/s. The number of CV cycles continued until the recorded peak current density was less than 10 μA/cm^2. This type of test is referred to as accelerated stress test. Figure 7 shows the first cycle of CV tests carried out using FTO in different pH conditions at a scan rate of 10 mV/s. The cathodic sweep showed strong current waves corresponding to hydrogen evolution and reduction of SnO_2 to Sn as well. The reverse sweep showed anodic peaks prior to oxygen evolution. These anodic peaks were attributed to oxidation of Sn to tin oxide or oxyhydroxide. It is noted that these anodic peaks were predominant in HCl solution as well as in the alkaline solutions implying the oxidation of a significant amount of Sn. On the other hand, no such predominant oxidation peaks were recorded in the NaCl solution. These observations indicated that reduction of SnO_2 was not significant during cathodic polarizations in the neutral solution.

5.2 *Evaluation of Localized Corrosion*

Cyclic voltammetry has been employed to investigate the localized corrosion behavior of metallic materials in chloride containing solutions. Dayeb and Keera [21] investigated pitting behavior of carbon steel in simulated offshore environments by using cyclic voltammetry (CV). The CV was carried out by scanning the potential from −2.0 to −0.1 V(SCE) in 0.1–0.3 M NaCl solution with the addition of 0.01–0.04 M Na_2S, 0.02–0.1 M Na_2SO_4 or 0.05–0.2 M $NaHCO_3$ at 30 °C. The scan rate was 10 mV/s. Figure 8 schematically illustrates the cyclic voltammogram. An anodic peak current (A1) was observed at around -0.8 VSCE during the anodic scan which was attributed to oxidation of Fe to Fe(II). A second anodic current wave

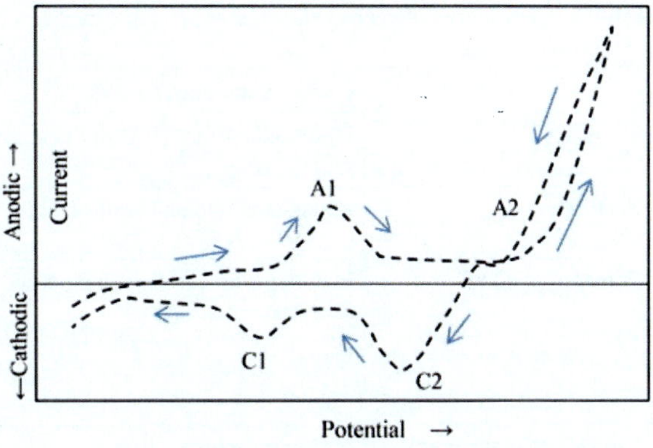

Fig. 8 Schematic cyclic voltammogram of carbon steel in chloride containing solution to evaluate pitting corrosion

(A2) initiated at potentials more positive than −0.4 which was attributed to initiation of pitting. When the potential scan continued beyond the pitting potential, the current increased monotonically indicating growth of pits. Reversal of potential scan revealed hysteresis indicating slower passivation of pits, and two cathodic current peaks were reported [21]. The peak C2 indicated reduction of corrosion products at the pits, and the cahodic peak C1 was attributed to the reduction of iron oxide. Observation two cathodic peaks during the reverse scan implied occurrence of pitting. In the absence of pitting, only the C1 peak was recorded.

It should be noted that the cyclic polarization studies are generally employed to evaluate the localized (pitting) corrosion of stainless alloys. ASTM standard G61 specifies the procedure for conducting cyclic polarization. Cyclic voltammetry (CV) differs from cyclic polarization (CP) in several ways. Table 1 lists the differences between CV and CP.

5.3 *Evaluation of Passivation and Oxide Films*

Cyclic voltammetry has been widely employed to study the passivation behavior of metals and alloys [22–24]. Li et al., investigated the passivation behavior of low carbon steel in acidic solutions containing nitrite as an inhibitor [22]. In addition to potentiodynamic polarization technique, cyclic voltammetry studies were carried out. The CV results delineated the oxidation reaction ($Fe(II) \rightarrow Fe(III) + e^-$), and cathodic reactions ($2NO_2^- + 6H^+ + 4e^- \rightarrow N_2O + 3H_2O$; $Fe(III) + e^- \rightarrow Fe(II)$; $Fe(II) + 2e\text{-} \rightarrow Fe$; and $2H^+ + 2e^- \rightarrow H_2$). Multiple cycling caused either decrease or increase in the anodic peak currents depending on thickening of the passive layer or breakdown of the passive layer. Eliyan and Alfantazi [23] evaluated passivation of high strength pipe steel (API-X100) in carbonate/bicarbonate containing solution using cyclic voltammetry. The potential was scanned between −1.2 to 1.2 V_{SCE} at a slow scan rate. Two anodic peaks were recorded at −0.7 and −0.35 V_{SCE} during the forward scan and a single cathodic peak at −0.7 V_{SCE}. The first anodic peak was

Table 1 Difference between cyclic voltammetry and cyclic polarization tests for evaluating pitting corrosion behavior

	Cyclic voltammetry (CV)	Cyclic polarization
V-I plot	Potential–current (V-I) in linear scale	Semi-log scale (V-log I)
Scan rate	Faster scan rate and varies from 1 to 100 mV/s	Slow scan rate (0.166 mV/s)
Potential window	Includes both cathodic and anodic potentials	The scan starts from OCP and goes in the anodic direction. A small window of cathodic polarization may be included to remove the air formed film
Number of scans (cycles)	Multiple cycles	One cycle
Pertinent data obtained	Oxidation and reduction currents, peak potentials, pitting potential	Passivation current, film breakdown potential, and pitting protection potential No reduction current is recorded
Materials evaluated	Steel, Al, and Mg alloys	Alloys that form a stable passive film in chloride solutions such as stainless steels, nickel base alloys

associated with dissolution suppression by passive films containing $FeCO_3$. Increase in number of cycles of CV did not affect the first peak but the magnitude of second anodic peak and the cathodic peak increased [23]. Thickness of passive film and its reactivity could be evaluated based on the peak current values.

CV is employed for evaluating the electron density of passive films [25]. Chemical capacitance (C_μ) of the passive film can be calculated from the CV results using the relation:

$$C_\mu = \frac{\int I.dt}{\Delta V} = \frac{I}{\left({}^{dV}/_{dt}\right)} \tag{49}$$

where, I = current recorded during the CV, and dV/dt = scan rate. The distribution of electronic states in the passive film or density of states (DOS), g(E), can be related to the chemical potential by the relation [26]:

$$C_\mu = q^2 \cdot g(E) \tag{50}$$

where, C_μ = capacitance (farad/cm^2), q = elementary charge expressed (coulomb), $g(E)$ = density of states (cm^{-2} eV^{-1}). The shape of the V- I profile of the cyclic voltammogram could represent the distribution of the electronic states as a function of energy level.

5.4 Evaluation of Coating Performance

The CV has been used for coating of thin films on metallic substrates for corrosion protection [27, 28], and evaluating performance of the coatings as well [29]. Wang and Lin [28] studied the performance of nanostructured TiN coating on stainless steel substrate as durable electrode for Electro-Fenton process systems. Linear sweep voltammetry was used to determine the kinetics parameters of oxygen reduction reaction. The rate of reduction or oxidation of metallic ions in the electrochemical system of the Electro-Fenton process was studied by the cyclic voltammetry.

Several types of organic coatings are used for corrosion protection. Electrochemical stability of the organic coatings for various applications could be related to the electronic orbital energy levels of the organic materials. The highest occupied molecular orbital (HOMO) represents the highest energy level of the electron or energy required to extract an electron from a molecule. The lowest unoccupied molecular orbital (LUMO) is the minimum energy level available in the unoccupied orbital or energy necessary to inject an electron to a molecule. The LUMO and HOMO energy levels can be determined from the CV results based on the take-off potentials of reduction or oxidation waves, respectively [30]. The difference between the incipient potentials of oxidation and reduction waves gives an estimate about the optical energy band gap of the material. Since most of the coating materials will have an energy band gap larger than 1.23 V, aqueous solutions may not be used for evaluating the HOMO–LUMO energy levels because water will decompose before oxidation or reduction of the organic molecule takes place. Non-aqueous electrolytes with a wide stability window need to be used.

6 Photo Spectroscopy

Photo spectroscopic techniques are used along with electrochemical methods to understand the corrosion behavior of a material including dissolution, passivation and inhibition. The photo spectroscopic techniques can be used in-situ or ex-situ to analyze the surface of the specimens or corrosion products. These techniques include photo absorption spectroscopy (using ultraviolet–visible (UV–Vis) and infrared light sources), and vibrational spectroscopy such as Raman Spectroscopy and Fourier Transformed Infrared (FTIR) spectroscopy. In this chapter, the discussion will focus on UV–Vis photo absorption spectroscopy.

6.1 UV–Vis Absorption Spectroscopy

In photo absorption spectroscopy, light beam of power ‘P’ passes through a liquid solution with a thickness of b cm. The solution has absorbing species of concentration

'C'. Due to the interaction between photons and species, the light beam is attenuated from P_o to P as seen in Fig. 9. The transmittance 'T' of the solution is considered as the ratio between transmitted power 'P' and the incident power P_o, given as $\boldsymbol{T} = P/P_o$. Transmittance is expressed as percentage. The absorbance '$\boldsymbol{A}$' of a solution is given by the relation: $\boldsymbol{A} = -\log_{10}\boldsymbol{T} = \log P_o/P$. Further, the absorbance can be related to the thickness (*b* in cm) of solution through which the light passes, and the concentration '*c*' of the absorbing species in the liquid and which is given by the relation: $\boldsymbol{A} = abc$, where, a = a proportionality constant denoted as 'absorptivity'. If the absorptivity has a unit of $\text{Lcm}^{-1}\ \text{mol}^{-1}$, it is termed as molar absorptivity and denoted as 'E' and $\boldsymbol{A} = \varepsilon bc$.

The absorbance occurs by excitation of electrons to higher energy levels when the electromagnetic radiation interacts with the absorbing species. The types of electronic transitions occurring during absorption spectroscopy include $\sigma \rightarrow \sigma^*$, $\text{n} \rightarrow \sigma^*$, and $\text{n} \rightarrow \pi^*$ as schematically shown in Fig. 10. In addition to the electron transition from bonding orbitals to antibonding orbitals or non-bonding orbitals to anti bonding orbitals, which usually occurs in organic molecules, other absorption involves electronic transitions of *4f* and *5f* electrons in case of lanthanides and actinides, *3d* and *4d* electrons in the first and second transition metals, and finally the charge transfer absorption.

The valence electrons of organic or inorganic compounds excited to higher energy levels by absorbing electromagnetic energy and therefore will respond to photo spectroscopy. The electrons participating in formation of single bonds require high energy for excitation in the vacuum ultraviolet region ($\lambda < 185$ nm). The $\sigma \rightarrow \sigma^*$ transition also requires a large energy as schematically noted in Fig. 9. The $\sigma \rightarrow \sigma^*$ absorption peaks of organic compounds such as methane and ethane occurs at 125 and 135 nm

Fig. 9 Schematic illustration of light of power Po passing through a solution and getting attenuated to P

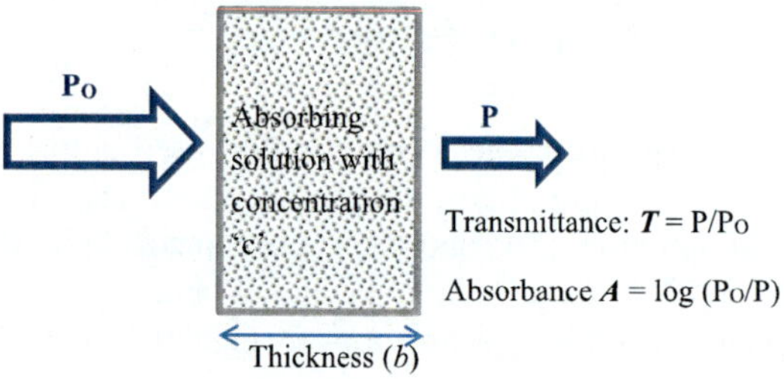

Fig. 10 Electron transitions during optical absorbance

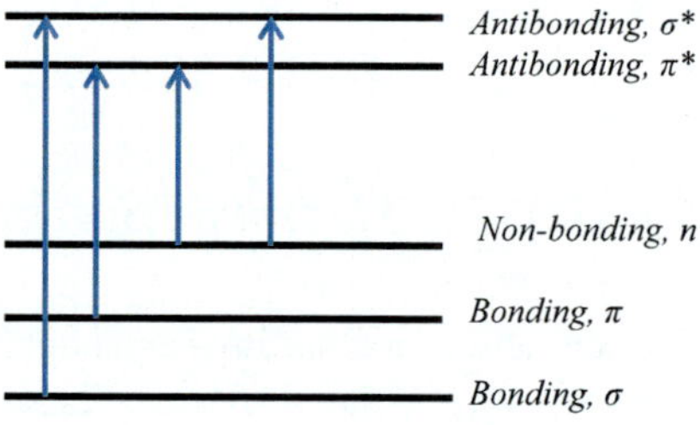

wavelengths, respectively. The $\sigma \rightarrow \sigma_o^*$ transition does not occur in the ordinary ultraviolet wavelengths.

The n $\rightarrow \sigma^*$ transition occurs by transition of unshared electron pairs (non-bonding electrons) to antibonding orbitals by absorbing photons with wavelength of 150–250 nm. Type of atomic bond and structure of the molecule determine the energy required for n $\rightarrow \sigma^*$ transition. The molar absorptivity based on this transition will be in the range of 100–3000 Lcm^{-1} mol^{-1}.

The n $\rightarrow \pi^*$ and $\pi \rightarrow \pi^*$ transitions require lower energy than that of n $\rightarrow \sigma^*$. The energy required is in the range of 200–700 nm within the UV–Vis spectral region. The molar absorptivity for the n $\rightarrow \pi^*$ is in the range of 10 -100 L cm^{-1} mol^{-1} whereas the $\pi \rightarrow \pi^*$ transition results in very high molar absorptivity in the range of 1000–10,000. The π orbitals are available through unsaturated functional groups of organic compounds.

The absorption peaks of transition metal ions are broad in contrast to sharp peaks observed for lanthanides or actinides. The absorption maxima associated with d-d transitions of transition metal ions occur in the wavelengths of 290–1370 nm. The wavelength depends on the ligand field strength. The wavelength decreases with increase in the ligand field strength in the order $I^- > Br^- > Cl^- > F^- > OH^- > H_2O > SCN^- > NH_3 > CN^-$.

The molar absorptivity is larger (>10,000 Lcm^{-1} mol^{-1}) when charge transfer absorption is operating. Fe(III)/thiocyanate complex exhibits charge transfer absorption through electron transfer from thiocyanate ion to Fe(III) ion, forming a neutral SCN radical and Fe(II).

6.2 *Vibrational Spectroscopy*

Other spectroscopic techniques such as FTIR and Raman differ from the UV–V is spectroscopy even though all of these techniques utilize photons are probes. In the vibrational spectroscopy (including FTIR and Raman), the electromagnetic radiation interacts with the vibration of the atoms and molecules that causes either absorption (FTIR) or scattering (Raman Spectroscopy). These techniques also use low energy photons in the wavelengths of visible to infrared region.

Atoms and/ or molecules in a material vibrate at their equilibrium positions when the temperature is higher than absolute zero. The vibration energy of molecules can be modeled using a mass less spring model, and the frequency of molecular vibration is given by the relation:

$$f = \frac{1}{2\pi}\sqrt{\frac{K}{\mu}} \tag{51}$$

where K = bond strength, μ = equivalent mass of the atoms under consideration

$$\mu = \frac{m_1 m_2}{m_1 + m_2} \tag{52}$$

In vibrational spectroscopy, the energy is expressed in terms of wave number ($\overline{\nu}$) using the relations:

$$\mathrm{E} = \mathrm{h}\nu\frac{hc}{\lambda} = \mathrm{hc}\overline{\nu} \tag{53}$$

where h is Planck's constant 6.626 × 10^{-34} J.s, c = velocity of light 3 × 10^{10} cm/s, the term 'hc' can be considered as a conversion constant ≈2 × 10^{-23} J.cm.s.

Typically, the vibrational spectroscopy is carried out at wave numbers ranging from 200 to 4000 cm^{-1}. The crystalline solids exhibit lattice vibrations in the range of 20–300 cm^{-1} that are sensitive to temperature. Molecular vibrations are not as sensitive to temperature changes as lattice vibrations. In order for an effective interaction the frequency the photon should match with the frequency of molecular vibration. In such case, the energy of the photon is absorbed by the molecular vibration and the molecule is excited to the next vibrational quantum number. In order to be infrared active, the vibration of molecules should result in non-zero net dipole moment which is mathematically represented as:

$$\left(\frac{\partial \mu}{\partial x}\right)_{x=0} \neq 0 \tag{54}$$

where $\frac{\partial \mu}{\partial x}$ is gradient of dipole moment across the vibrational displacement, $\mu = q.l$, and *dx* is vibrational displacement. The dipole moment can be induced by the electric field of the interacting light which is an electromagnetic wave. Therefore, the major difference between absorption spectroscopy and infrared vibrational spectroscopy is the absorbance mechanism. In both the techniques the photon energy is absorbed. In UV–Vis spectroscopy the absorbance results in electron transition from on orbital to another (anti bonding) orbital. On the other hand, in IR spectroscopy the absorbance results in vibrational excitation when the net dipole moment is nonzero.

6.2.1 Raman Spectroscopy

Raman Spectroscopy is based on the inelastic scattering of the electromagnetic radiation by the molecules. This technique employs a single wavelength light source. The scattered light has a frequency change which is called Raman Shift caused by excited molecular vibration. In order to be Raman active, the vibration of a molecule should = cause changes in the polarizability is related to the electric dipole moment by the relation:

$$\mu = \alpha.E \tag{55}$$

where, α = polarizability, E = electric field.

Polarizability can be defined as the ability to induce a dipole moment or tendency to move charge by an electric field. Large molecules are more polarizable covalently bonded material is more polarizable than ionically bonded ones. On a similar note, metals are not polarizable by the electric field available through the electromagnetic radiation. Here, the electric field of the electromagnetic radiation participates in the electric dipole moment. The condition for Raman activity can be expressed as:

$$\left(\frac{\partial \alpha}{\partial x}\right)_{x=0} \neq 0 \quad (56)$$

7 UV–Vis Spectroscopy in Corrosion Research

In order to understand the surface chemistry of the FTO during electrochemical conditioning, *in-situ* UV–Vis photo absorbance spectroscopy was carried out (Spectro 115E, 350–1050 nm, Gamry Instruments). The experimental arrangement is shown in Fig. 11. The electrochemical cell consisted of 0.8 cm wide FTO as the working electrode, a Pt-wire loop counter electrode, and a miniature Ag/AgCl reference electrode. These electrodes were assembled in a 1 cm square quartz cuvette having a volume of 10 ml. The photo absorbance spectra were collected by conditioning the FTO at

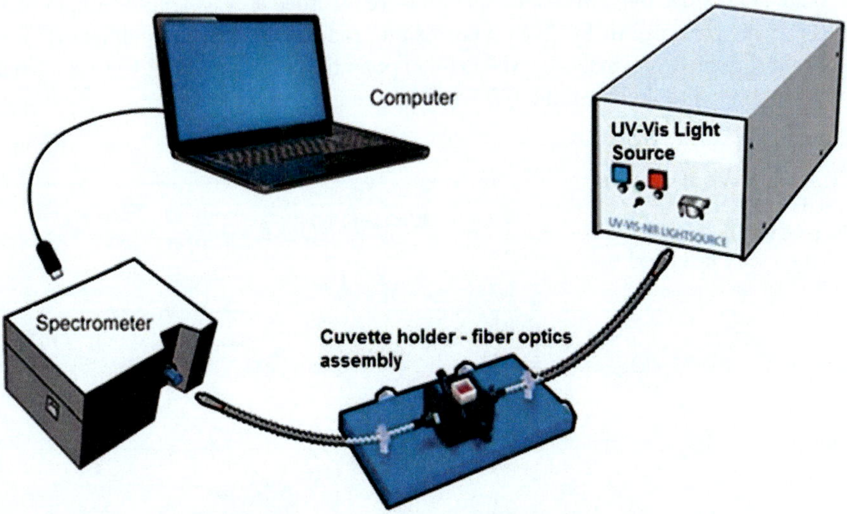

Fig. 11 Experimental setup of UV–Vis spectroscopy in which the clear quartz cuvette is configured to also be an electrochemical cell. The electrochemical cell is created in such a way that the beam entering the cell will interact with the quartz front layer, solution, solution/working electrode (FTO) interface, and through the quartz back layer

different potentials determined from the cyclic voltammetry (CV). During the CV, first the potential was scanned in the cathodic direction from OCP to −2.0 $V_{Ag/AgCl}$ (pH 1.1) or −2.8 VAg/AgCl (pH 13.1) at a scan rate of 10 mV/s. After reaching the cathodic potential limit, the potential scanning continued in the anodic direction to a maximum potential of 2.5 $V_{Ag/AgCl}$ (pH 1.1), or 1.8 $_{VAg/AgCl}$ (pH 13.1).The cyclic voltammetry data allowed us to gather information regarding potential windows and polarization potentials for the potentiostatic *in-situ* UV–Vis experiments. The potentials of potentiostatic (PS) experiments were determined from the cyclic voltammetry (CV) peak currents and shoulders. The *in-situ* UV–Vis absorbance spectra were collected by conditioning the FTO specimen at different constant potentials. Each experiment started by applying the highest cathodic potential and moving to the next desired potential in the anodic direction. Each potential was applied for 120 s and optical absorbance spectra were collected at every 30 s. Typical results are shown in Fig. 11.

Functional groups in a molecule can be identified from the wavelengths of absorption peaks by correlating with the type of bond. In this investigation, the optical path includes the electrolyte, FTO/electrolyte interface, and the bulk of FTO. The absorption spectra contain information of all these components. However, the variation in the spectra between different potentiostatic conditions of a given electrolyte system could be attributed to the changes occurring at the FTO/electrolyte interface. Figure 12 shows the absorption spectra of FTO in different pH solutions under cathodic polarization. The spectra were obtained under reducing conditions. The optical absorbance decreased as the wavelength of light increased. The UV absorbance peaks of $SnCl_2$ were reported to occur at 236 and 323 nm with a shoulder tapering off at around 350 nm [31]. Since the absorbance spectra were recorded at wavelengths longer than 350 nm, the presence of $SnCl_2$ as a corrosion product could not be confirmed. The lower absorbance than that observed in other potential conditions could be associated with cathodic dissolution of the FTO that resulted in a porous film.

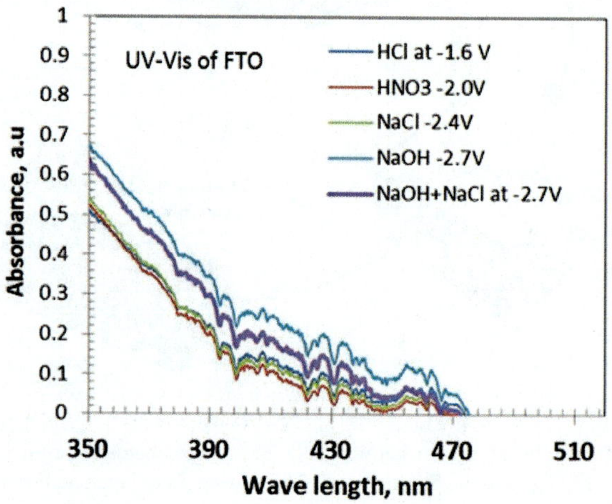

Fig. 12 UV–Vis absorbance spectra of FTO specimens exposed to different pH solutions and conditioned at cathodic potentials

The absorbance of FTO in nitric acid solution was similar to that observed in the HCl solution. This result suggests that polarization of FTO in cathodic direction in the nitric acid solution caused significant dissolution. The UV absorption of NO_3^- ion occurs at 200 nm (owing to $\pi \rightarrow \pi^*$ transition) and near 300 nm ($n \rightarrow \pi^*$ transition) [32]. These absorption peaks are not shown here due to limitation of the instrument used in this study. The absorbance spectra collected in NaCl solution are also similar to those observed in the nitric acid solution. The absorbance spectra collected in 0.1 M NaOH solution at -2.7 $V_{Ag/AgCl}$ were similar to that of acidic pH condition. At this cathodic potential a significant reduction current was recorded in the CV. The reduction of SnO_2 could occur in two steps: first step as Sn(IV) → Sn(II); and second step Sn(II) → Sn(0) [33]. Conversion of Sn to SnH_4 could also occur at low potentials. The high absorbance observed at this potential indicates that no significant change occurred with the specimen in 120 s. When the potential moved to -1.9 $V_{Ag/AgCl}$ (shoulder of the cathodic wave) a significant reduction in the absorbance was observed implying degradation of the specimen. Application of anodic potentials did not change the absorbance behavior significantly from that observed in -1.9 $V_{Ag/AgCl}$. The UV–Vis absorbance peak for SnO in alcohol solutions was reported at around 330 nm [34]. In this investigation, the absorbance data were recorded for wavelengths >350 nm. Therefore, it is not possible to assign the absorbance peaks unequivocally to SnO. However, presence of SnO after electrochemical conditioning could not be ruled out. The absorbance results of FTO tested in 0.1 M NaOH +0.1 M NaCl solution are similar to those observed in the NaOH solution.

UV–Vis spectroscopy was utilized to investigate different mechanism of several complex metallic alloys based on Al-Cr-Fe system. The photo wavelength used were in the range of 190–1200 nm. The calibration was performed using Fe(II) and Fe(III) salts of known concentrations in the range of 0.1 to 10 μmol/L [35]. UV–Vis absorption spectra of Al species could not be observed because Al does not form complexes with H_2O or Cl^- which absorb light in the visible region. On the other hand, Cr–H_2O and Cr-Cl^- complexes could form but have low molar absorptivity. Fe(III) shows an absorption peak at about 340 nm and a steep absorption shoulder at 280 nm with a maximum at 220 nm. On the other hand, Fe(II) did not have a strong finger print as that of Fe(III), however a minor absorption peak could be noted at about 350 nm and another significant peak at about 250 nm. UV–Vis spectroscopy was used for monitoring residues of corrosion inhibitors in oilfields [36]. Not only the structural characterization of inhibitors but also the absorption of organic inhibitors on the corroding surface is usually analyzed by UV–Vis spectroscopy [37].

References

1. R.G. Compton, C.E. Banks, *Understanding Voltammetry*, 2nd edn. (Imperial College Press, 2014), pp. 35–100
2. L. Eberson, *Electron Transfer Reactions in Organic Chemistry* (Springer, Berlin, 1987)
3. J. Franck, Elementary processes of photochemical reactions. Trans. Faraday Soc. **21**, 536–542 (1926)

4. E. Condon, A theory of intensity distribution in band systems. Phys. Rev. **28**, 1182–1201 (1926)
5. L. Redey, D.R. Vissers, Reference electrode development for molten LiCI-KCI eutectic electrolyte. J. Electrochem. Soc. **128**, 2705 (1981)
6. R. Greef, R. Peat, L.M. Peter, D. Pletcher, J. Robinson, *Instrumental Methods in Electrochemistry* (Ellis Horwood, London, 1990), pp. 42–228
7. H. Matsuda, Y. Ayabe, Z. Eectrochem **59**, 494 (1955)
8. J.E.B. Randles, Trans Faraday Soc. **44**, 327 (1948)
9. A. Sevcik, Coll. Czech. Chem. Comm. **13**, 349 (1958)
10. R.S. Nicholson, I. Shain, Anal. Chem. **37**, 178–190 (1965)
11. A.J. Bard, L.R. Faulkner, *Electrochemical Methods: Fundamentals and Applications* (Wiley, New York, 1980)
12. E.J. Laviron, General Expression of the LSP voltammogram in the case of diffusionless electrochemical systems. J. Electroanal. Chem. **101**, 19–28 (1979)
13. J. Osteryoung, J.J. O'Dea, J. Electroanal. Chem. **14**, 209 (1986)
14. J.E. Ramon et al., Cem. Concr. Compos. **110**, 103590 (2020)
15. M. Goldman, C. Tully, J.J. Noël, D.W. Shoesmith, Corros. Sci. **169**, 108607 (2020)
16. M.A. Dayeb, S.S. Abd El-Rehim, H.H. Hassan, A.H. Shaltot, J. Alloys. Compd. **820**, 153428 (2020)
17. J.D. Tucker et al., Met. Mater. Trans A **50A**, 3388 (2019)
18. J. Pelle, N. Grut, B. Gwinner, M.L. Schlegal, V. Vivier, Electrochim Acta **335**, 135578 (2020)
19. A. Korjenic, K.S. Raja, J. Electrochem. Soc. **166**(6), C169–C184 (2019)
20. J. Xu, S. Huang, Z. Wang, Solid State Commun. **149**, 527 (2009)
21. M.A. Deyab, S.T. Keera, Egypt. J. Pet. **21**, 31–36 (2012)
22. X. Li, P. Zhang, H. Huang, X. Hu, Y. Zhou, F. Yan, RSC Adv. **9**, 39055–39063 (2019)
23. F.F. Eliyan, A. Alfantazi, Metall. Mater. Trans. B **48**(5), 2615–2619 (2017)
24. N.A. Al-Mobarak, Chem. Technol. Fuels Oils **48**(4), 321–329 (2012)
25. V.G. Kytin, J. Bisquert, I. Abayev, A. Zaban, Phys. Rev. B **70**, 193304 (2004)
26. B. Klahr, S. Gimenez, F. Fabregat-Santiago, T. Hamann, J. Bisquert, J. Am. Chem. Soc. **134**, 4294 (2012)
27. S.T. Doslü, B.D. Mert, B. Yazıcı, Arab. J. Chem. **11**, 1–13 (2018)
28. Y. Wang, Y. Lin, Nanomaterials **8**, 494 (2018)
29. S. Abaci, B. Nessark, J. Coat. Technol. Res. **12**(1), 107–120 (2015)
30. R.E.M. Willems, C.H.L. Weijtens, X. de Vries, R. Coehoorn, R.A.J. Janssen, Adv. Energy Mater. **9**, 1803677 (2019)
31. L. Jiang, G. Sun, Z. Zhou, S. Sun, Q.I. Wang, S. Yan, J. Phys. Chem. B, **109**, 8774(2005)
32. D. Mayerstein, A. Treinin, Trans. Faraday Soc. **57**, 2104 (1961)
33. E. Matveeva, J. Electrochem. Soc. **152**, H138 (2005)
34. M. Marikkannan, V. Vishnukanthan, A. Vijayshankar, J. Mayandi, J. M. Pearce, AIP Advan. **5**, 027122 (2015)
35. A. Beni et al., Electrochim. Acta **179**, 411–422 (2015)
36. D. Dolezal, T. Bolanca, S.C. Stefanovic, Mat.-wiss.u.werkstofftech, 42 (2011) 229
37. A.P. Samide, I. Bibicu, Surf. Interface Anal. **40**, 944–952 (2008)

The Utilization of Scanning Electrochemical Microscopic (SECM) Technique in Corrosion Monitoring

A. Madhan Kumar

Abstract Various conventional electrochemical corrosion measurements have been deployed to evaluate the corrosion phenomenon on various metallic materials. However, these techniques have been inadequate to elucidate the corrosion process on the metal/solution interface at the micrometer scale with early interpretation. Hence, we are in need to discover the appropriate technique with a longitudinal determination in the microscopic level that might specify the early steps of surface degradation in the actual phase. Recently, several methods, including scanning reference electrode (SRET) techniques, scanning vibrating electrode (SVET), localized electrochemical impedance spectroscopy (LEIS), and scanning electrochemical microscopy (SECM) have been established to overcome the issues with conventional electrochemical techniques. Amongst the methods, SECM has been presented to assess the corrosion phenomenon in early stages at the micrometer scale, as it delivers the electrochemical behavior and surface topographic evidence about the electrochemical reactions at the interface in solution. Owing to their higher spatial resolution and electrochemical sensitivity, the application of SECM has enabled inspection on the metal/electrolyte interface. This method has previously delivered exact exciting evidence regarding the beginning of localized corrosion of numerous metallic materials, the formation of anodized layers on metallic samples, and the illustration of thin inhibitor films on various metallic materials and galvanic corrosion of Zn–Fe cell. Moreover, the usage of SECM to investigate the degradation of protective coatings on various metallic substrates is very promising in investigating the multifaceted reactions happening in the aggressive environment. This chapter intends to demonstrate the fundamentals, different modes briefly, and the feasibility of utilizing the SECM technique in corrosion monitoring research.

Keywords SECM · Corrosion · Coatings · Inhibitors · Passivation

---

A. Madhan Kumar (✉)
Interdisciplinary Research Center for Advanced Materials, King Fahd University of Petroleum & Minerals, Dhahran, Saudi Arabia
e-mail: madhankumar@kfupm.edu.sa

I. Toor (ed.), *Recent Developments in Analytical Techniques for Corrosion Research*,
https://doi.org/10.1007/978-3-030-89101-5_4

1 Introduction

Assessment of electrochemical behavior of liquid/solid, liquid/liquid, and liquid/gas interfaces has been providing significant understandings on the mechanisms and kinetics of various interfacial reactions over the electrode surfaces. Characterizing the electrochemical interfaces has become one of the most interesting prospects in basic and applied research, which is a result of remarkable importance in the growth of wide-ranging technologies and elucidates the complexity of the most fundamental processes. Interminable reflection of the primary stages of the complete electrochemical process and the observation of their connection with the systematic features of the resulting electrochemical interfaces are research areas of abundant concern. Thus, electrochemical techniques discover extensive applications and offer valuable understanding of the mechanism and kinetics of interfacial electrochemical reactions across solid–liquid interfaces. Hence, an abundant effort is dedicated to exploring these concerns to enlighten the understanding of the necessary information about the electrochemical interfaces, which is a vital indication for the development of effective electrochemistry platforms. However, the conventional electrochemical techniques have been accomplished at the regular dimension and offering an approximate spatial electrochemical activity over the surface of the materials. Such evidence delivers slight information about how electroactivity is distributed across a surface and is often inherent in exposing local surface features in micron and nanoscale, especially becoming further integrated into the modern world. Therefore, the need for a highly confined analytical tool proficient for instantaneously monitoring the precise electrochemical activity in the nanoscale dimension is extremely desired [1].

Among the techniques utilized to probe the local electrochemical phenomenon, scanning electrochemical microscopy (SECM) has established to be a useful analytical tool for examining an extensive sort of interfacial phenomenon along with temporal and high spatial resolution [2, 3]. As a scanning probe tool, SECM is constructed according to the changes in the faradaic current when the scanning tip or probe moves on the electrode surface. Engaging scanning tips in the nearby vicinity to the inspected substrate surface permits high-resolution data about electrochemical reactivity or ion fluxes arising at the liquid–liquid and solid–liquid interface to be achieved [4]. Moreover, the preference to get steady-state currents without the typical difficulties, including iR drop, charging current, and convection impacts, permit the ultra-microelectrode (UME) as a scanning tip to explore the full range of electrochemical phenomenon over the surfaces [5]. Thus, SECM is well recognized as a vital tool in numerous fields of fundamental electrochemical research. SECM can deliver quantitative information about the topographic and electrochemical features of a surface, though the ultra-microelectrode's dimension indeed governs the accuracy of the information. SECM has pointedly advanced in a range of modes, and probe types since its first inception in 1989 by Bard groups, and in the variety of applications, and noticeable improvements have been accomplished in the visualizing the distributions of electroactivity through potential or current mapping [6, 7].

This chapter will firstly provide an overview of the fundamentals and working principle with the operations of SECM modes, then the utilization of SECM procedure in the field of corrosion research and, finally, updates the advances and challenges of applying the SECM technique in the intended field.

2 Fundamentals of SECM Technique

2.1 *Basic Principle*

SECM technique is based on the monitoring of the current using a UME tip while it is scanned in nearby vicinity on the surface of a sample that is valuable in gaining information about surface topographic and electrochemical reaction kinetics [8]. In the case of the conductive disk of radius, *a*, in an insulating sheath, if the UME is pointedly away from a substrate, the steady-state diffusion-controlled current is assumed as (1):

$$i_{\mathrm{T},\infty} = 4\mathrm{nFDca} \tag{1}$$

where D denotes the diffusion coefficient of species, c represents concentration, n denotes the number of electrons passed per mole, and F represents the Faraday's constant. Generally, the substrate perturbs the UME probe's electrochemical activity at the vicinity of a surface, which offers facts about the substrate's electrochemical features.

2.2 *Instrumentation*

Figure 1 illustrates the schematic representation of the SECM technique comprising six components, such as UME tip, electrolytic cell, piezoelectric controller, and bipotentiostat attached to the computer. The electrolytic cell has consisted of the horizontally positioned substrate with an auxiliary and reference electrodes, and these are kept in an electrolytic medium with the UME probe placed overhead the substrate. A bipotentiostat accurately quantify and regulate the potential and current at the UME tip and substrate. A highly sensitive piezoelectric controller system contains piezoelectric motors and stepper that can able to regulate the fine and coarse scanning of the UME tip [9]. Further, x and y directions are considered on the substrate's plane, when the z-direction is represented as the general direction to the substrate. The UME can scan in x, y, and z directions, and the current is monitored concerning their coordinates.

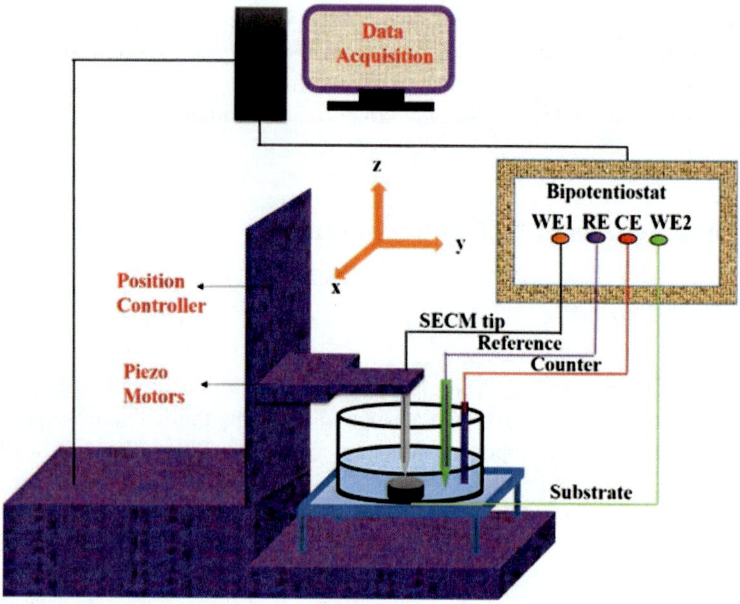

Fig. 1 Schematic diagram of SECM instrumentation

2.3 *Ultra Microelectrodes*

SECM ultra microelectrodes are electrodes with a diameter of the tip in the range 1–25 μm, which consists of attractive features including negligible ohmic drops, less double-layer charging and, high mass transport rates [10]. In general, UMEs are fabricated by sealing noble metal (platinum, gold, and carbon) wires into glass or quartz and polishing a cross-section of the tip to achieve a disk-shaped UME. For a characteristic disk-shaped UME, hemispherical diffusion between the electrode surface and the solution bulk completely controls the mass transport at the interface. The active UME tip's area and shape strongly influence the diffusion process jointly across the UME tip. Previous reports showed that the minor improvements in the SECM scheme and alterations in the UME tips are the vital adaptations depending on the applications and information needed [11, 12].

2.4 *Operation Modes of SECM*

SECM has been employed in a diverse array of applications through different modes of operations. These include the feedback (positive and negative) mode, the redox competition mode, the generation/collection mode, the penetration mode, and the potentiometric and the ion transfer mode. Further, the selection of SECM mode

Table 1 Summary of redox mediators and the relevant parameters

Metal/Alloy	Redox mediators	Mode of SECM
Carbon steel	FcMeOH, Fe^{2+}, H^+ and $Fe(CN)_6{}^{4-}$	Feedback, SG/TC and Redox competition
Stainless steel	FcMeOH, DO, OH^-, Cl^-, Fe^{2+}, H^+ and $Fe(CN)_6{}^{4-}$	Feedback, SG/TC, TG/SC and Redox competition Potentiometric
Magnesium	H_2, FcMeOH, DO, Mg^{2+} and OH^-	Feedback, SG/TC, redox competition and AC-SECM
Titanium	FcMeOH, $Fe(CN)_6{}^{4-}$ and Ti^{3+}	Feedback, SG/TC
Nickel	Ni^{2+}, Fe^{2+}, OH^- and Cr^{2+}	Feedback, SG/TC, redox competition and AC-SECM
Zinc	Zn^{2+}, DO	SG-TC, redox competition, Potentiometric
Copper	Cu^{2+}	SG/TC, feedback
Aluminum	FcMeOH, DMAFc, $Fe(CN)_6{}^{4-}$, I^-, Quinones(C), NB, TCNQ, H_2, DO, OH^- and Cl^-	Feedback, SG-TC, redox competition and potentiometric
Zirconium	FcMeOH	Feedback

depends mainly on the source of the electrochemical species contributing to the redox reactions accountable for the faradaic currents detected at the UME tip. Table 1 summarized the utilized redox mediators and the potential values to achieve their diffusion-limited regimes.

2.4.1 Feedback Mode

Frequently utilized SECM mode is the feedback mode in which only the UME tip current is measured. During the feedback mode, the current monitored at the UME tip is disturbed by the existence of a substrate at nearby vicinity by obstructing the diffusion of solution species towards the UME tip (negative feedback) and by regenerating the O at the substrate's surface (positive feedback). This phenomenon allows us to describe both conducting and insulating substrates and elucidating the electrochemical reactivity and the potential imaging of surfaces.

While the UME probe is positioned nearby the conducting substrate, the formed O species are diffused towards the substrate and be reduced to R. After the reduction reaction, the flux of R towards the UME probe is increased, and $i_T > i_{T,\infty}$, which designated as "positive feedback" (Fig. 2). In the insulating substrate, the tip produces species, O, does not respond, $i_T < i_{T,\infty}$. The observed reduction in current is defined as "negative feedback" (Fig. 2).

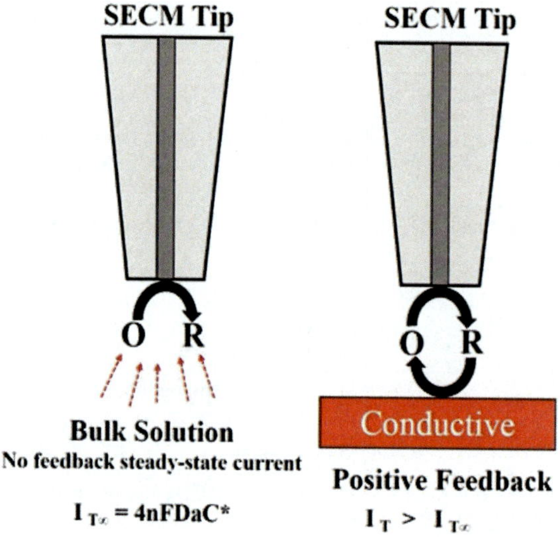

Fig. 2 Feedback modes of SECM

2.4.2 Generation/Collection Mode

Generation and collection (G/C) mode is another standard operating mode of SECM and its own more sensitivity than other modes of the SECM. In the G/C mode, both the UME probe and substrate are utilized as a working electrode; one produces electrochemically active species that can be consumed by the other electrode.

This mode is usually performed in two different versions, such as the tip generation/substrate collection (TG/SC) mode and the substrate generation/tip collection (SG/TC) mode, displayed in Fig. 3. In SG/TC mode, the substrate produces the electroactive species, which can be consumed by the SECM probe. In this

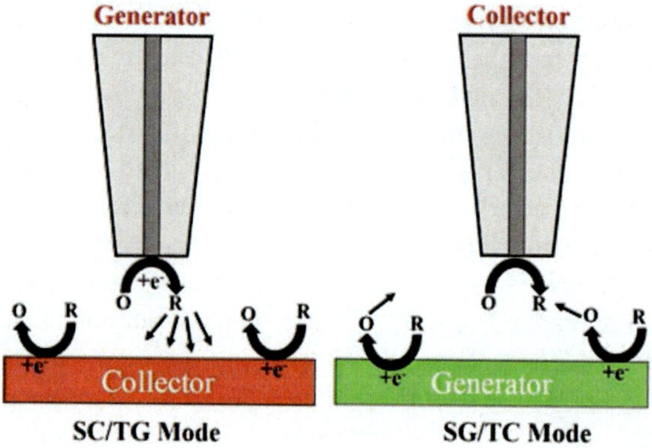

Fig. 3 Generation/collection mode of SECM

mode, the testing electrolyte primarily has redox species that are not yet measurable. When these redox species were oxidized/reduced by the substrate, then the oxidized/reduced forms can be monitored by the biased SECM probe. Whereas, in the case of TG/SC mode, the UME produces electroactive species, which can be identified at the substrate [13, 14]. In general, SG/TC mode is predominantly employed in monitoring the chemical flux or concentration profiles from a sample whereas, the TG/SC is utilized to elucidate the reaction kinetics or achieve alterations to the substrate. Despite the practical usages in enzymatic and corrosion research, this G/C model has few minor shortcomings, including less collection efficiency, absence of steady-state at the large substrate, and interference between the substrate and tip reaction [15, 16].

2.4.3 Redox Competition Mode

In the RC mode (Fig. 4), the substrate and the UME tip are polarized at the specific potential to make competition for the same redox species in solution, and current is only recorded at the UME tip.

While the SECM probe is moving at constant height over the substrate, the measuring tip current is always constant above the inactive area. However, it reduces if the probe is over the active area because of the consumption of redox species at both the SECM probe and the sample [1]. The decrease in current is directly associated with the surface reactivity of the substrate. Though RC mode is less common than other modes, it still plays a vital role in studies of corrosion, surface catalytic, and cell respiration activity. Compared with the G/C mode, the RC mode has proper lateral resolution and higher sensitivity due to the diminished background current without restrictions on the substrate size and the activity of the support [11].

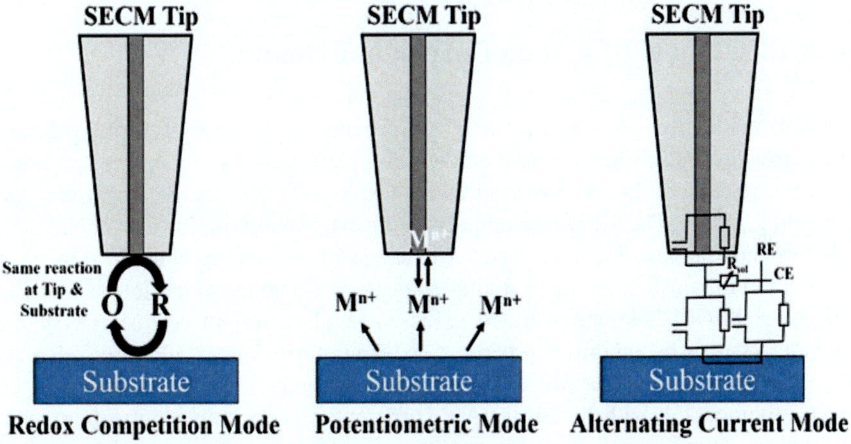

Fig. 4 SECM mode of redox competition, potentiometric and alternating current

2.4.4 Potentiometric Mode

In contrast to the modes mentioned above (instead of measuring current), potentiometric mode detects a local potential between an ion-selective electrode and a reference electrode. Analogous to the SG/TC mode, as the substrates generated electrochemically active species (Fig. 4), a redox reaction does not occur at the UME tip to produce a faradaic current [15]. This mode's essential benefits include higher selectivity, sensitivity, and capability to identify non-electroactive species. Since there are no faradaic reactions involved, the species' concentration with their oxidation state continues to be the same during the scanning. Besides, the detected local potential is directly associated with analytic activity. Moreover, the distance between the probe and substrate slightly influences the resultant signals, and hence the distance is less significant in comparison with other SECM modes [16].

3 Applications of SECM

Almost three decades after the first introduction by Bard et al., the SECM technique has found significant achievements for a wide range of applications [17, 18]. Though the basic theoretic and experimental principles of SECM were chiefly discovered by Bard and co-workers, between 1986 and 1994, the first commercial SECM instrumental setup was commercialized in 1999 by CH instruments. SECM can monitor the chemical reactivity of almost any interface leading to an extensive range of applications. Up to this date, the number of publications associated with SECM has considerably raised, which witnessed its widespread applications in solar cells, water splitting, batteries, and biological fields [15].

4 Utilization of SECM on Corrosion Research

Local variances in electrochemical activity associated with microcells forming during the corrosion process have been predicted by in situ scanning micro electrochemical measurements, hence providing enhanced spatial information regarding the corroding system. The SECM technique is being extensively employed in elucidating the corrosion process due to its capacity to assess localized corrosion phenomena at the micron dimension. As the initial attempt to elucidate the corrosion of metallic materials, SECM has demonstrated extensive applications in corrosive systems, including corrosion inhibitors, coatings, galvanic corrosion, etc. Figure 5 depicts the various applications of SECM on corrosion research. Though SECM cannot be directly comparable with the SEM or AFM observations concerning high topographic resolution, it can be a valuable analytical technique for observing the variety of electrochemical reactivity of the metallic samples on a scale ranging from sub-micron to nanometer dimension recently [11]. Moreover, vital evidence regarding earlier

Fig. 5 Schematic representation of applications of SECM on corrosion research

stages of the corrosion phenomenon is obtained by altering the operation modes of SECM experimental approaches.

To eliminate the possibility of interference of utilized redox mediators with the corrosion reactions during the SECM monitoring, it is suggested to utilize the available ionic species in the investigated system as redox mediators. For instance, the released Fe (II) ions from Fe-based alloys could be monitored at the SECM probe, and variation in the corrosion process of steel substrate can be examined [19]. Notably, the released Fe (II) ions from the anodic reaction of steel substrates could be sensed at the SECM tip by their oxidation to iron (III). The mentioned route can be positively engaged to monitor the localized corrosion on iron-based materials, and the breakdown of passive oxide layers and consequent metastable pitting of stainless steel [20–22]. Consecutively, it is also recommended to utilize another electrochemically-active species such as dissolved oxygen as a redox mediator in the testing electrolyte. In this situation, a competition exists between the SECM probe and the substrate to consume the DO at their cathodic regions. Hence, the necessity for the inclusion of external redox mediator is eliminated.

4.1 *Corrosion of Metals and Alloys*

SECM has been utilized in potentiometric mode to evaluate Mg alloys' corrosion behavior by monitoring the Mg^{2+} ions and pH, and also to assess the homogeneity of surface modifications using feedback mode (Fig. 6). SECM technique can provide qualitative information about the hydrogen evolution reaction (HER), though it is not considered a diffusion-controlled reaction [23]. The utilization of the SECM technique to evaluate Mg's corrosion behavior is limited due to the evolution of H_2 bubbles, which make it challenging to place the SECM tip near the sample surface. In general, the SECM approach curve could be used for regulating the distance between

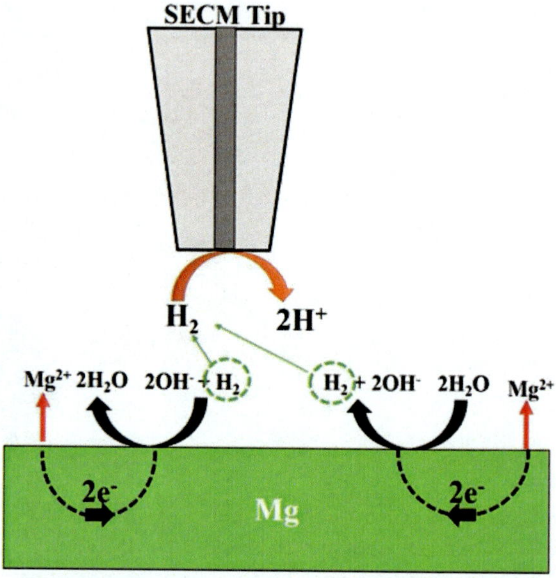

Fig. 6 Hydrogen evolution measurements using the SG/TC mode of SECM

the tip and substrate. However, approach curves cannot be efficiently employed in positioning tip due to accumulated H_2 gas over the corroding Mg surface. Tefashe et al. [24] evaluated the hydrogen evolution reactions on the corroding Mg alloy by the SG/TC mode and reported that quantitative SECM approach curves could be achieved by adequate controlling of corroding media and exposure time. Dauphin-Ducharme et al. [25] compared the elemental composition and surface morphology of sand case AM50 substrates with the local HE from the obtained SECM results.

Jamali et al. [26] examined the SECM monitoring of hydrogen evolution and its consequences for Mg's localized corrosion in an orderly environment. They first assessed the monitoring of H_2 on platinum to verify the method and detect the impact of the localized corrosion process, including surface alkalization, on the hydrogen evolution (Fig. 7). The negative difference effect (NDE) designates the accumulation of H_2 from the corroded Mg substrate under the anodically polarized condition, and it can be understood with the higher current related to hydrogen evolution in Fig. 7b compared to Fig. 7a. It can also be observed, as displayed in Fig. 7c that the increase in current reveals the increase in the production of H_2 from the Mg in a cathodically polarized state. The profilometry result (Fig. 7d) indicates that the actual areas of H_2 evolution can be linked with areas of both negative (white arrow) and positive (black arrow) profiles with the corrosion pits being related with the negative profiles. Their results revealed that the SECM tip's sensitivity diminished with the high pH in nearby the substrates surface due to the alteration in H_2 electro-oxidation reactions. Further, their studies also suggested that combining the SECM results with another surface analytical tool, including profilometry or EDX, probably provides more accurate information on H_2 evolution and localized corrosion on Mg alloys.

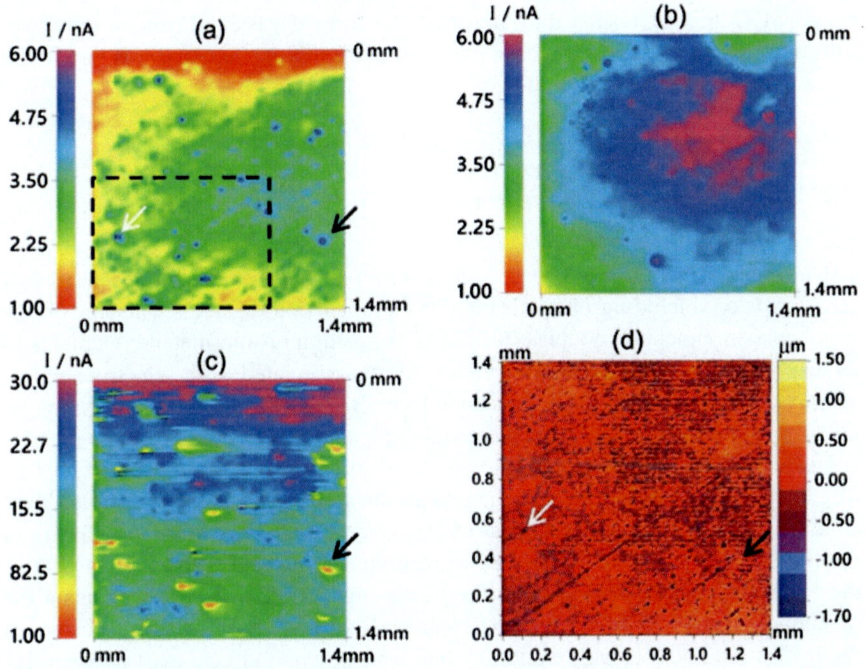

Fig. 7 Hydrogen evolution (HE) maps **a–c** and the associated surface profile **d** of AZ31 immersed in 10 times diluted SBF. HE maps were acquired by SG/TC mode of SECM using a 25 mm Pt tip biased at 0.0 V. The potential of AZ31 substrate was **a** unbiased (OCP), **b** polarized anodically relative to OCP at −1.5 V, and **c** polarized cathodically relative to OCP at −1.8 V. Note the different scales between (**a**) and (**c**)

Electrochemical behavior of titanium and its alloys have also been described using the SECM technique to examine their local surface electrochemical activity, passive film formation, and resultant localized corrosion such as pitting corrosion. Izquierdo et al. [27] have utilized the SECM technique to evaluate the formation of passive films on Ti and NiTi surface through substrates polarization in an artificial physiological medium. Their obtained SECM results revealed that the tendency of the electron transfer reaction on their surfaces is not homogeneously distributed and effectively influenced by the semiconductive nature of the passive films, the period of exposure, and applied potential. Izquierdo et al. [28] have also evaluated the influence of surface modification on the corrosion protection behavior of ZrTi alloys in an acidified physiological solution using the SECM and found that heat-induced oxidation at 500 °C formed more compact and electrically insulating oxide films which provided the enhanced corrosion-resistant behavior in the tested medium.

Bolat et al. [29] have evaluated the influence of pH of testing electrolytes and the amount of Mo on the TiMo alloys using the SECM technique. The passive layer produced in neutral and acidic mediums efficiently suppresses the anodic dissolution and breakdown of the passive film was not found in the potential region up to +1.00 V versus SCE. SECM results indicated that the formed passive layer on the TiMo substrate show dielectric features to charge the transfer process, and was monitored as insulators by SECM technique.

4.2 *Inspection on Localized Corrosion in Early Stages*

SECM measurements have been effectively employed in characterizing the localized corrosion, including pit nucleation and propagation on steels, galvanic corrosion between bimetallic couples, localized dissolution of intermetallic elements in Al alloys, etc., It can able to offer vital information related to the electrochemical activity of the substrates at a microscopic level. In general, nucleation and propagation of pits on the steel surface releases the high amount of Fe^{2+} ions in the pit sites compared to the remaining sites, thus by monitoring the tip current over the pit site, SECM results could provide the changes in the pitting corrosion with time. Yuan et al. [30] have evaluated the impact of H_2 on the pitting corrosion of X70 carbon steel in NaCl solution using the SECM technique. Nucleation and propagation of the pits were monitored by measuring the amount of released Fe^{+2} ions using the magnitude of the SECM tip current. Their results found that hydrogen accelerates the nucleation and the proliferation of pits, and synergic effects exist between H^+ and Cl^- ions on pitting corrosion. Regardless of this approach, other reports deal with the localized production of Cl^- ions at the SECM probe at the vicinity of the studied substrate [31–33]. Released Cl^- ions from the redox reaction happening at the SECM probe sufficiently can make able to initiate a single pit. This methodology has been effectively used to generate pits on iron, SS, aluminum, and copper samples [34]. Limitations of produced pits in terms of their dimensions and locations can be monitored by regulating the proportions of the SECM tips engaged in forming an adequately higher local concentration of chloride ions. Besides, SECM has also been employed to detect the active species that can cause the pitting initiation on steel substrates. The existence of sulfide inclusions as MnS in steel substrates is one of the main reasons for nucleating the pits by destroying and dissolving the passivation layer in an acidic media, producing the aggressive species thiosulphates. Using the iodide–triiodide redox mediator, SECM feedback mode was utilized to detect such inclusions and evaluate their contributions in the pitting process. During the positive feedback SECM operations, sulfide inclusions reduce triiodide (generated at the SECM tip) into iodide at the SECM tip, whereas the passive surface film experiences the negative feedback.

On the other hand, the breakdown of passive films on metallic substrates can also be identified through the SECM feedback mode. During this investigation, the substrate is generally polarized to convert iodide into triiodide to be monitored at the SECM probe. It can be utilized to sense the pitting precursor spots with the function of exposure period [35] since these disruptions in the passive layer exposed small areas of the reactive surface, which can further perform as pitting nucleation spots [36].

Evaluating the galvanic corrosion between the iron and zinc has received significant attention due to the galvanization of steel in which we deposit the protective zinc layer on the steel surface to enhance the corrosion resistance of steels components. While the damage occurred on the galvanized layer, it will become a sacrificial anode and encountered the galvanic corrosion. Using the SECM amperometric or potentiometric modes, the released Zn ions can be monitored from the galvanic couple's anodic dissolution. Further monitoring of the oxygen consumption or change in the pH, the oxygen reduction over the Fe surface can also be monitored [37, 38]. Apart from the general anodic dissolution of the Zn anode and cathodic reduction of oxygen on the Fe cathode, SECM helps to elucidate the corrosion mechanisms through high spatial resolution and chemical specificity.

SECM technique can detect the low concentrations of species, which can be utilized to assess oxygen reduction at the Zn anode, spatially localized on edge utmost distanced from the Zn cathode corroborated with a change in the local pH [39]. Raj and Nishimura et al. have quantified the anodic dissolution of Fe in galvanic couples and found the increased release of Fe with enlarged spacing between the electrodes in galvanic couples [40]. The obtained information revealed that mass transport in the solution could significantly influence the corrosion process at enlarged spaces. The distance between the Fe and Zn in a galvanic couple can also experience the additional impact, such as the deposition of Zn on the Fe surface [41]. However, Marques et al. [38] exhibited that the behavior mentioned above causes a negligible decrease in the oxygen reduction by observing the consumption of DO with SECM redox competition mode on a Fe/Zn galvanic couple.

Simoes et al. [42] have described the galvanic corrosion occurring in a Fe/Zn couple exposure in NaCl medium using the SECM and SVET techniques. SECM mapping of ionic currents monitored above the surface with the functions of the distances is shown in Fig. 8. The anodic and cathodic activities are identified evidently and found on zinc and iron, respectively, as expected that Zn sacrificially undergoes oxidizing by inhibiting the corrosion of Fe in a galvanic couple. While the two metals (Fe and Zn) are coupled, the potential changed to around −1.040 V, which is marginally positive than the potential of Zn in the same solution (−1.060 V). On this potential, Fe is not able to oxidize, and thus, the oxidation of Zn and the cathodic reduction of oxygen are the main reactions that occurred at the Fe surface. Due to the higher diffusion of Zn^{2+} ions from the galvanic couple, the measured ionic currents turn into less localized and less intense with increasing space between Fe and Zn in a galvanic couple. The mentioned phenomenon is indicated by obtaining the sharp signals near the surface, turning into more full and lesser lengthier distances, eventually disappearing at adequately lengthier distances. The obtained results revealed that SECM has a benefit of spatial resolution by the vicinity of the SECM tip to sample surface, on the other hand, the anodic reactivity of Zn can be monitored with high sensitivity using the SVET technique. Izquierdo et al. [43] have utilized the SECM and SVET techniques to predict the breakdown of the passive film and resultant localized corrosion on 304 and 316 stainless steels in HCl solution under open circuit potential and applied polarization. SECM results of monitoring the free Fe ions revealed that nucleation and propagation of pits occurred only with the

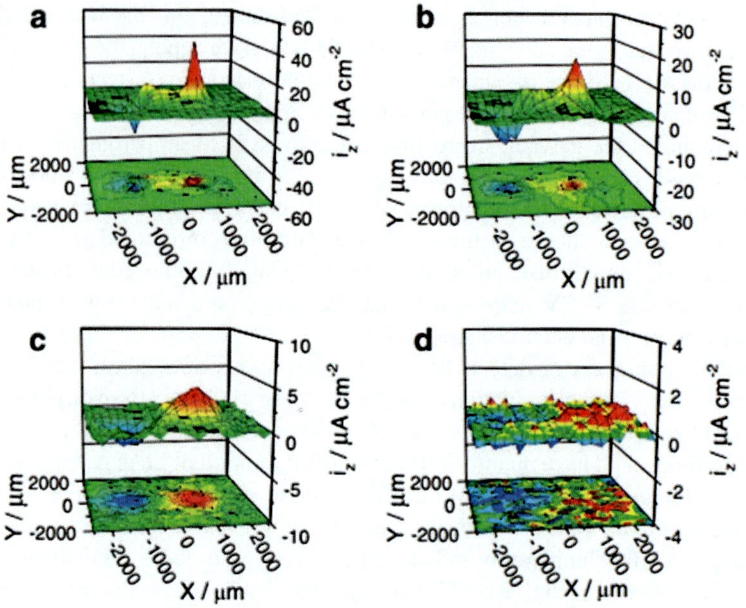

Fig. 8 Ionic current mapping of a galvanic couple (pure Fe and pure Zn) in 0.1 M NaCl obtained at various distances from the surface: **a** 100 μm, **b** 300 μm, **c** 700 μm and **d** 2000 μm

release of Fe (II) ions. The reported data inferred significant variance in the electrochemical activity of both steel substrates. Besides, by galvanic coupling 304 and 316 substrates, variances in the corrosion resistance concerning their proneness to localized corrosion were monitored using the SECM results.

Meng et al. [44] have utilized the SECM to evaluate the stress corrosion on the scratches of Alloy 690 substrates in NaOH solution, and the results revealed that the tip current increased at the scratch faster that the remained surface, indicating the higher corrosion rate on the scratched sites due to the high deformation energy by deformed scratch groove and scratch banks store. The strong localized corrosion facilitates the initiation of crack at the scratch groove's bottom. Volker et al. [45] also reported that the significant increase of tip current at scratch in the tin-coated steel substrate due to the dissolution of Fe. Zhu and Luo [46] concurrently inspected the influence of compression and tensile stress on the electrochemical activity of the Alloy 800 substrates in a ferrocene methanol solution containing thiosulfate, and their results revealed that both the compression and tensile stress increased the localized surface reactivity on the stressed areas, demonstrating the susceptibility of localized corrosion on Alloy 800 under stress. Bing et al. [47] have evaluated the influence of pH on the corrosion behavior of 800 alloy substrates by in situ SECM technique. Positive feedback is obtained in acidic medium, and negative feedback is occurred in both neutral and alkali medium, validating the formation of a passivated layer in neutral and alkaline mediums. Increased SECM tip current with the application of positive potential further validated the electrochemical activity of the passive films generated on the Alloy 800.

4.3 Evaluation of Passive Films

Formation of oxide/passive layer on metallic substrates provides the improved corrosion resistance, and further local breakdown of the passive layer at some conditions causes the localized corrosion. Differences in the local chemical reactivity caused by the formation of passive films on metallic substrates can be monitored by micro electrochemical measurements, which provide essential information related to the passivating system. Herein, SECM is an effective procedure for exploring the electrochemical phenomenon happening at the metal/solution interface of the passivating metallic surface [43]. Bard et al. [7] first utilized this technique to study passive layers on titanium and chromium substrates, and their results revealed that the investigated passive films owe insulating features. Evaluation of the stages of breakdown of passive layers on SS was also done using SECM by Wipf et al. [48]. Further, the non-homogeneity of the thickness or imperfect structure of a passive layer on a pure iron was explored by monitoring the alteration in redox activity through SECM techniques by Fushimi et al. [49].

Effect of alloying constituents and contaminations on the passivation phenomenon of Ti alloys can also be evaluated using the SECM technique as it can able to image the electron transfer reactions at metal/metal-oxide interfaces. Fushimi et al. [50] evaluated the electrochemical reactivity of Fe, Ti, and FeAlCr substrates through SECM, and their results revealed that grain boundaries and crystallographic orientation are strongly influenced the variations on the electrochemical reactivity based on the crystallographic grains. Indira and Nishimura [51] evaluated the influence of Cr addition on the localized corrosion behavior of low-alloy steels using the SECM measurements. Their results revealed that the measured SECM current at the anodic regions diminished with a Cr content in the steel substrate in NaCl solution. Yanagisawa et al. [52] evaluated the passivation tendency of dual-phase steel in buffer medium through the SECM TG/SC mode. Their results indicated that the passive layer produced on the martensitic phase exhibited higher electronic conductivity in comparison with the passive layer produced on the ferrite phase.

The thickness of the oxide films on metallic materials is usually varied, ranging from nano to micron scale, yielding significant alterations in their chemical nature, crystallinity, and electronic features. The conventional electrochemical techniques could not obtain information about the surface heterogeneity and defects on the passive films. However, local micro defects and minor changes in the thickness can able to give a significant impact on the electron transfer reactions on the metallic surfaces. SECM technique can monitor the mentioned surface heterogeneity and defects as well as changes in the properties of thin passivation films. Bolat et al. [53] have investigated the passive layer's localized reactivity produced on the three

ZrTi alloys substrates in a physiological medium. SECM results revealed that the surface reactivity increased when the substrates were anodically polarized, although the nature of the metallic material can mainly govern this feature's magnitude.

Ramirez-Cano et al. [54] examined the distribution of pH on 2-mercaptobenzothiazole (2MBT) treated Cu substrates in NaCl medium using the insitu SECM in potentiometric mode. Figure 9a, b displayed the SECM line scan and mapping results, in which the SECM map illustrates the group of lines demonstrating various locations of the substrate monitored. The 2MBT treated Cu substrate acts as the cathode, and thus cathodic reduction of DO undergoes over its surface, which changes the pH of its proximity significantly alkaline compared to that of the remaining surface. The 2MBT treated Cu exhibits an alkaline pH, related to the depletion of DO at the cathode, while the non-treated surface shows an acid pH (4.5–5.5), as Cu suffers corrosion. The obtained results revealed that the formed thin

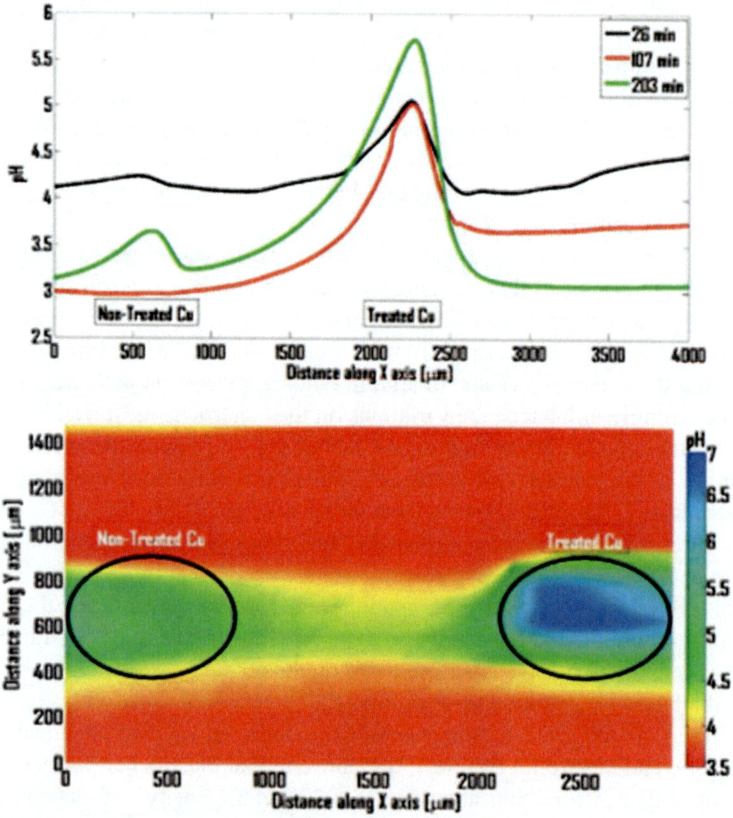

Fig. 9 Potentiometric SECM line scan and map showing the pH distribution in a plane parallel to the surface of the Cu wires after immersion in 1 mM NaCl for 1 h. Sensing tip: Sb disc, 30 μm diameter; tip–substrate distance: 30 μm; scan rate: 60 μm s^{-1}

inhibitor film by 2MBT over the Cu substrate could shift the pH to more alkaline regions compared with the untreated Cu substrate.

Gao et al. [55] have evaluated the electrochemical reactivity of stressed passive films by in situ SECM and point defect model (PDM). The obtained SECM current distribution images reveal that both tensile and compressed stress reduce the stability of the passive layer, further increasing the surface reactivity and dissolution rate due to the reduction in binding force between atoms. PDM results suggest that the stability of the passive layer decreased due to the increase in the rate constant of defect reactions that exist in the passive film.

Asserghine et al. [56] have utilized the SECM to evaluate the passivation tendency of nitinol in the acidic and neutral medium through feedback mode. SECM results validated that nitinol substrate is in a passive state in neutral medium and rupture of the passive protective film with the release of Ni^{2+} ions occurred in acidic medium. Besides, SECM was also utilized to check the self-healing tendency of the formed TiO_2 film on the nitinol substrate, and the results validated that longer deposition time is desired to generate a compacted passive layer on the nitinol substrate.

4.4 Characterization of Thin Inhibitors Films

Evaluating the thin inhibitor film growth and its protection against corrosion is another potential utilization of SECM in corrosion science. Consumption of chemical compounds as corrosion inhibitors is one of the most efficient corrosion control strategies worldwide. Organic inhibitors' efficiency is directly connected with the degree of adsorption and uniform coverage on the metallic base surface. The effectiveness of inhibitors is generally inspected through conventional gravimetric and electrochemical approaches. Most conventional approaches consistently deliver ample evidence without adequate spatial resolution, which is needed to elucidate the mechanisms and thoroughly identify the tendency of the interactions and phenomenon to be accountable for the corrosion inhibition. This is the main reason behind the utilization of SECM on the corrosion inhibition research.

Mansikkamaki et al. [57] have explored the potential to evaluate the formation of benzotriazole (BTA) film on Cu surface and examined the formed BTA layer with the function of applied potential and exposure period validates that SECM can be a valuable analytical technique for corrosion inhibition research. In this investigation, SECM images have not obtained for inhibited Cu samples or to evaluate the inhibitor efficiency on Cu samples without application of potential. Izquierdo et al. [58] utilized the SECM feedback mode to examine the formation of BTA inhibitor films on Cu substrates at their open circuit potential using the ferrocene–methanol as redox mediator. They found that the SECM approach curves were changed in shape during the formation of inhibitor film on the Cu surface. Their results revealed that conducting the nature of uninhibited Cu surface towards ferrocenium reoxidation could change into insulating behavior while the Cu surface covered with the BTA films. Izquierdo et al. [59] have also evaluated BTA films' formation on the Fe/Cu galvanic couples in NaCl solution using the sensing probe SECM antimony tips to be operated in both potentiometric and amperometric modes. They found that the formed BTA films on Cu surface made the Cu, electrochemically inactive compared with Fe surface, and cathodic regions developed over less elevated Fe surface and the anodic regions. Their results validated the application of the SECM technique to assess the spatial distributions of redox species, pH, and ionic currents related to the phenomenon's corrosion inhibition. Izquierdo et al. [60] have also explored the utilization of SECM to evaluate the formation of 2-mercaptobenzimidazole (MBT) films on Cu surface with the function of exposure time. Their results validated that the nature of the electrolyte could significantly influence the efficiency of organic inhibitors, hence chloride ions stimulating the growth of denser and highly insulating films than those formed in Na_2SO_4 medium.

Schuhmann et al. [61, 62] have described the utilization of frequency-dependent AC-SECM to evaluate the inhibition performance of several organic compounds on Cu substrates in Na_2SO_4 medium. The efficiency of corrosion inhibition can be examined by reducing the local electrochemical activity of the substrate due to the formed inhibitor layers. From the obtained results, they revealed that AC-SECM is an effective procedure to elucidate the corrosion inhibition phenomenon. Singh et al. [63] have evaluated the corrosion inhibition performance of plant-derived cationic dye on Al 7075 alloy in NaCl, and their SECM results revealed that steady transition of Al surface from a conducting to an insulating nature occurred in the presence of inhibitors in testing electrolyte. Li et al. [64] have described the applicability of SECM on the corrosion inhibition of mixed self-assembled monolayers (SAM_{mix}) produced by 1-dodecanethiol (DT), and 2-(Pyridin-2-yliminomethyl)-phenol (HL) compounds over Cu substrates in NaCl medium. $SAM_{HL/DT}$ exhibited more excellent stability and compactness than the $SAM_{DT/HL}$, demonstrating that the assembly arrangement played a significant role in improving the performance of SAM_{mix}. The adequate surface coverage of DT provided the SAM layer without any defects, significantly improving the corrosion inhibition efficiency of Cu substrate against chloride ions. The obtained results validated the advantage of SECM compared with the conventional experiments by monitoring transitory information regarding the nucleation and distribution of the pits on SAM_{mix}.

Esmaeili et al. [65] have assessed the corrosion inhibition of aryl-triazino-benzimidazole-2-thiones (ATBT) molecules on carbon steel in 2M HCl solution by SG/TG mode of SECM technique using the Fe^{2+}/Fe^{3+} as redox mediator. By applying the SG/TC mode, carbon steel corrosion in acidic medium was directly monitored through the amperometric measurement of Fe^{2+} ions. Their SECM results and local concentration of Fe^{2+} ions corroborated that corrosion current is reduced in the presence of inhibitors by forming the homogenous inhibitor film over the steel substrate. The obtained local concentrations of Fe^{2+} ion and SECM tip current were compared with the inhibition efficiency of the investigated ATBT compound.

Singh et al. [66] studied the electrochemical activity of the hexamethylenediamine functionalized glucose (DHA) thin inhibitor films on Cu substrates in NaCl medium using the SECM redox competition mode to scan the amount of DO as a cathodic precursor. SECM tip current in the blank solution (Fig. 10) was found to

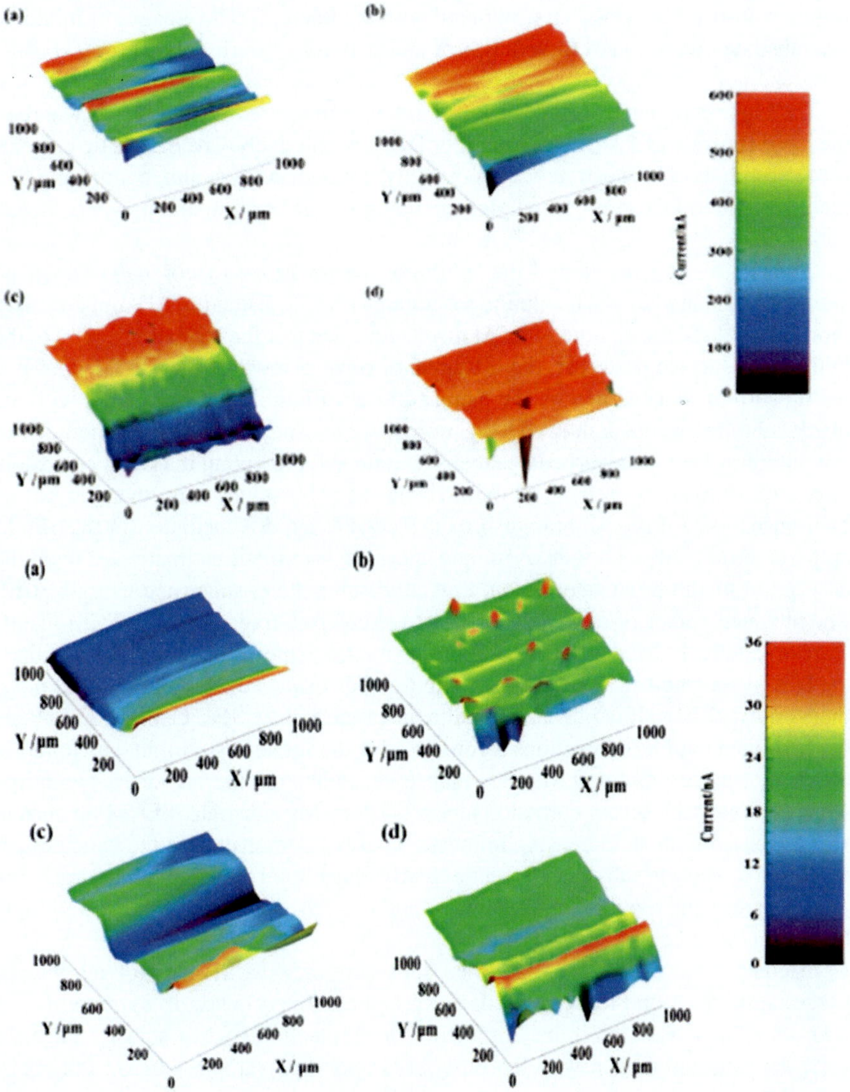

Fig. 10 SECM mapping results for copper substrates in test solution without and with the addition of DHA **a** 1 h; **b** 4 h; **c** 8 h; and **d** 24 h

be higher due to a reduced quantity of DO, revealing that the DO is utilized in the cathodic reaction. SECM tip current increased with the exposure period by showing propagated corrosion zones (Fig. 10), suggesting that the Cu substrates affected by strong corrosive attack after exposure of 24 h in NaCl medium. The distribution of the SECM current density on Cu substrate with inhibitor was nearly unchanged (Fig. 10), and only minor current peaks were observed on the Cu substrates during initial exposure time (Fig. 9). Compared with the blank, SECM images of inhibited Cu substrates were found to be uniform and featureless during the entire exposure time.

The highest tip current of 5.79×10^2 nA was measured for the blank, whereas, the lowest current of 32 nA was found for DHA, respectively. The SECM tip currents for the inhibited Cu substrate were pointedly lower than the blank, confirming that the investigated DHA compound makes the thin inhibitor film and prevents the Cu corrosion.

Wang et al. [67] evaluated the inhibition performance of corydalis yanhusuo extract on J55 steel in NaCl solution saturated with CO_2 using the AC-amperometry mode of SECM. The obtained SECM results revealed that the insulating nature of the inhibitored surface perturbed the diffusion of oxygen towards the SECM tip. Thus the tip current reduced nearby the metal surface with inhibitor. Compared with the blank, inhibited samples showed lower currents, further confirming the generation of thin inhibitor film on adsorbed on steel substrate and protecting it against corrosion. Varvara et al. [68] assessed the inhibition behavior of horse-chestnut ethanolic extract on bronze in simulated acid rain using Fc-MeOH as a redox mediator for the SECM amperometric feedback mode. A unique approach for kinetic estimation of dynamic adsorption of inhibitive species on a metallic substrate is suggested using SECM. The obtained values of rate constant related to electron transfer from SECM results were productively fixed into the Langmuir isotherm. Umoren et al. [69] have utilized the SECM technique to monitor the inhibition performance of several natural polymers on AZ31 Mg alloy in NaCl medium using the SECM SG/TC mode to evaluate the hydrogen evolution reactions. From Fig. 11, the higher tip current was found in the case of an un-inhibited Mg alloy substrate, indicating the higher H_2 evolution, which revealed the severe corrosion attack over the Mg substrate. SECM tip current was increased with an increasing immersion period, demonstrating the harshness of corrosion on the Mg substrate. It can be clearly understood that a significant variation on tip current and the maximum value of 300 nA was observed on the un-inhibited Mg substrate after exposure of 24 h.

In comparison with the uninhibited substrate, the SECM maps of inhibitored substrates were found to be more uniform and exhibited lower tip currents (0−15 nA) after 24 h NaCl medium, indicating the protection of Mg substrate against corrosion. In comparison with the pure ALG and HEC samples, ALG- and HEC-formulations were pointedly reduced the SECM tip current, revealing the efficiency of the formulations in suppressing the corrosion rate of Mg substrates. It was summarized that adding the natural polymers, mainly their formulations, efficiently inhibited the corrosion of AZ31 Mg alloy by adsorption and production of thin inhibitor layers over the Mg surface.

Table 2 summarized the previous reports in the past decade that dealt with the numerous SECM models operating in the different modes to elucidate the corrosion inhibition mechanism of different organic compounds on many metallic materials in different environments [57–76].

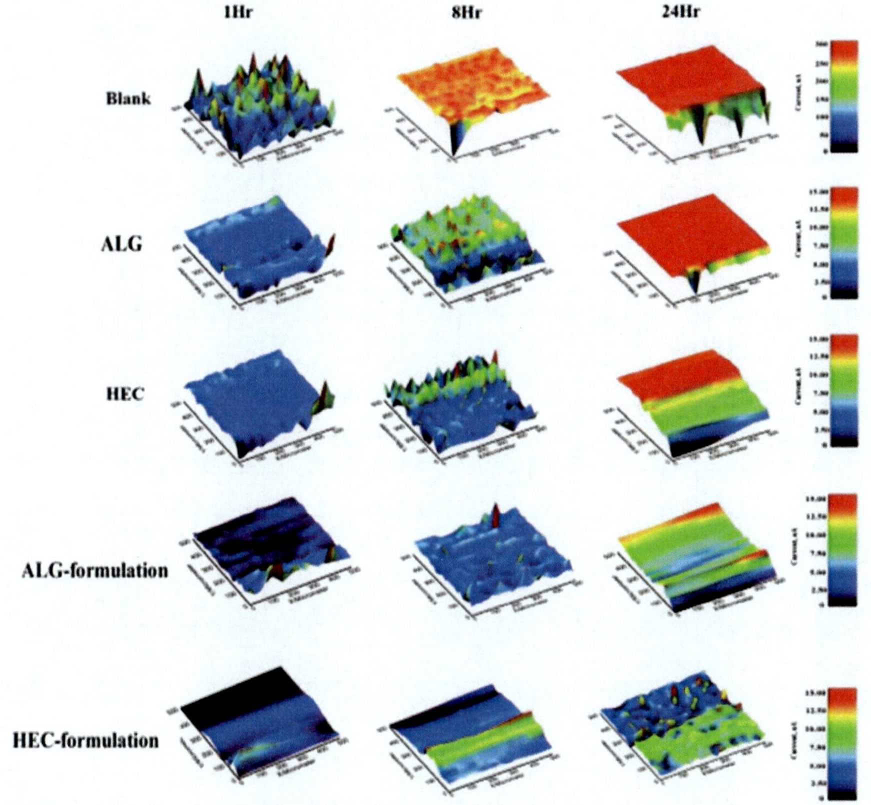

Fig. 11 SECM mapping images of AZ 31 Mg substrates with and without inhibitor as function of immersion time

4.5 Evaluation of Degradation of Protective Coatings

From the past few decades, many micro electrochemical techniques have been effectively utilized to obtained vital information related to coating degradation at the early stages. The utilization of SECM to assess coated metallic substrates' anticorrosion behavior is a promising strategy in describing the complex phenomenon at metal/coating systems in aggressive environments. The formation of the blister on the coating surface due to the water absorption can easily be monitored using this technique through the difference in the redox activity of tip and substrate. The observed phenomenon can make deviations in the current response at the tip with time, while the SECM probe is scanned at a constant height from the substrate [77].

Using the SECM feedback mode, early stages of coating deterioration at short exposure periods can be examined by the coatings swelling and resultant blister formation. By monitoring the SECM tip current through feedback mode, local interference in the diffusion-restricted transference of a redox mediator can directly be

Table 2 Summary of SECM application on corrosion inhibition research along with redox mediators and the relevant parameters

Inhibitor	Metallic components	Medium	SECM mode	Redox mediator
2-Mercapto benzothiazole	Cu	NaCl	Potentiometric mode	NIL [54]
Benzotriazole	Cu	NaCl	Feedback	DO [57]
Benzotriazole	Cu	Na_2SO_4, KCl	Feedback	Ferrocene–methanol [58]
Benzotriazole	Fe/Cu couple	Na_2SO_4	Amperometric	Ferrocene-methanol [59]
2-mercaptobenzimidazole	Cu	Na_2SO_4	Feedback	Ferrocene-methanol [60]
Benzotriazole derivatives	Cu	Na_2SO_4	AC-SECM Feedback	NIL [61]
Benzotriazole, 5-methyl-benzotriazole, 2-mercaptobenzimidazole, and ethyl xanthate	Cu	Na_2SO_4	AC-SECM Feedback	NIL [62]
Plant derived cationic dye	7075 Al	NaCl	Feedback	DO [63]
SAM from 2-(Pyridin-2-yliminomethyl)-phenol (HL) and 1-dodecanethiol (DT) molecules	Cu	NaCl	Feedback	Ferrocene-methanol [64]
Aryl-triazinobenzimidazole-2-thiones	Carbon steel	HCl	SG/TC	$FeSO_4(NH_4)_2SO_4$ [65]
Porphyrin derivatives	J55 Steel	CO_2 saturated NaCl	Feedback	DO [70]
Porphyrin derivatives	N80 Steel	CO_2 saturated NaCl	Feedback	DO [71]
Pyridine derivatives	N80 steel	15% HCl	Feedback	DO [72]
2-(Pyridin-2-yliminomethyl)-phenol	Cu	NaCl	Feedback	Ferrocene-methanol [73]
Hexamethylene diamine functionalized glucose	Cu	NaCl	Redox competition	DO [66]
Piperazine based polymer	N80 steel	NaCl	Feedback	DO [74]
Corydalis yanhusuo extract	J55	CO_2 saturated NaCl	AC-Amperometry	NIL [67]
Horse-chestnut extract	Bronze	Simulated acid rain	Amperometric feedback	Ferrocene-methanol [68]

(continued)

Table 2 (continued)

Inhibitor	Metallic components	Medium	SECM mode	Redox mediator
Polygonatum fructan	Mild steel	HCl	Feedback	$FeSO_4(NH_4)_2SO_4$ [75]
Sodium alginate and hydroxyethyl cellulose	AZ31 Mg	3.5% NaCl	Redox competition	DO [69]
Imidazole derivatives	J55 steel	CO_2 saturated brine	Feedback	DO [76]

connected to the topographic alterations of insulating surfaces originated from the coated metallic surface [78, 79]. Usage of dissolved oxygen (DO) as a redox mediator eliminator the issues of freely diffusing redox mediator and DO is needed for the corrosion reactions occurring at the metal/coatings interfaces. Souto et al. [80] have used the SECM feedback mode to assess the effect of the composition of electrolytes on coated steel substrates' corrosion behavior and revealed that the formation of blister quickly occurs in the chloride medium after immediate exposure (Fig. 12). Simultaneously, sulfate or nitrate solutions do not cause blisters/bulges on the coating's surface even after 24 h of exposure. Moreover, monitoring of topographic changes using the DO as a redox mediator eliminates the probability of ferrocene–methanol to provide the needed information through a straight reaction with the surface. This study benefits the utilization of SECM as an innovative analytical tool for the faster examination of coated metallic components.

Along with the SECM feedback mode, the redox competition and G/C modes have also been utilized to obtain vital information about the coating degradation at the early stages and their protection efficiency towards the base metallic substrates. Gonzalez-Garcia et al. [81] have utilized the different mode of SECM including feedback, redox competition, and SG/TC modes to get the spatial information regarding the surface topography, water inclusion with the resultant blistering and the expulsion of Fe (II) ions from the anodic regions and the consumption of DO at the cathodic regions in the coated metallic steel with the artificial scratches. By adjusting the SECM tip potential, the released Fe (II) ions from the metal corrosion at the coating

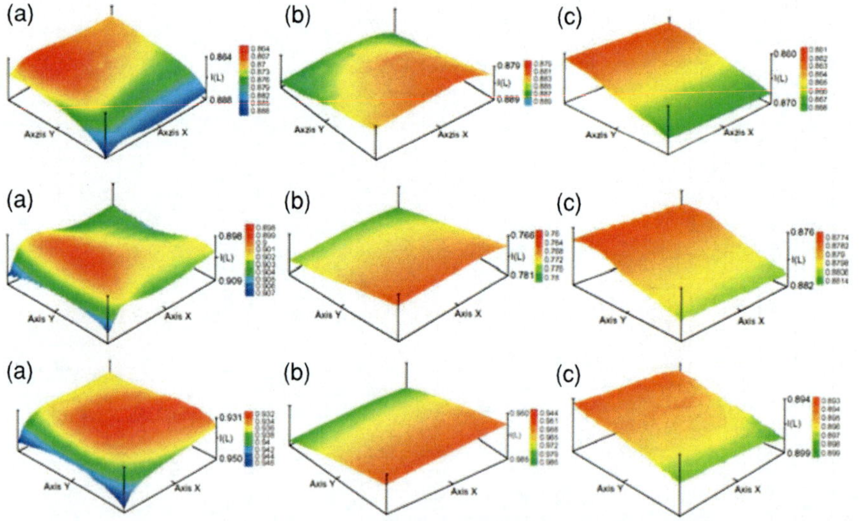

Fig. 12 SECM images generated by SECM of polyester-coated CCS specimens immersed in: **a** 0.1 M KCl; **b** 0.1 M K_2SO_4, and **c** 0.1 M KNO_3 solutions. The images were taken after 30 min, 4 h and 9 h of immersion and represent 200 μm × 200 μm in X and Y directions. The Z scale is the tip current in nA. The time origin is the beginning of the scan. Tip potential, +0.50 V versus the Ag/AgCl/KCl (saturated) reference electrode

defects could be measured by detecting the tip current. Similarly, the tip potential is adequately polarized to sense the DO in the solution at the vicinity of coated substrates since the competition existing between the tip and substrates as the DO is directly consumed by the cathodic reactions at the metal/coating interface. Santana et al. [82] have used the SECM technique to assess the anti-corrosion performance of steel coated with an epoxy-polyamine with glass flakes anti-corrosive pigments and also evaluated the influence of the composition of electrolyte on coating's performance. During the SECM experiments, DO was employed as a redox mediator, eliminating the possible interference with the electrochemical corrosion reactions at the metal/coating system. Their results revealed that negative feedback is maintained when the tip is scanned over the intact coating irrespective of the electrolytes, whereas redox competition mode exists at the coating's defected sites. The amount of DO gets reduced due to the consumption of DO in cathodic regions of the corroding system.

Utilizing the redox competition mode, the self-healing efficiency of polymer coatings has also evaluated by detecting the variation in the DO on the defected regions. Gonzalez-Garcia et al. [83] have explored the possibility of redox competition mode to elucidate the mechanism of self-healing tendency of shape memory polyurethane (SMPU) coatings on defected sites by monitoring the magnitude of tip current and found that decreased tip current due to the diminishing the DO content at the defected sites before self-healing of SMPU, suggesting the regain of the barrier feature and suppressing the DO reduction on cathodic sites. Similarly, Pilbath et al. [84] also evaluated the self-healing features of epoxy coatings incorporating micro capsulated linseed oil on steel samples. They found that the DO level is increased at the scratched regions, which validated the improved corrosion protection performance offered by the self-healing coating.

Yang et al. [85] prepared the benzimidazole loaded pH-sensitive zeolite imidazole framework (ZIF) nanoparticles and inserted them into the polymeric coating to evaluate their self-healing tendency using the SECM technique. The variation of SECM tip current nearby the scratch of the self-healing coating was monitored using the exposure period (Fig. 13). During the initial exposure time, the distribution of tip current on the scratch regions of both coated substrates was decreased, representing that the aggressive species from the electrolyte started localized corrosion on the defected sites. From Fig. 13a, the tip current density of the pure coatings was prompt decreased during 50 h of exposure, and the lowest current density of 2.0×10^{-7} $\mu A/cm^2$ was found to be in the scratched region after 50 h of exposure. Whereas, in the case of the coatings with ZIF-7, increasing the tip current with exposure time indicated the lower consumption of H^+ ions, further validated the self-healing features of ZIF-7 incorporated coatings. The penetration of the pure coating was found to be about 200 μm on both sides of the scratch after 50 h of immersion. However, the coatings incorporated with nano ZIF 7 exhibited self-healing and anti-corrosion performance on the scratch sites. The acid environment facilitates the discharge of an adequate quantity of the BI from the ZIF-7 nanoparticles to reduce the consumption of H^+ by the corrosion process, further increase in the monitored tip current. From Fig. 13b, the largest width of the scratch site, was found to be only 150 μm after 50 h

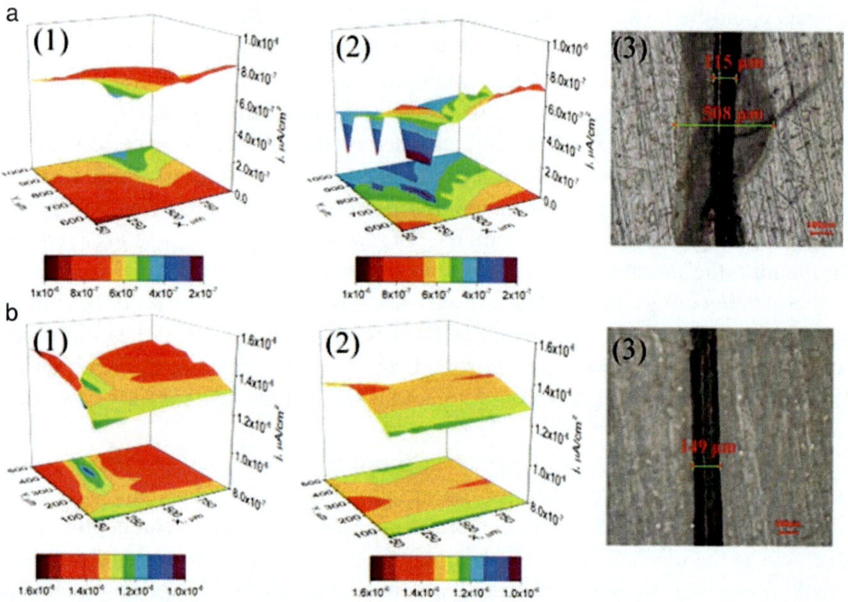

Fig. 13 Three-dimensional (3D) SECM current density maps of (**a**) the control coating sample (**b**) and the epoxy coating sample containing ZIF-7 after immersion for (1) 4 and (2) 50 h and (3) optical microscopy images taken of the scratched surface of the coatings after 50 h of immersion in 0.1 M HCl

of exposure time, which was much smaller compared with the pure coatings (about 500 μm).

Wang et al. [86] synthesized the graphene oxide–mesoporous silicon layer–nanosphere incorporated with tannic acid (GSLNTA) and incorporated into polymeric coating to evaluate the self-healing performance under marine alternating hydrostatic pressure (AHP) using in-situ AC-SECM technique. The monitoring of various electrochemical parameters of |Z| and –phase elucidate the self-healing mechanism by the local variations of GSLNTA coatings in the AHP environment. Wang et al. [87] also evaluated the self-healing performance of epoxy coatings with the addition of hexamethylene diisocyanate trimer (HDIt) microcapsules on Al alloy surface through SECM and AC-SECM techniques. SECM mapping results in Fig. 14 illustrate the distributions of tip current on the scratched surface of the coating in 0.1 M KCl solution. The variance on the colors of SECM images was produced by the variation in the surface topography and electrochemical activity of substrate, environmental noise, and UME movement during the SECM test. After exposure of 48 h, the scratched region of the epoxy coating with HDIt microcapsules was almost covered with the products of self-healing polyurethane polymerized by HDIt and water. By comparing the results obtained from both SECM and AC-SECM, they revealed that AC-SECM provided significant local electrochemical topographies of

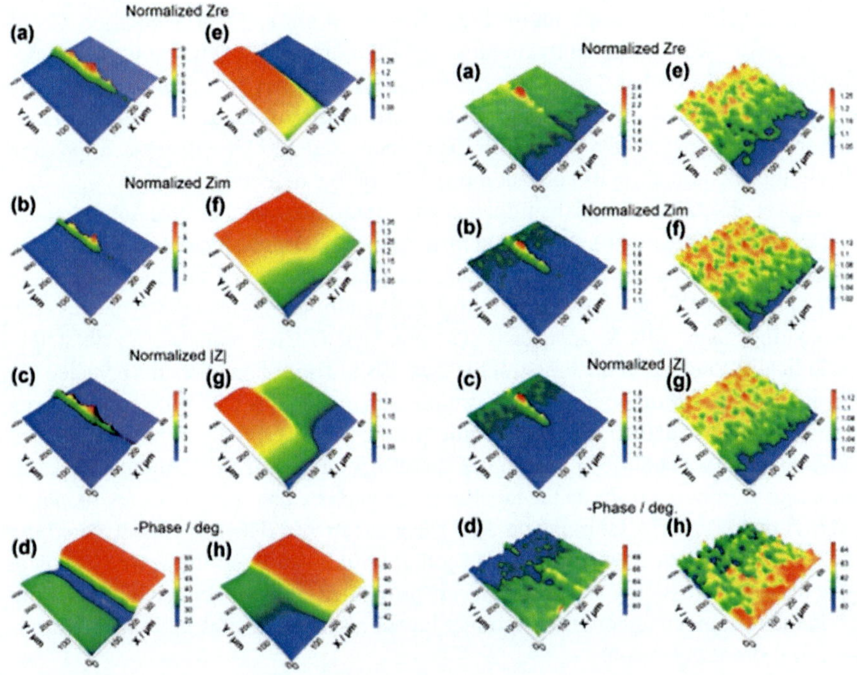

Fig. 14 SECM images of a self-healing coated surface immersed in a solution of 0.6 M NaCl. DC potential: 0 V. Amplitude of the AC potential signal: 10 mV, and frequency of 60 000 Hz. Tip diameter: 25 mm; tip-substrate distance: 15 mm.(**a–d**) Sample with scratch. (**e–h**) Healed sample. Scanning step of the UME tip was 12.5 mm

the self-healing surface at various AC frequencies with improved spatial resolution and was less provoked by the scanning step size and the purity of the probe surface.

Niteen et al. [88] evaluated the anticorrosion behavior of epoxy coating with the inclusion of tungstate and vanadate doped polypyrrole/aluminum flake composite materials AA2024-T3 substrates through feedback and redox competition modes of SECM. Their results revealed that the consumption of DO at the scratched sites of epoxy composite coatings increases with exposure time, however less corrosion attack/product formed nearby the scratched areas. Simoes et al. [89] have described the mechanism of cathodic protection tendency of Mg-rich coating on Al substrate by monitoring the cathodic activity using the SECM technique and revealed that the cathodic protection effectively suppressed the nucleation and the growth of pit by Mg to Al substrates.

Wang et al. [90] inspected the delamination of epoxy coated X80 steel under alternating stray current using the feedback SECM mode. The shape of the obtained tip current in SECM mapping results revealed the corrosion pits and delamination sites of epoxy coatings, which increased severe with stray AC densities and exposure periods. Xia et al. [91] evaluated the surface reactivity at the organic coating defective sites by relating the SECM results with the electrochemical noise (ECN) technique. Corrosion intensity was evaluated by monitoring both noise and spectral noise resistance. SECM results revealed that the tip current at the defected sites is pointedly higher compared to that of the entire site. SECM probe approach curves result indicated that the rate constant of the defect initially increased and then decreased, possibly due to the accumulation of corrosion products on the defective sites.

Vega et al. [92] evaluated the influence of Ti interlayers on the corrosion resistance of TiN PVD coatings using the potassium hexa cyano ferrate and Prussian blue during the SECM experiments. The same group has also investigated the open porosity of CrN and TiN PVD coatings by monitoring the released electroactive species from the coatings using SECM area scans. The obtained SECM results were validated by periodically conducting the topographical analysis. The obtained results revealed that the anodic dissolution is significantly suppressed in coatings with thicker interlayers. Conversely, the coating with a thin interlayer is highly prone to noticeable local anodic spots. Ma et al. [93] have utilized the SECM and LEIS techniques to assess the effect of imperfections on the localized corrosion behavior of micro-arc oxidation (MAO) processed HA coatings on AZ31 Mg substrates. Their results revealed that the MAO HA coating with simulated scratch initially suffered the severe corrosion with the exposure period and progressively prolonged from the scratch to the coating. Then the accumulated corrosion product layer on scratched sites inhibits the further propagation of corrosion.

Lacombre et al. [94] have also utilized a similar setup to evaluate the influence of TiO_2 particles and temperature on the water absorption and resultant swelling in the epoxy coatings in NaCl solution. Water absorption results from gravimetric and EIS analyses were correlated with the measured swelling from the insitu SECM experiments. The water absorption and swelling of coatings are affected by the aging temperature due to the local breakdown of the epoxy/TiO_2 interphase and internal stresses exiting among the pigments in the epoxy matrix. Trinh et al. [95] have utilized the ionic liquids during in situ SECM experiments to differentiate the influences of water and temperature on the swelling of organic coating. The swelling of coated substrates in different temperatures was monitored using SECM in the saline and 1-Ethyl-3-methylimidiazolium ethyl sulfate (RTIL) solution. SECM results revealed that temperature stimulated swelling in the saline medium was observed to be higher than the swelling in RTIL, validating the central role of water in entering the free volume of the polymeric network. Additional swelling was also occurred, independent of the water and temperature, possibly due to a relaxation of internal stresses within the polymer network during hygrothermal aging.

Elkebir et al. [96] have evaluated the influence of physical aging (PA) on the water absorption of epoxy coatings using the insitu monitored SECM probe approach curves. The obtained results revealed that water absorption followed by swelling was lower in the case of coatings with PA compared to that of coatings without PA. Further analysis of diffusion factors indicated that the PA enhanced the densification of the epoxy matrix due to the stronger interactions among polar groups, which reduced the available polar groups for water absorption.

The influence of nanomaterials on the anti-corrosion performance of organic coatings has also been evaluated using the SECM approach. Kumar et al. [97] have evaluated the impact of the addition of silicon (Si) nanoparticles on the anti-corrosion performance of epoxy-coatings on carbon steel substrates in NaCl solution. Figure 15 displays the SECM mapping results of scratched epoxy coated steel without and with Si nanoparticles in different immersion periods. The tip current for the scratched epoxy coating without Si nanoparticles (Fig. 15a) was high and reduced nearby the scratched sites. With prolonged immersion periods, the reduction in tip current was observed at the scratched site by spreading red color, signifying the depletion of DO by anodic dissolution in the scratched site. Whereas, the tip current of the scratched epoxy coatings with Si nanoparticles was significantly lesser, which is accredited to the depletion of DO by the anodic dissolution of Si nanoparticles from the epoxy coating. Furthermore, the lower current was continued for a prolonged exposure period (8 h). Thus, anodic dissolution of Si can suppress the cathodic reduction of oxygen and the dissolution of Fe (anodic reaction), which suggested the improved, enhanced anti-corrosion behavior of the epoxy coatings with the incorporation of Si nanoparticles.

Suleiman et al. [98] have evaluated the impact of metal oxides on the anti-corrosion performance of sol–gel coatings on steel samples in NaCl solution through the SECM SG/TC and redox competition modes. The release of Fe^{2+} ions at coated steel samples and the depletion of DO at the tip was monitored by applying tip potential of +0.60 V and −0.70 V, respectively. Therefore, the observation of both local changes in the

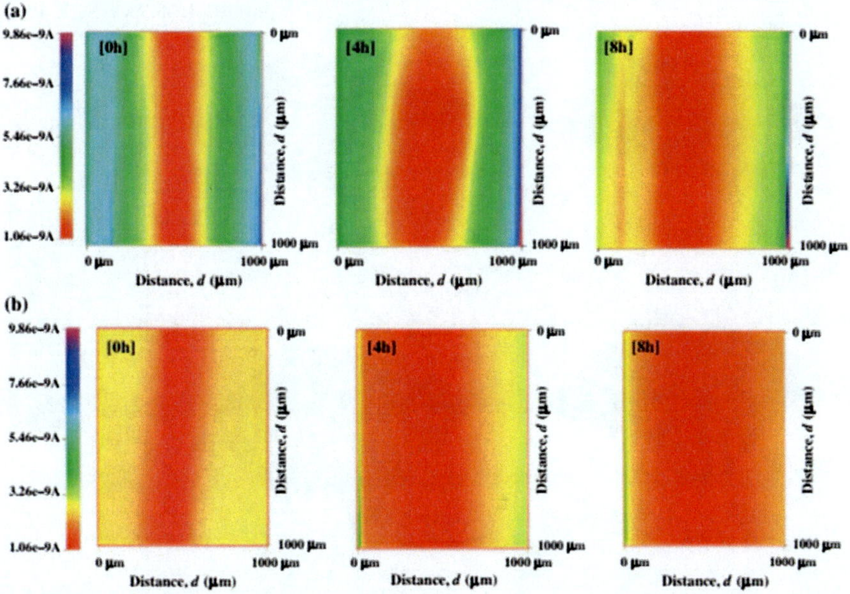

Fig. 15 SECM topographic image of **a** scratched sample without and **b** with Si nanoparticles in 0.1 M NaCl solution

quantity of Fe^{2+} ions and a reduction in the DO is a straightforward approach to assess the magnitude of corrosion at the metal/coating interface. The SECM mapping results of the sol–gel coated MS substrates without metal oxide (Fig. 16a) indicated that the corrosion originated only after 4 h of immersion and underwent a strong corrosive

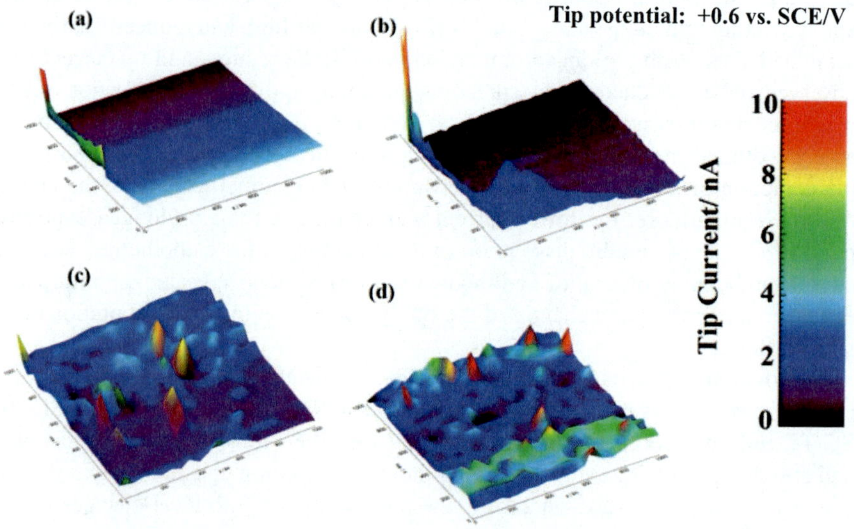

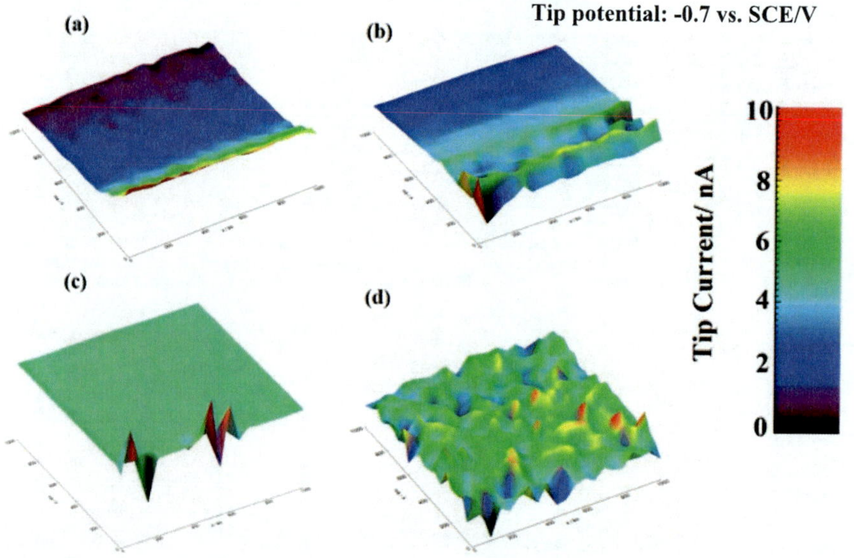

Fig. 16 SECM mapping images of sol–gel coated MS substrates at the Tip potential of −0.7 SCE/V during the immersion in 3.5 wt.% NaCl Solution **a** 1 h, **b** 4 h, **c** 8 h and **d** 24 h (tip diameter: 15 μm, tip substrate distance: 15 μm, scan rate: 50 μm s^{-1})

attack after 24 h of exposure. The acquired results inferred that the base hybrid polymeric coating (CT) possibly owns the micro defects such as pores and micro-cracks in which violent species from the electrolyte could penetrate the coating and the consequent corrosion started at the metal/coatings interface. Instead, the SECM mapping images of the coated substrate with metal oxide were nearly identical and undistinguished during the exposure time, while the SECM tip current was pointedly lower than that of the bare and sol–gel coated MS without metal oxide. The obtained SECM results confirmed that the incorporation of metal oxide could fill the pores and micro-cracks in the coatings improving the barrier protection.

5 Integration with Other Techniques

In general, the interconnected results can be assisted to understand the electrochemical activity of the corroding system completely and also aid in elucidating the precise mechanism, thus improving the corrosion mitigation strategies. Smyrl and co-workers [99] used SECM linked with near-field scanning optical microscopy (NSOM) to examine cathodic activity at AA2024. Their comparison of NSOM topographic maps and SECM images showed that localized cathodic activity is associated with the second-phase particles. Recently, the scanning electrochemical impedance microscopy (SEIM) technique has been dedicated to examining the variation of topographical and morphological with the difference in the electrochemical properties of surfaces. SEIM is the updated version of AC-SECM, in which local impedance response can be received through multi-frequency AC-SECM at each analyzed site, and the generated data can be systematically studied through modeling and spectral fitting. SEIM can be utilized to examine self-assembled monolayers' performance, ultrathin epoxy coatings, and corrosion inhibitors [100–102].

Gabrielli et al. [103] have utilized the SECM Ag/AgCl microelectrode and an electrochemical quartz crystal microbalance (EQCM) to detect the nucleation of pits and the resultant pit propagation by mass variations. Aouina et al. [104] also used these strategies to examine the nucleation and propagation of pits on 316L SS and successively investigate the resultant chemical species by Raman and atomic emission spectroscopic techniques. The integration of an EQCM can permit to regulate the distance between the tip and substrate since contact between the UME probe and the surface triggered a significant spike in the frequency.

By joining the potentiometric and amperometric modes of SECM techniques, the essential advancements in corroding systems' electrochemical properties are adequately studied. Dual amperometric/potentiometric tips made by few transition metal oxides such as antimony permit regular SECM operation to detect pH in the same experiment. Also, ion-selective microelectrodes can be utilized as SECM tips for appropriate sensing of related chemical species. Hence, integrations of SECM modes can be employed for more sensitive applications of SECM imaging in corrosion science.

Izquierdo et al. [34] combined the SECM with the atomic force microscopy (AFM) to examine the initiation and propagation of the corrosion pits on the passivating iron substrate. The AFM–SECM probe produces HNO_3 in the vicinity during the oxidation of nitrite ions with the release of protons at specific locations on the passivated metallic surface. SECM TG/SC mode of operation was performed using the nitrite ions as a redox mediator to achieve the enhanced spatial resolution. Further, the redesigned AFM tip delivers the surface topographic mapping of the pit nucleation with the resultant propagation stages.

Since the intermetallic secondary particles accumulated in the alloy matrix and possibly clustered, quarantined investigations of electrochemical activity can be complicated using the conventional techniques. Senoz et al. [105] combined the use of SECM with the scanning Kelvin probe force microscopy (SKPFM) technique to evaluate the galvanic corrosion on Al substrates with the secondary phases of copper in micrometer dimension. SKFPM was used to find the potential regions for corrosion, and the redox competition mode of SECM was used to detect the corroded sites after immersion. Consumption of DO at the SECM tip competed with the cathodic sites of the Al-Cu galvanic couples, causing the reduction in the cathodic current on the copper secondary phases. The correlation between the electrochemical activity and contact potential was achieved from their results. Similarly, Davoodi et al. [106] investigated the cathodic activity over the Fe-rich secondary phases of 3xxx series Al alloys and observed the reduction in cathodic current on the Fe- rich sites through a hybrid SECM-AFM arrangement linked with SKPFM technique.

The crystallographic orientation of a metallic surface on its preferential dissolution and localized corrosion has been examined by combining the SECM with the Electron Backscatter Diffraction (EBSD) technique. EBSD-SECM is considered a competent integration since the spatially resolved sensitive SECM tip could adequate to inspect within the grains. The previous report related to EBSD-SECM has compared the SECM reactivity mapping results with the EBSD orientation map [50, 107]. However, the obtained results are continued to be qualitative since the approach curves could not be used to measure the thermodynamic and kinetic parameters on crystallographic orientations. Besides, limited studies have found in the literature with industrial alloys [50, 108, 109]. Recently, Verchere et al. [110] have utilized the coupled EBSD-SECM technique to quantitatively evaluate the deviations of electrochemical activity based on the crystallographic orientation of the passive nickel-based alloy. The obtained SECM results were quantitatively compared with the proper crystallographic orientation, indicating the actual relation between the grain's crystallographic plane and the electrochemical activity of the passive layer. The reported information validates that combining these techniques is optimistic in predicting the quantifiable kinetic or corrosion models.

Nanoscale SECM is an efficient analytical tool that facilitates highly spatial resolution monitoring of the electrochemical phenomenon on the materials in the nano dimension. However, it prerequisites to precisely characterize the nanoelectrode tip to diagnose SECM nanoscale images quantitatively. Chen et al. [111] established an effective integration of TEM of a nanopipette with SECM of a nanopipette-facilitated ITIES (interfaces between two immiscible electrolyte solutions) tip to develop quantifiable SECM scanning at the nanometer scale. TEM results of nanotips and nanopores related to their dimension (size and geometry) were vital for the consistent exploration of SECM images, which identified the elongation of rounded nanopores alongside the direction of the tip. It is recommended that the relation between the spatial resolution of SECM and the precise dimension of a projected tip should be described to facilitate high-resolution SECM imaging at a few nanometer scales.

Deboss et al. [112] prepared the conductive polydopamine coated AFM-SECM probes through the pulsed electrochemical deposition to perform the electrochemical force spectroscopic experiments. The distinctive integration of the SECM and AFM approach offers the information related to electrochemical reactivity at the micro and nanoscale, whereas nanomechanical features, including adhesion forces, stiffness and, elastic moduli are concurrently gained. It is concluded from their results that the prepared probes can able to attain information on the adhesion properties of bacteria, cells, and tissues and may be designed to the particular desired application by selecting suitable electrochemically controlled polymer.

6 Future Challenges and Opportunities

Though SECM technology is a comparatively new research area, broad range of corrosion systems have been analyzed ranging from the observation of anticorrosion coatings, growth of thin inhibitor films and passive layers along with the basic mechanism behind general corrosion to localized corrosion. The discussed findings in SECM validates the remarkable resolving potential of this strategy and the outcomes of these technology offer an micro meter scale basis for establishing the electrochemical kinetics of interfaces at the corroding systems. The arena of SECM is predicted to expand considerably as novel methodological progresses in SECM tips and hybrid modes and dimensions materialize. Utilization of innovative modifications on SECM technology will impact research in numerous disciplines, such as physics, biology, chemistry, materials science, and medicine.

One of the main boundaries of the SECM over conventional SPMs is the attained resolution which is regulated by the dimension of the SECM tip and by the distance between tip and substrate. Current research works on the SECM have been targeted to resolve this limitations and several studies are now engaged concerning the improvement of innovative approaches for the construction of conductive SECM tips in nanometer-scale and also on the integration of the SECM with other SPMs techniques. Along with this, some developments on the instrumental design and procedure and also in the predictable applications could be recommended, including (a) A range of UME tips fabricated by micromachining to improve the time resolution for SECM mapping; (b) Precise locating of the UME tip above an imperfect regions (c) Heterogeneous corrosion reactions at grain boundaries or on individual grains; and (d) SECM analysis of localized corrosion attack on metallic materials in the presence of stress or strain in actual time.

7 Conclusions and Perspective

Beginning from the discovery, SECM techniques have been employed in investigating the corroding systems concerning various metals and alloys. Because of the many SECM modes, analysis of surface electrochemical activity using various modes delivers the most significant evidence compared to the other techniques. SECM experiments need marginal substrate preparation in comparison with the other spectroscopic techniques. SECM technique can perform under steady-state circumstances, removing the double layer charging current and a resistive potential drop in electrolytes, and few complications that are often related to the other conventional electrochemical tests. The additional benefit of the SECM is the view of arithmetical simulating of the image, which further permits the acquisition of evidence and exploration of the followed reaction kinetics. Resulting in a summary of the SECM strategies, this book chapter has offered a wide-ranging summary of the experimental details, comprising all of the SECM probes, electrochemical redox mediators, electrolytes, and relevant parameters. Comprised analysis and information of these specifics can permit researchers to understand the theories and essential constituents to incorporate SECM procedure into the corrosion science and technology.

In conclusion, a promising option for the improvement of SECM sustain within combinational techniques. Numerous illustrations have been designated in the chapter. However, novel progress has continuously being described, such as the combination of SECM with SVET, AFM, Raman, etc. The novel combined SECM strategies with the advanced procedures transfer toward the multifunctional systematic stages appropriate to elucidate corrosion phenomenon in different environments and much high acceptance by the corrosion science community.

Acknowledgements We thank King Fahd University of Petroleum and Minerals (KFUPM, Saudi Arabia) for providing all support to prepare this book chapter.

References

1. X. Shi, W. Qing, T. Marhaba, W. Zhang, Atomic force microscopy—scanning electrochemical microscopy (AFM-SECM) for nanoscale topographical and electrochemical characterization: principles, applications and perspectives. Electrochimica Acta **332**, 135472 (2020)
2. X. Lu, Q. Wang, X. Liu, Review: recent applications of scanning electrochemical microscopy to the study of charge transfer kinetics. Analytica Chimica Acta **601**, 10–25 (2007)
3. S. Bergner, P. Vatsyayan, F.-M. Matysik, Recent advances in high resolution scanning electrochemical microscopy of living cells—a review. Anal. Chim. Acta **775**, 1–13 (2013)
4. J. Izquierdo, P. Knittel, Christine Kranz scanning electrochemical microscopy: an analytical perspective. Anal. Bioanal. Chem. **410**, 307–324 (2018)
5. P. Bertoncello, Advances on scanning electrochemical microscopy (SECM) for energy. Energy Environ. Sci. **3**, 1620–1633 (2010)
6. A.J. Bard, F.-R. Fan, J. Kwak, O. Lev, Scanning electrochemical microscopy. Introduction and principles. Anal. Chem. **61**, 132–138 (1989)

7. C. Lee, A.J. Bard, Scanning electrochemical microscopy. Application to polymer and thin metal oxide films. Anal. Chem. **62**, 1906–1913 (1990)
8. G. Nagy, L. Nagy, Scanning electrochemical microscopy: a new way of making electrochemical experiments. Fresenius J. Anal. Chem. **366**, 735–744 (2000)
9. A.J. Bard, X. Li, W. Zhan, Chemically imaging living cells by scanning electrochemical microscopy. Biosens. Bioelectron. **22**, 461–472 (2006)
10. X. Zhang, M. Liu, W. Kong, H. Fan, Recent advances in solar cells and photo-electrochemical water splitting by scanning electrochemical microscopy. Front. Optoelectron. **11**, 333–347 (2018)
11. L. Huang, Z. Li, Y. Lou, F. Cao, D. Zhang, X. Li, Recent advances in scanning electrochemical microscopy for biological applications. Materials **11**, 1389 (2018)
12. W.S. Roberts, D.J. Lonsdale, J. Griffiths, S.P.J. Higson, Advances in the application of scanning electrochemical microscopy to bioanalytical systems. Biosens. Bioelectron. **23**, 301–318 (2007)
13. I. Morkvenaite-Vilkonciene, A. Ramanavicienea, A. Kisieliutea, V.B.A. Ramanavicius, Scanning electrochemical microscopy in the development of enzymatic sensors and immunosensors. Biosens. Bioelectron. **141**, 111411 (2019)
14. C.M. Sanchez-Sanchez, J. Rodnguez-Lopez, A.J. Bard, Scanning electrochemical microscopy. 60. Quantitative calibration of the SECM substrate generation/tip collection mode and its use for the study of the oxygen reduction mechanism. Anal. Chem. **80**, 3254–3260 (2008)
15. D. Polcari, P. Dauphin-Ducharme, J. Mauzeroll, Scanning electrochemical microscopy: a comprehensive review of experimental parameters from 1989 to 2015. Chem. Rev. **116**, 13234–13278 (2016)
16. J.L. Fernandez, A.J. Bard, Scanning electrochemical microscopy 50. Kinetic study of electrode reactions by the tip generation-substrate collection mode. Anal. Chem. **76**, 2281–2289 (2004)
17. A.J. Bard, Guydenuault, C. Lee, D. Mandler, D.O. Wipf, Scanning electrochemical microscopy: a new technique for the characterization and modification of surfaces. Acc. Chem. Res. **23**, 357–363 (1990)
18. D.O. Wipf, A.J. Bard, Scanning electrochemical microscopy, X. High resolution imaging of active sites on an electrode surface. J. Electrochem. Soc. **138**(5) (1991)
19. J. Izquierdo, L. Martín-Ruíz, B.M. Fernández-Pérez, L. Fernández-Mérida, J.J. Santana, R.M. Souto, Imaging local surface reactivity on stainless steels 304 and 316 in acid chloride solution using scanning electrochemical microscopy and the scanning vibrating electrode technique. Electrochimica Acta **134**, 167–175 (2014)
20. Y. Gonzalez-Garcıa, G.T. Burstein, S. Gonzalez, R.M. Souto, Imaging metastable pits on austenitic stainless steel in situ at the open-circuit corrosion potential. Electrochem. Commun. **6**, 637–642 (2004)
21. Y. Yin, L. Niu, M. Lu, W. Guo, S. Chen, In situ characterization of localized corrosion of stainless steel by scanning electrochemical microscope. Appl. Surface Sci. **255**, 9193–9199 (2009)
22. R. Leiva-García, R. Akid, D. Greenfield, J. Gittens, M.J. Munoz-Portero, J. García-Antón, Study of the sensitisation of a highly alloyed austenitic stainless steel, Alloy 926 (UNS N08926), by means of scanning electrochemical microscopy. Electrochim. Acta **70**, 105–111 (2012)
23. R.M. Souto András Kiss, J. Izquierdo, L. Nagy, I. Bitter, G. Nagy, Spatially-resolved imaging of concentration distributions on corroding magnesium-based materials exposed to aqueous environments by SECM. Electrochem. Commun. **26**, 25–28 (2013)
24. U.M. Tefashe, M.E. Snowden, P.D. Ducharme, M. Danaie, G.A. Botton, J. Mauzeroll, Local flux of hydrogen from magnesium alloy corrosion investigated by scanning electrochemical microscopy. J. Electroanal. Chem. **720–721**, 121–127 (2014)
25. P. Dauphin-Ducharme, R.M. Asmussen, U.M. Tefashe, M. Danaie, W.J. Binns, P. Jakupi et al., Local hydrogen fluxes correlated to microstructural features of a corroding sand cast AM50 magnesium alloy. J. Electrochem. Soc. **161**(12), C557–C564 (2014)

26. S.S. Jamali, S.E. Moulton, D.E. Tallman, M. Forsyth, J. Weber, G.G. Wallace, Evaluating the corrosion behavior of Magnesium alloy in simulated biological fluid by using SECM to detect hydrogen evolution. Electrochim. Acta **152**, 294–301 (2015)
27. J. Izquierdo, M.B. González-Marrero, M. Bozorg, B.M. Fernández-Pérez, H.C. Vasconcelos, J.J. Santanad, R.M. Souto, Multiscale electrochemical analysis of the corrosion of titanium and nitinol for implant applications. Electrochim. Acta **203**, 366–378 (2016)
28. J. Izquierdo, G. Bolat, D. Mareci, C. Munteanu, S. González, R.M. Souto, Electrochemical behavior of ZrTi alloys in artificial physiological solution simulating in vitro inflammatory conditions. Appl. Surf. Sci. **313**, 259–266 (2014)
29. G. Bolat, D. Mareci, R. Chelariu, J. Izquierdo, S. González, R.M. Souto, Investigation of the electrochemical behavior of TiMo alloys in simulated physiological solutions. Electrochim. Acta **113**, 470–480 (2013)
30. Y. Yuan, L. Li, C. Wang, Y. Zhu, Study of the effects of hydrogen on the pitting processes of X70 carbon steel with SECM. Electrochem. Commun. **12**, 1804–1807 (2010)
31. C. Gabrielli, S. Joiret, M. Keddam, H. Perrot, N. Portail, P. Rousseau, V. Vivier, A SECM assisted EQCM study of iron pitting. Electrochim. Acta **52**, 7706–7714 (2007)
32. N. Aouina, F. Balbaud-Célérier, F. Huet, S. Joiret, H. Perrot, F. Rouillard, V. Vivier, Single pit initiation on 316L austenitic stainless steel using scanning electrochemical microscopy. Electrochimica Acta **56**, 8589–8596 (2011)
33. F. Falkenberg, K. Fushimi, M. Seo, Study on initiation of localised corrosion on copper thin film electrode by combinational use of an EQCM with a liquid-phase ion gun. Corros. Sci. **45**, 2657–2670 (2003)
34. J. Izquierdo, A. Eifert, R.M. Souto, C. Kranz, Simultaneous pit generation and visualization of pit topography using combined atomic force–scanning electrochemical microscopy. Electrochem. Commun. **51**, 15–18 (2015)
35. T.E. Lister, P.J. Pinhero, Scanning electrochemical microscopy study of corrosion dynamics on type 304 stainless steel electrochem. Solid-State Lett. **5**(11), B33–B36 (2002)
36. T.E. Lister, P.J. Pinhero, Microelectrode array microscopy: investigation of dynamic behavior of localized corrosion at type 304 stainless steel surfaces. Anal. Chem. **77**(8), 2601–2607 (2005)
37. A.M. Simões, A.C. Bastos, M.G. Ferreira, Y. González-García, S. González, R.M. Souto, Corros. Sci. **49**(2), 726–739 (2007)
38. A.G. Marques, M.G. Taryba, A.S. Panão, S.V. Lamaka, A.M. Simões, Corros. Sci. **104**, 123–131 (2016)
39. J. Izquierdo, L. Nagy, S. González, J.J. Santana, G. Nagy, R.M. Souto, Electrochem. Commun. **27**, 50–53 (2013)
40. X.J. Raj, T. Nishimura, J. Mater. Eng. Perform. **25**(2), 474–548 (2016)
41. E. Tada, K. Sugawara, H. Kaneko, Electrochim. Acta **49**(7), 1019–1026 (2004)
42. A.M. Simoes, A.C. Bastos, M.G. Ferreira, Y. Gonzalez-Garcıa, S. Gonzalez, R.M. Souto, Use of SVET and SECM to study the galvanic corrosion of an iron–zinc cell. Corrosion Sci. **49**, 726–739 (2007)
43. J. Izquierdo, L. Martín-Ruíz, B.M. Fernández-Pérez, L. Fernández-Mérida, J.J. Santanac, R.M. Souto, Imaging local surface reactivity on stainless steels 304 and 316 in acid chloride solution using scanning electrochemical microscopy and the scanning vibrating electrode technique. Electrochimica Acta **134**, 167–175 (2014)
44. F. Meng, E.-H. Han, J. Wang, Z. Zhang, W. Ke, Localized corrosion behavior of scratches on nickel-base alloy 690TT. Electrochim. Acta **56**, 1781–1785 (2011)
45. E. Volker, C.G. Inchauspe, E.J. Calvo, Scanning electrochemical microscopy measurement of ferrous ion fluxes during localized corrosion of steel. Electrochem. Commun. **8**, 179–183 (2006)
46. R.K. Zhu, J.L. Luo, Investigation of stress-enhanced surface reactivity on alloy 800 using scanning electrochemical microscopy. Electrochem. Commun. **12**, 1752–1755 (2010)
47. L. Bing, X.I.A. Da Hai, Characterization of pH effect on corrosion resistance of nuclear steam generator tubing alloy by in-situ scanning electrochemical microscopy. Acta Physico Chimica Sinica **30**, 59–66 (2014)

48. J.W. Still, D.O. Wipf, Breakdown of the iron passive layer by use of the scanning electrochemical microscope. J. Electrochem. Soc. **144**, 2657 (1997)
49. K. Fushimi, K. Azumi, M. Seo, Evaluation of heterogeneity in thickness of passive films on pure iron by scanning electrochemical microscopy. ISIJ Int. **39**(4), 346–351 (1999)
50. K.A. Lill, K. Fushimi, M. Seo, A.W. Hassel, Reactivity imaging of a passive ferritic FeAlCr steel. J. Appl. Electrochem. **38**, 1339–1345 (2008)
51. K. Indira, T. Nishimura, SECM study of effect of chromium content on the localized corrosion behavior of low-alloy steels in chloride environment. JMEPEG **25**, 4157–4170 (2016)
52. K. Yanagisawa, T. Nakanishi, Y. Hasegawa, K. Fushimi, Passivity of dual-phase carbon steel with ferrite and martensite phases in ph 8.4 boric acid-borate buffer solution. J. Electrochem. Soc. **162**, C322–C326 (2015)
53. G. Bolat, J. Izquierdo, J.J. Santana, D. Mareci, R.M. Souto, Electrochemical characterization of ZrTi alloys for biomedical applications. Electrochim. Acta **88**, 447–456 (2013)
54. J.A. Ramírez-Cano, L. Veleva,, R.M. Souto, B.M. Fernández-Pérez, SECM study of the pH distribution over Cu samples treated with 2-mercaptobenzothiazole in NaCl solution. Electrochem. Commun. **78**, 60–63 (2017)
55. Z.-M. Gao, C. Wang, W.-H. Miao, R. Zhu, D.-H. Xia, Characterization of a stressed passive film using scanning electrochemical microscope and point defect model. Trans. Indian Inst. Metals **70**, 1337–1347 (2017)
56. A. Asserghine, M. Medvidović-Kosanović, L. Nagy, G. Nagy, In situ monitoring of the transpassivation and repassivation of the passive film on nitinol biomaterial by scanning electrochemical microscopy. Electrochem. Commun. **107**, 106539 (2019)
57. K. Mansikkamaki, P. Ahonen, G. Fabricius, L. Murtomai, K. Kontturi, Inhibitive effect of benzotriazole on copper surfaces studied by SECM. J. Electrochem. Soc. **152**(1), B12–B16 (2005)
58. J. Izquierdo, J.J. Santana, S. González, R.M. Souto, Uses of scanning electrochemical microscopy for the characterization of thin inhibitor films on reactive metals: The protection of copper surfaces by benzotriazole. Electrochimica Acta **55**, 8791–8800 (2010)
59. J. Izquierdo, L. Nagy, J.J. Santana, G. Nagy, R.M. Souto, A novel microelectrochemical strategy for the study of corrosion inhibitors employing the scanning vibrating electrode technique and dual potentiometric/amperometric operation in scanning electrochemical microscopy: application to the study of the cathodic inhibition by benzotriazole of the galvanic corrosion of copper coupled to iron. Electrochim. Acta **58**, 707–716 (2011)
60. J. Izquierdo, J.J. Santana, S. González, R.M. Souto, Scanning micro electrochemical characterization of the anti-corrosion performance of inhibitor films formed by 2-mercaptobenzimidazole on copper. Progr. Org. Coat. **74**, 526–533 (2012)
61. M. Pahler, J.J. Santana, W. Schuhmann, R.M. Souto, Application of AC-SECM in corrosion science: local visualization of inhibitor films on active metals for corrosion protection. Chem. Eur. J. **17**, 905–911 (2011)
62. J.J. Santana, M. Pahler, W. Schuhmann, R.M. Souto, Investigation of copper corrosion inhibition with frequency-dependent alternating-current scanning electrochemical microscopy. Chem. Plus Chem. 77, 707–712 (2012)
63. A. Singh, Y. Lin, W. Liu, S. Yu, J. Pan, C. Ren, D. Kuanhai, Plant derived cationic dye as an effective corrosion inhibitor for 7075 aluminum alloy in 3.5% NaCl solution. J. Industr. Eng. Chem. **20**, 4276–4285 (2014)
64. C. Li, L. Li, C. Wang, Study of the inhibitive effect of mixed self-assembled monolayers on copper with SECM. Electrochim. Acta **115**, 531–536 (2014)
65. N. Esmaeili, J. Neshati, I. Yavari, Scanning electrochemical microscopy for the investigation of corrosion inhibition of triazinobenzimidazole-2-thiones in hydrochloric acid solution. Res. Chem. Intermed. **42**, 5339–5351 (2016)
66. D.S. Chauhan, A. Madhan Kumar, M.A. Quraishi, Hexamethylenediamine functionalized glucose as a new and environmentally benign corrosion inhibitor for copper. Chem. Eng. Res. Des. **150**, 99–115 (2019)

67. J. Wang, Y. Lin, Z. Sun, X. Luo, W. Liu, X. Deng, Corydalis yanhusuo extract as a green inhibitor for J55 steel in 3.5%NaCl solution saturated with CO_2. Green Chem. Lett. Rev. **12**(3), 353–363 (2019)
68. S. Varvara, G. Caniglia, J. Izquierdo, R. Bostan, L. Găină, O. Bobis, R.M. Souto, Multiscale electrochemical analysis of the corrosion control of bronze in simulated acid rain by horse-chestnut (Aesculus hippocastanum L.) extract as green inhibitor. Corrosion Sci. **165**, 108381 (2020)
69. S.A. Umoren, M.M. Solomon, A. Madhankumar, I.B. Obot, Exploration of natural polymers for use as green corrosion inhibitors for AZ31 magnesium alloy in saline environment. Carbohydrate Polym. **230**, 115466 (2020)
70. A. Singh, Y. Lin, K.R. Ansari, M.A. Quraishi, E.E. Ebenso, S.S. Chen, W. Liu, Electrochemical and surface studies of some Porphines as corrosion inhibitor for J55 steel in sweet corrosion environment. Appl. Surf. Sci. **359**, 331–339 (2015)
71. A. Singh, Y. Lin, M.A. Quraishi, L.O. Olasunkanmi, O.E. Fayemi, Y. Sasikumar, B. Ramaganthan, I. Bahadur, I.B. Obot, A.S. Adekunle, M.M. Kabanda, E.E. Ebenso, Porphyrins as corrosion inhibitors for N80 steel in 3.5% NaCl solution: Electrochem. Quantum Chem. QSAR Monte Carlo Simul. Stud. Mol. **20**, 15122–15146 (2015)
72. K.R. Ansari, M.A. Quraishi, A. Singh, Pyridine derivatives as corrosion inhibitors for N80 steel in 15% HCl: electrochemical, surface and quantum chemical studies. Measurement **76**, 136–147 (2015)
73. C. Li, L. Li, C. Wang, Y. Zhu, W. Zhang, Study of the protection performance of self-assembled monolayers on copper with the scanning electrochemical microscope. Corros. Sci. **80**, 511–516 (2014)
74. A. Singh, N. Soni, Y. Deyuan, A. Kumar, A combined electrochemical and theoretical analysis of environmentally benign polymer for corrosion protection of N80 steel in sweet corrosive environment. Results Phys. **13**, 102116 (2019)
75. W. Zhang, H.-J. Li, L. Chen, S. Zhang, Y. Ma, C. Ye, Y. Zhou, B. Pang, Y.-C. Wu, Fructan from polygonatum cyrtonema Hua as an eco-friendly corrosion inhibitor for mild steel in HCl media. Carbohydrate Polym. **238**, 116216 (2020)
76. A. Singh, K.R. Ansari, A. Kumar, W. Liu, C. Songsong, Y. Lin, Electrochemical, surface and quantum chemical studies of novel imidazole derivatives as corrosion inhibitors for J55 steel in sweet corrosive environment. J. Alloys Compounds **712**, 121–133 (2017)
77. R.M. Souto, Y. Gonzalez-Garcıa, S. Gonzalez, G.T. Burstein, Damage to paint coatings caused by electrolyte immersion as observed in situ by scanning electrochemical microscopy. Corros. Sci. **46**, 2621–2628 (2004)
78. R.M. Souto, Y. González-Garcí, S. González, Evaluation of the corrosion performance of coil-coated steel sheet as studied by scanning electrochemical microscopy. Corros. Sci. **50**, 1637–1643 (2008)
79. J.J. Santana, M. Pähler, R.M. Souto, W. Schuhmann, Direct evidence of early blister formation in polymer-coated metals from exposure to chloride-containing electrolytes by alternating-current scanning electrochemical microscopy. Electrochimica Acta **77**, 60–64 (2012)
80. R.M. Souto, Y. González-García, S. González, Characterization of coating systems by scanning electrochemical microscopy: surface topology and blistering. Progr. Org. Coat. **65**, 435–439 (2009)
81. Y. Gonzalez-Garcia, J.J. Santana, J. Gonzalez-Guzman, J. Izquierdo, S. Gonzalez, R.M. Souto, Scanning electrochemical microscopy for the investigation of localized degradation processes in coated metals. Prog. Org. Coat. **69**, 110–117 (2010)
82. J.J. Santana, J. González-Guzman, L. Fernández-Mérida, S. González, R.M. Souto, Visualization of local degradation processes in coated metals by means of scanning electrochemical microscopy in the redox competition mode. Electrochim. Acta **55**, 4488–4494 (2010)
83. Y. González-García, J.M.C Mol, T. Muselle, I. De Graeve, G. Van Assche, G. Scheltjens, B. Van Mele, H. Terryn, SECM study of defect repair in self-healing polymer coatings on metals. Electrochem. Commun. **13**, 169–173 (2011)

84. A. Pilbáth, T. Szabo, J. Telegdi, L. Nyikos, SECM study of steel corrosion under scratched microencapsulated epoxy resin. Prog. Org. Coat. **75**, 480–485 (2012)
85. S. Yang, J. Wang, W. Mao, D. Zhang, Y. Guo, Y. Song, J.-P. Wang, T. Qi, Guo Liang Li, pH-responsive zeolitic imidazole framework nanoparticles with high active inhibitor content for self-healing anticorrosion coatings. Colloids Surf. A **555**, 18–26 (2018)
86. W. Wang, H. Wang, J. Zhao, X. Wang, C. Xiong, L. Song, R. Ding, P. Han, W. Li, Self-healing performance and corrosion resistance of graphene oxide–mesoporous silicon layer–nanosphere structure coating under marine alternating hydrostatic pressure. Chem. Eng. J. **361**, 792–804 (2019)
87. W. Wang, X. Likun, H. Sun, X. Li, S. Zhao, W. Zhang, Spatial resolution comparison of AC-SECM with SECM and their characterization of self-healing performance of hexamethylene diisocyanate trimer microcapsule coatings. J. Mater. Chem. A **3**, 5599–5607 (2015)
88. N. Jadhav, M.B. Jensen, V. Gelling, Tungstate and vanadate-doped polypyrrole/aluminum flake composite coatings for the corrosion protection of aluminum 2024–T3. J. Coat. Technol. Res. **12**, 259–276 (2015)
89. A.M. Simoes, D. Battocchi, D.E. Tallman, G.P. Bierwagen, SVET and SECM imaging of cathodic protection of aluminum by a Mg-rich coating. Corros. Sci. **49**, 3838–3849 (2007)
90. X. Wang, Q. Liu, Y. Chun, Y. Li, Z. Wang, Evaluation of delamination of X80 pipeline steel coating under alternating stray current via scanning electrochemical microscopy. JMEPEG **27**, 3060–3071 (2018)
91. D.-H. Xia, J. Wang, W. Zhong, Z. Qin, X. Likun, H. Wenbin, Y. Behnamiand, J.-L. Luo, Sensing corrosion within an artificial defect in organic coating using SECM. Sens. Actuators B Chem. **280**, 235–242 (2019)
92. J. Vega, H. Scheerer, G. Andersohn, M. Oechsner, Experimental studies of the effect of Ti interlayers on the corrosion resistance of TiN PVD coatings by using electrochemical methods. Corros. Sci. **133**, 240–250 (2018)
93. H. Ma, Y. Gua, S. Liu, J. Che, D. Yang, Local corrosion behavior and model of micro-arc oxidation HA coating on AZ31 magnesium alloy. Surf. Coat. Technol. **331**, 179–188 (2017)
94. C. Vosgien Lacombre, G. Bouvet, D. Trinh, S. Mallarino, S. Touzain, Effect of pigment and temperature onto swelling and water uptake during organic coating ageing. Progr. Org. Coat. **124**, 249–255 (2018)
95. D. Trinh, C. Vosgien-Lacombre, G. Bouvet, S. Mallarino, S. Touzain, Use of ionic liquids in SECM experiments to distinguish effects of temperature and water in organic coating swelling. Progr. Org. Coat. **139**, 105438 (2020)
96. Y. Elkebir, S. Mallarino, D. Trinh, S. Touzain, Effect of physical ageing onto the water uptake in epoxy coatings. Electrochimica Acta **337**, 135766 (2020)
97. A. Madhankumar, N. Rajendran, T. Nishimura, Influence of Si nanoparticles on the electrochemical behavior of organic coatings on carbon steel in chloride environment. J. Coat. Technol. Res. **9**, 609–620 (2012)
98. R.K. Suleiman, A. Madhan Kumar, A.Y. Adesina, F.A. Al-Badour, M.H. Meliani, T.A. Saleh, Hybrid organosilicon-metal oxide composites and their corrosion protection performance for mild steel in 3.5% NaCl solution. Corrosion Sci. **169**, 108637 (2020)
99. N. Casillas, P. James, P.W.H. Smyrl, Novel approach to combine scanning electrochemical microscopy and scanning photoelectrochemical microscopy. J. Electrochem. Soc. **142**, L16 (1995)
100. V. Kuznetsov, A. Maljusch, R.M. Souto, A.S. Bandarenka, W. Schuhmann, Characterization of localized corrosion processes using scanning electrochemical impedance microscopy. Electrochem. Commun. **44**, 38–41 (2014)
101. A. Estrada-Vargas, A. Bandarenk, V. Kuznetsov, W. Schuhmann, In situ characterization of ultrathin films by scanning electrochemical impedance microscopy. Anal. Chem. **88**, 3354–3362 (2016)
102. A.S. Bandarenka, A. Maljusch, V. Kuznetsov, K. Eckhard, W. Schuhmann, Localized impedance measurements for electrochemical surface science. J. Phys. Chem. C **118**, 8952–8959 (2014)

103. C. Gabrielli, S. Joiret, M. Keddam, H. Perrot, N. Portail, P. Rousseau, V. Vivier, Development of a coupled SECM-EQCM technique for the study of pitting corrosion on iron. J. Electrochem. Soc. **153** (2006)
104. N. Aouina, F. Balbaud-Célérier, F. Huet, S. Joiret, H. Perrot, F. Rouillar, V. Vivier, Single pit initiation on 316L austenitic stainless steel using scanning electrochemical microscopy. Electrochim. Acta **56**, 8589–8596 (2011)
105. C. Sençz, A. Maljusch, M. Rohwerder, W. Schuhmann, SECM and SKPFM studies of the local corrosion mechanism of Al alloys—a pathway to an integrated SKP-SECM. Electroanalysis **24**(2), 239–245 (2012)
106. A. Davoodi, J. Pan, C. Leygraf, S. Norgren, The role of intermetallic particles in localized corrosion of an aluminum alloy studied by SKPFM and integrated AFM/SECM. J. Electrochem. Soc. **155** (2008)
107. L.C. Yule, C.L. Bentley, G. West, B.A. Shollock, P.R. Unwin, Scanning electrochemical cell microscopy: a versatile method for highly localized corrosion related measurements on metal surfaces. Electrochim. Acta **298**, 80–88 (2019)
108. E. Martinez-Lombardia, Y. Gonzalez-Garcia, L. Lapeire, I. De Graeve, K. Verbeken, L. Kestens, J.M.C. Mol, H. Terryn, Scanning electrochemical microscopy to study the effect of crystallographic orientation on the electrochemical activity of pure copper. Electrochim. Acta **116**, 89–96 (2014)
109. K. Fushimi, Y. Takabatake, T. Nakanishi, Y. Hasegawa, Microelectrode techniques for corrosion research of iron. Electrochim. Acta **113**, 741–747 (2013)
110. L. Verchere, I. Aubert, O. Devos, Influence of the crystallographic orientation on the electrochemical reactivity measured by scanning electrochemical microscopy on nickel-based alloy 600. Electrochim. Acta **313**, 292–302 (2019)
111. R. Chen, R.J. Balla, A. Lima, S. Amemiya, Characterization of nanopipet-supported ITIES tips for scanning electrochemical microscopy of single solid-state nanopores. Anal. Chem. **89**, 9946–9952 (2017)
112. S. Daboss, J. Lin, M. Godejohann, C. Kranz, Redox switchable polydopamine-modified AFM-SECM probes: a probe for electrochemical force spectroscopy. Anal. Chem. **92**, 8404–8413 (2020)

Electrochemical Impedance Spectroscopy: A Useful Tool for monitoring the Performance of Corrosion Inhibitors

Saman Zehra, Ruby Aslam, and Mohammad Mobin

Abstract This chapter intends to present the fundamental application of Electrochemical Impedance Spectroscopy (EIS) in the assessment of the efficiency of corrosion inhibitors. EIS is an impressive technique that employs the utilization of a small amplitude AC signal to inquire about the characteristics of the impedance of the cell by scanning over an extensive range of frequencies, which generates the spectrum for the electrochemical cell under the examination. It is an efficient approach that evaluates the persistency of the films formed by the corrosion inhibitors. It is merely used in order to explain the formation and destruction of the film of corrosion inhibitors thus formed on the metal substrate. It also helps in the prediction of the mechanism of corrosion protection by the studied inhibitors. So, this chapter will concisely cover the application of EIS as a key tool in investigating their capability to act as corrosion inhibitors.

Keywords Corrosion · Corrosion inhibitors · Electrochemical technique · EIS · Equivalent circuits

1 Introduction

Electrochemical Impedance Spectroscopy (EIS) is an effective, flexible, highly efficient and fast, non-destructive technique and is commonly used in various applications, such as corrosion research [1], biomedical devices [2], batteries [3], semiconductors and solid state devices [4], sensors [5], fuel cells [4], etc. It is a transfer function method applied to the electrochemical processes.

Electrochemical techniques have already been successfully used in the field of corrosion inhibition and protection like in the studies of the coatings, the analysis of the passive layers, and the evaluation of the corrosion inhibitors [6, 7]. EIS holds a special spot amongst the various electrochemical techniques, as it already has shown its applicability in the provision of corrosion and protection mechanism details. It

S. Zehra (✉) · R. Aslam · M. Mobin
Corrosion Research Laboratory, Department of Applied Chemistry, Faculty of Engineering and Technology, Aligarh Muslim University, Aligarh 202 002, India

I. Toor (ed.), *Recent Developments in Analytical Techniques for Corrosion Research*,
https://doi.org/10.1007/978-3-030-89101-5_5

provides useful information when an applied coating or adsorbed film is present on the wide range of materials in different aggressive media. From the EIS data mechanistic information, surface properties and electrode kinetics can be predicted [8].

In the field of corrosion inhibitors, EIS is a helpful technique for evaluating the inhibitors' film persistency, and their performance and has been proven a powerful tool for predicting the mechanics of the electrochemical processes that are probably to occur at the interface of metal/aggressive media, which is useful for the determination of the rate of the corrosion, already proven by the literature [9–11]. In the past few years, the application of EIS has dramatically increased. The potential of EIS to elucidate the plethora of electronic and physical features of electrochemical structures such as the diffusion coefficient, adsorption processes, resistances to charge transfer, capacitances, etc. The data obtained from the EIS data is used to calculate the corrosion related electrochemical parameters and proved to be a very significant tool for evaluation of the mechanisms of inhibitory action.

This chapter will cover concisely the fundamental principles and application of EIS as a key tool in the investigation of corrosion inhibitors.

2 Basics of EIS

Alternating electrochemical perturbation (e.g., ac-potential) is applied to the system being examined during the EIS experiments. The perturbation is characterized as the potential or the current in sine wave at a certain frequency or a germination of some different frequency sine waves [7]. Electrical resistance (R) is defined, according to Ohm's law, as the voltage (V) to current (I) ratio, i.e. R = V/I, and is described as the conductor's propensity to resist the passage of electrical current through it. But this law is limited to the single circuit element, the ideal resistors [6]. However the circuit elements in the real show much more complex behavior. In the real world there are two more mechanism which impede or retards the electrical current in the electrical circuits, which are:

(a) Due to the magnetic field of the electric current, called the inductance indicated by the symbol L, the induction of voltages in the conductors is self-induced and the voltage between the conductors triggers electrostatic charge storage, referred to as the capacitance indicated by the symbol C.
(b) Both inductance and capacitance together are called as the reactance (X). In the AC circuits, the concept of resistance has been replaced by the quantity called as Impedance (Z). Likewise the resistance, Impedance is also the measure of the tendency of a circuit to impede the flow of the electrical current. The only difference between both of them is, the ideal resistance have only magnitude but impedance is defined by both the magnitude and phase angle, and represented by complex number.

Generally, impedance constitutes of a real part i.e. resistive (resistance) and an imaginary part i.e. the capacitive or the inductive (reactance). The phase angle of the complex impedance, meanwhile, is the phase change from which the voltage is ahead of the current. By adding AC potential to an electrochemical cell, the value of the parameter impedance is calculated. The current through the cell or a minor disruption in the fixed frequency potential or current is then measured. The response is measured and the impedance at each frequency is computed by [8]:

$$-Z\omega = E\omega / I\omega \tag{1}$$

where, $E\omega$ = Frequency-dependent potential and $I\omega$ = Frequency-dependent current. Now in the ohm's law R represents the impedance at the limit of zero frequency.

In the domain of corrosion, the actual component of the impedance is usually associated with the electrical resistance of the electrolytic solution, the charge transfer resistance of the corrosion reactions and the Ohmic resistance of the surface film of an electrochemical interface of the corrosion system like a metal dipped into the electrolyte. The imaginary component, also influenced by the surface film, is correlated with double layer capacitance at the metal/electrolytic solution interface and adsorption/desorption processes at the surface of the metals. A parameter known as polarization resistance is often used as a corrosion resistance measure in corrosion science and engineering. It is known as the resistance to corrosion reactions under open circuit conditions and largely depends on the resistance to charge transfer, the Ohmic resistance of the electrolyte and the surface film, and the resistance to diffusion in the case of corrosion regulated by mass transport. In reality, some of these variables are often dominant in the resistance of polarization.

3 Representation of the EIS Data

The most common ways in which the EIS spectra are interpreted and the methods from which the most useful knowledge can be extracted from the EIS spectra are now very useful to know at this point. Data from the EIS may be interpreted either as vector quantities (i.e. impedance magnitude and phase angle) or complex quantities (Ztotal = Zreal + Zimag), as seen in Fig. 1. Both are distinct types of impedance representation, but are mathematically equivalent.

Bode diagram and the Nyquist plots are commonly used to represent the impedance data.

(a) **BODE PLOTS**

In Fig. 2a, the phase diagram helps to represent the magnitude of the impedance vector as and the phase angle, θ, of the complex impedance [12]. In terms of real and imaginary elements, the magnitude of the impedance (Z) can be represented by the following equation [8]:

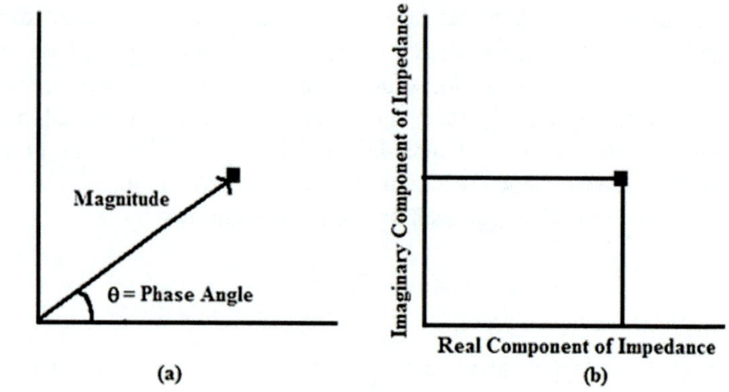

Fig. 1 **a** Vector Plane plot, and **b** Complex Plane plot, for the representation of EIS data

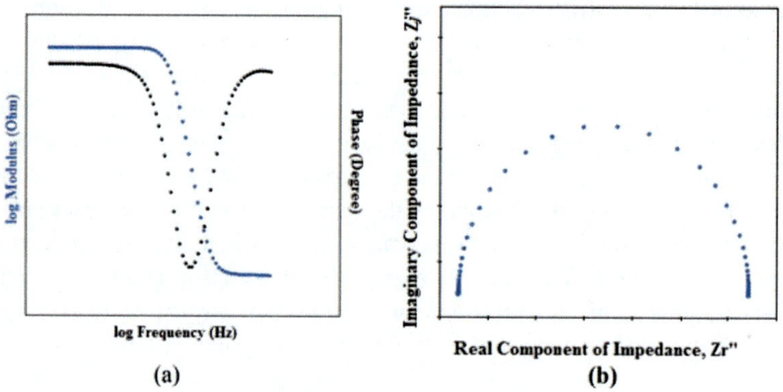

Fig. 2 General representation of **a** Bode Plot, and **b** Nyquist plot

$$z = \sqrt{Zr^2 + Zj^2} \tag{2}$$

where, Zr = the real component of the impedance and Zj = the imaginary component of the impedance.

The phase angle (θ) can be estimated by utilizing the following equation [8]:

$$\theta = tan^{-1}(\frac{Zj}{Zr}) \tag{3}$$

Bode diagrams are plotted on the log scales for quick estimation of small values owing to the wide variety of values frequently found in and f. Again, the high-frequency limit of yields Rs, and Rs + Rr yields the low-frequency limit. In the imaginary part, the slope of the transition area between the two asymptotic limits provides knowledge about the force of frequency dependence. The frequency at which $\varphi = -45^{\circ}$ can give the feature's characteristic frequency.

(b) **NYQUIST PLOTS**

The X-axis, on which the real element of impedance is taken, is plotted against the y-axis, on which the imaginary portion of impedance is taken [13], given in Fig. 2b. In the Nyquist map, also known as complex plane plots or Cole-Cole plots. The corresponding impedance at a given frequency is defined by any point in the map. In the electrochemistry group, it is a common convention to plot −Zimag (also written as −Z"or −Zj") on the y-axis such that the detail falls into the first quadrant of a graph. Although the Nyquist plots are useful for extracting details about the system's characteristic characteristics, all the frequency information is inherently lost. In the Nyquist map, there is one big drawback and the frequency cannot be calculated by merely visualizing the graphs. By annotating the frequencies of main data points such as high and low real axis intercepts and the characteristic frequency of the arc [14], this constraint has been solved.

4 Electrical Circuits in EIS

For the analysis of the data obtained from the EIS, a suitable equivalent electrical circuit model is fitted. Most of the circuit elements in the model are common electrical elements such as resistors, inductors and capacitors [15]. Equivalent electrical circuit model can be constructed by utilizing the passive elements, given in the Fig. 3, connected in network either in series and/or parallel combinations.

The impedance of the ideal resistors, capacitors inductors and Constant phase elements are given by the following equations [16]:

$$Z_R = R \tag{4}$$

$$Z_C = \frac{-j}{\omega.C} \tag{5}$$

$$Z_L = j\omega L \tag{6}$$

The impedance of an ideal resistor will be purely real and the current across the resistor will always be in phase with the voltage through it, i.e. indicating the phase change of 0 (in phase), whereas the impedance will be purely imaginary of an ideal inductor and capacitor. The impedance of an inductor would be opposite to that

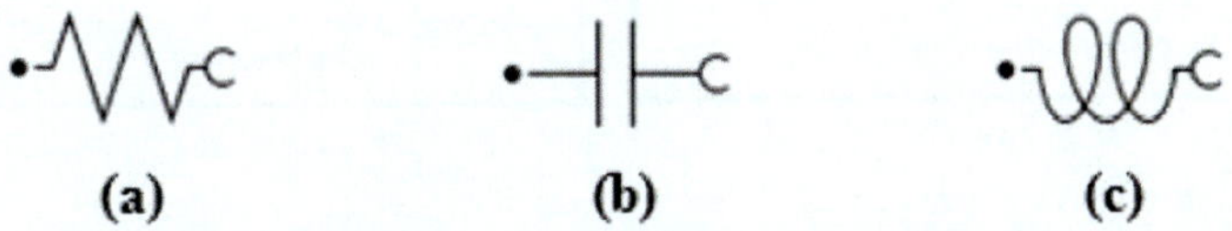

Fig. 3 Passive elements of the Equivalent electrical circuits: **a** Resistor, **b** Capacitor, and **c** Inductor

of a capacitor versus frequency response. The impedance of an inductor increases as the frequency is increased, while the impedance of the capacitor decreases with an increase in frequency. The capacitors show a phase change of −90 'while the inductors show a phase shift of 90' (out of phase) [17].

A simplified Randle's circuit, illustrated in Fig. 4, is one of the most commonly used cell models and a starting point for other complex models [8, 18].

The equivalent circuit of the simplified Randle consists of a solution resistance, a double layer capacitance and a charge transfer or a polarization resistance. The capacitance of the double layer (C*dl*) is parallel with the resistance of charge transfer (R*ct*) and this parallel combination is in series with the resistance of the solution (R*ct*). There is a semicircular arc in the accompanying Nyquist plot (Fig. 5a) intersecting the real axis (Z') at two points. The value of solution resistance (R*s*) is given by the intercept nearest to the origin and the value of overall resistance (R*s* + R*ct*) is given by the intercept farthest from the origin. Similarly, the high-frequency limit of yields R*s* and the low-frequency limit returns R*s* + R*ct* [19] in the Bode plot.

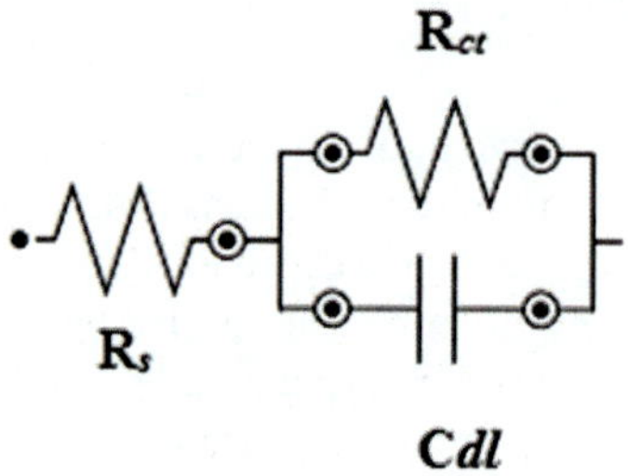

Fig. 4 Simple Randles equivalent circuit used for the fitting of EIS data for a corrosion system under charge transfer control, with one time constant feature

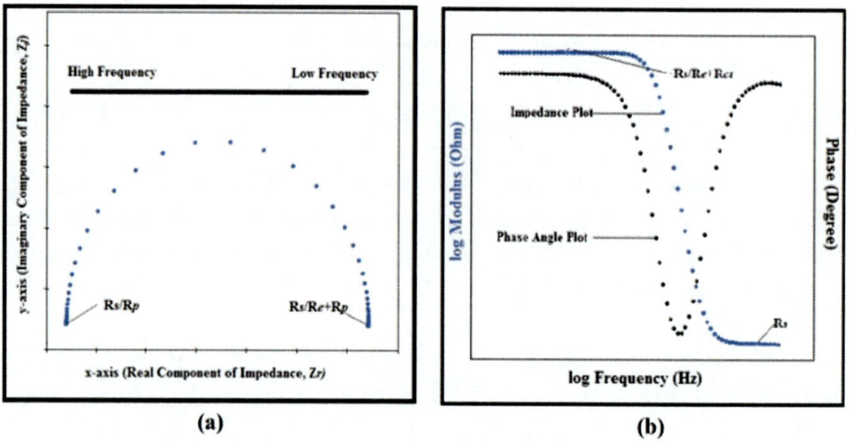

Fig. 5 **a** Nyquist and **b** Bode plot representing the solution resistance and charge transfer resistance

Based on the physical characterization of the electrode, more complicated models can be created, such as partly blocked electrode, electrode coated with an inert porous layer, completely porous electrode, electrode coated with several layers of inert porous layers, etc. For a heavily nested electrical circuit, the corresponding impedance is determined by lumping the basic circuits together. Some other modeling elements like Warburg impedance (W) and constant phase elements (CPE), given in Fig. 6 are also sometimes incorporated into the electrochemical circuit for much complex systems. W is the general impedance which is associated to the mass transfer processes i.e. diffusion control reactions, and causes the phase shift of 45°. On the other hand CPE is a very general element used to model the non-ideal or the imperfect capacitors. It causes phase shift of 80–90°. The impedance of the Warburg impedance and Constant phase elements are given by the following equations [20]:

$$Z_w = \frac{1}{Y_0\sqrt{j\omega}} \tag{7}$$

$$Z_Q = \frac{1}{Y_0(j\omega)^n} \tag{8}$$

The specific electrochemical interface on the microscopic level is not something that is-presumed smooth, uniform, and non-defective surface which is typical for a pure capacitor [21]. It comprises a substantial range of structural characteristics including porosity, surface inhomogeneities, and surface roughness. Correspondingly, the semicircle appears flattened in the Nyquist plot, as can be seen in Table 1 (point 2). Constant phase elements (CPE) are picked to substitute the double-layer capacitance (and other capacitances) in the equivalent circuits due to the extreme non-uniform distribution of current flow originating by surface inhomogeneity. The CPE's impedance is identified as:

$$\mathrm{Z_{CPE}} = 1/\mathrm{Y}(\mathrm{j}\omega)\alpha \tag{9}$$

Where Y is an electrode capacitance-related parameter and α is a coefficient between 0 and 1. A straight line with a slope of 45° is identified at the lower frequencies in the Nyquist plot [22] when the corrosion mechanism is controlled

Fig. 6 **a** Warburg impedance, **b** Constant phase Element, other circuit elements used for the fitting of EIS data

Table 1 A general representation of shapes of Nyquist and Bode plots and corresponding equivalent circuits utilized: (1) Simple Randles equivalent circuit, (2) Randles equivalent circuit modified with a constant phase element, (3) Randles equivalent circuit modified with a semi-infinite Warburg diffusional impedance element, (4) Randles equivalent circuit modified with an inductor and a resistance, (5) equivalent circuit with two time constants (6) equivalent circuit with three time constants

S. no	Nyquist plots	Bode plots	Equivalent circuit
1	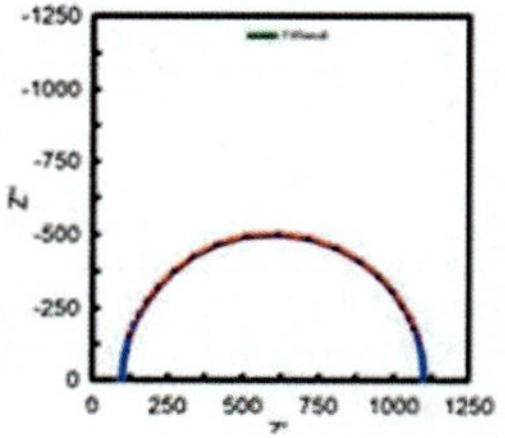	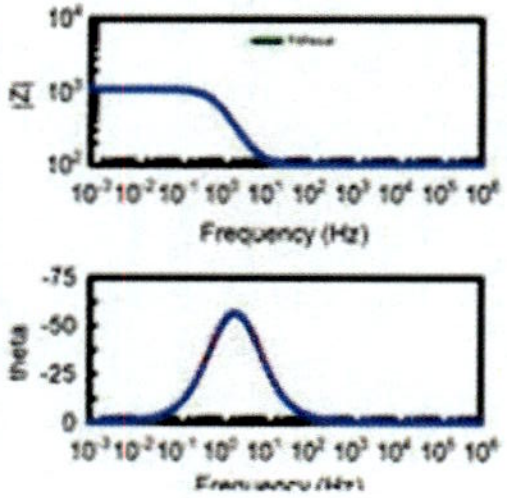	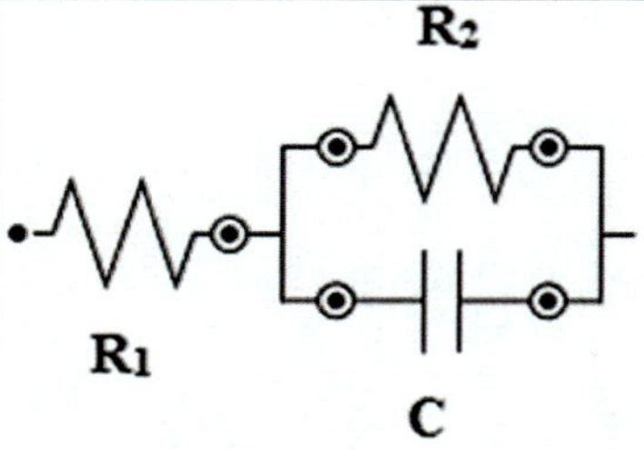

(continued)

Table 1 (continued)

S. no	Nyquist plots	Bode plots	Equivalent circuit
2	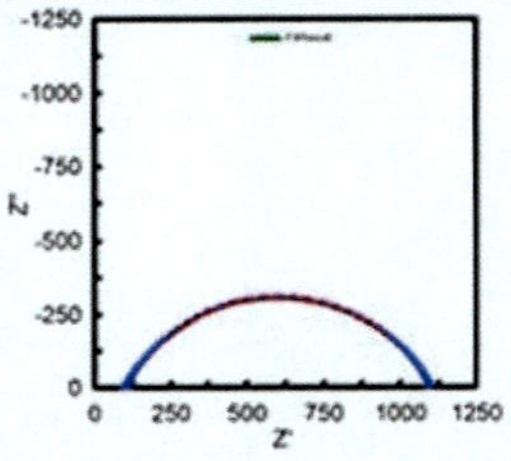	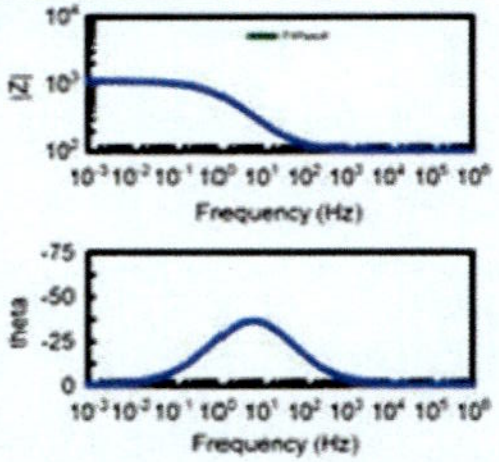	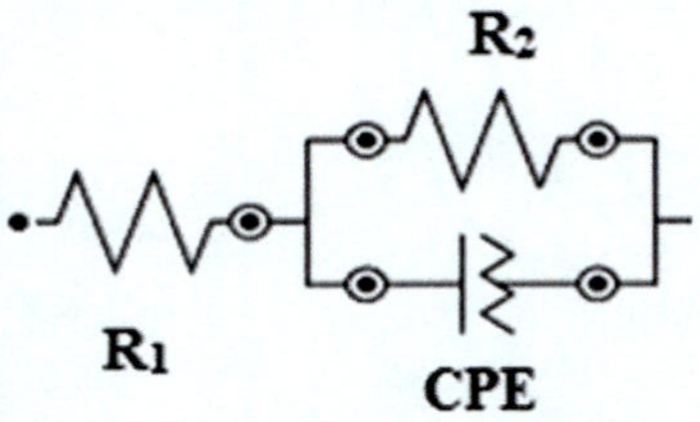
3	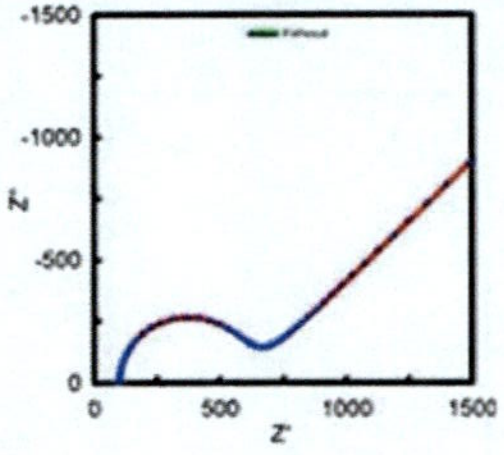	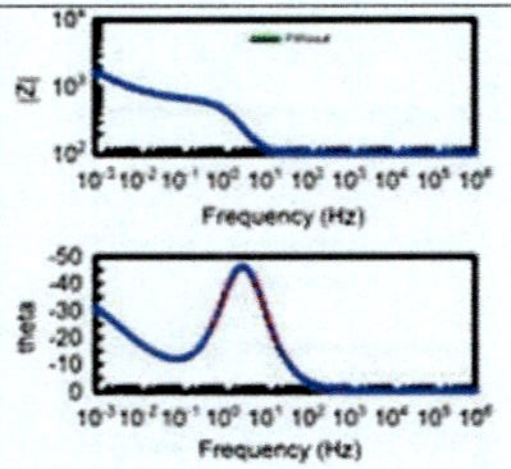	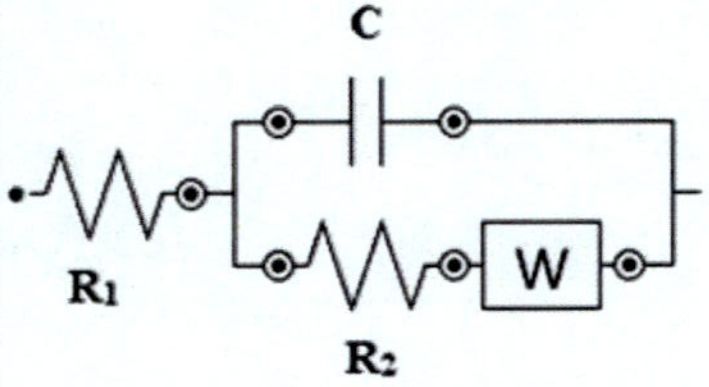

(continued)

Table 1 (continued)

S. no	Nyquist plots	Bode plots	Equivalent circuit
4	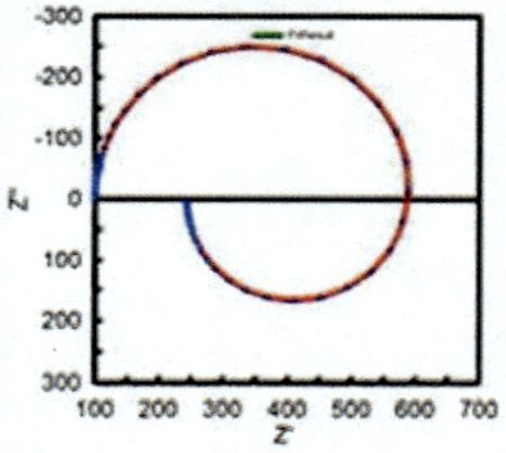	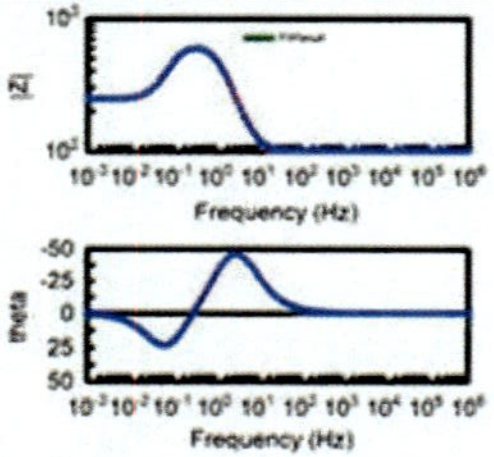	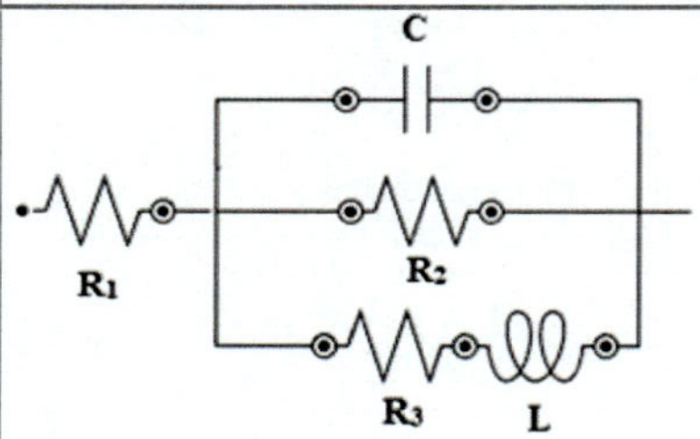
5	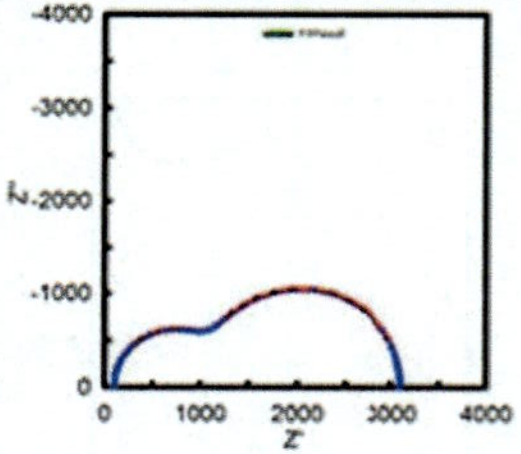	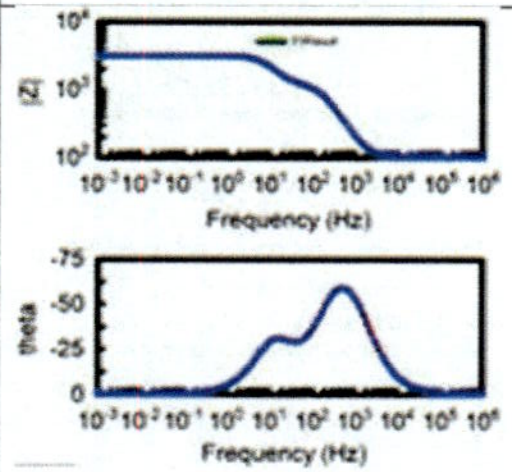	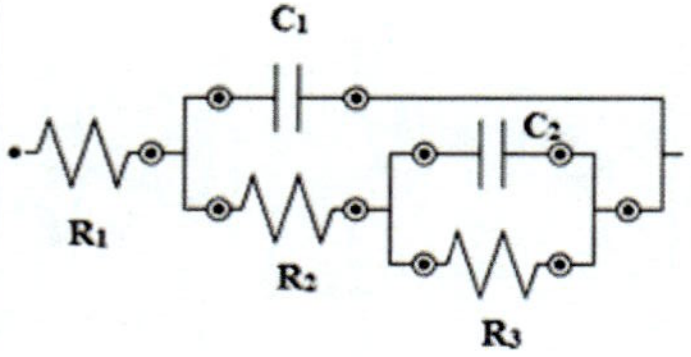

(continued)

Table 1 (continued)

S. no	Nyquist plots	Bode plots	Equivalent circuit
6	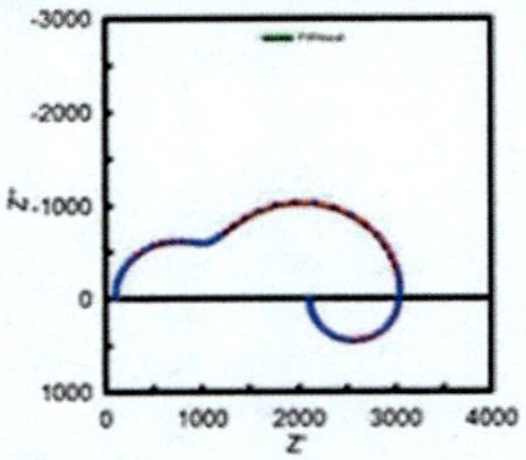	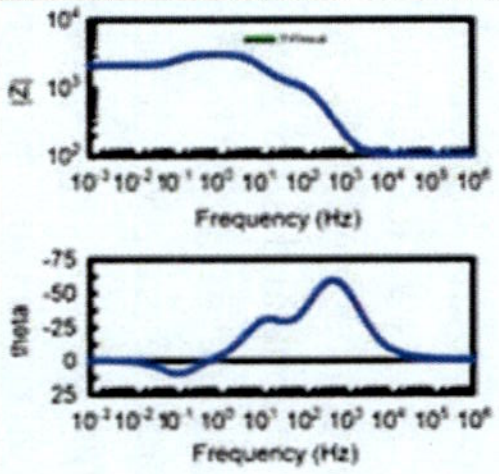	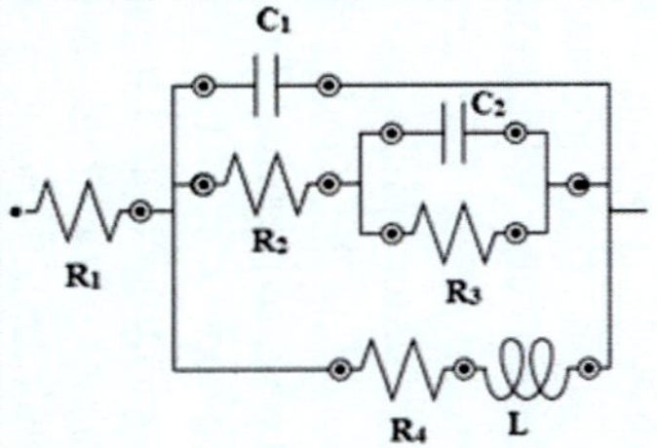

not only by a charge-transfer process but also by the diffusion of charged species through the corrosion products (Table 1 point 3). Thus, to simulate the EIS results, a transformed Randles-type circuit model with a semi-infinite Warburg diffusional impedance element (W) is being used. Such low-frequency tails may have been anticipated to be closely linked to the electroactive species' diffusion coefficient thru the corrosion layers. To the author's awareness, comparatively few corrosion EIS studies have recorded Nyquist plots characterised by a single high frequency capacitive loop and a low frequency inductive loop (Table 1 point 4). Pitting corrosion and associated adsorption/desorption of intermediates on the electrode surface or due to rapid anodic dissolution may be the cause for the existence of the inductive ring. As the corrosion protective layer is formed on the surface, the inductive loop could vanish, whereas larger inductive loops suggest more serious pitting corrosion. A resistance (R3) and an inductor (L) should be introduced into the standard Randle circuit in sight of the inductive response encountered at low frequencies. The development of protective corrosion component layers or the appearance of coatings on the surface of metals or alloys may also be observed by electrochemical impedance experiments. In such instances, the EIS continuum consists of two high- and low-frequency capacitive loops and there is no inductive loop (Table 1, point 5). And in equivalent circuit, the resistance and capacitance attributes of the protective corrosion layer or coating are R1 and C1, the charge transfer resistance is R2, and the capacitance of the electric double layer is marked by C2. It is common to see very complicated impedance diagrams, displaying a capacitive arc at high frequencies, a second capacitive arc at medium frequencies, and an inductive loop at the lowest frequencies (Table 1 point 6) throughout the literature discussing the usage of the EIS technique to research corrosion. However, a contentious and ambiguous issue is the understanding of the two capacitive arcs at high and medium frequencies, respectively. Some research demonstrates that the capacitive arc of high frequency could be attributable to the processes of charge transfer and the capacity effect of the electrochemical double layer, while the capacitive arc of medium frequency can be aligned with mass transport processes related to the diffusion of metal ions via the layer of corrosion products. Other works, on the other hand, relate the capacitive arc of high frequency to the resistance of passivating films or to layers of corrosion materials, while the capacitive arc of medium frequency has been attributed to the reaction of charge transfer [23].

5 Application of EIS in the Field of Corrosion Inhibitors

Corrosion Inhibition by inhibitor formulation is the most cost-effective and versatile means of corrosion protection to protect metallic surfaces in closed structures from corrosion in aggressive media. By being adsorbed onto the surface, the corrosion inhibitor covers the metallic surface, thereby creating a shield or film. Film formation minimizes the interaction between the metal surface and the hostile atmosphere around it. Corrosion inhibitor formulations have been used in numerous industries, such as steel part acid pickling, water recirculation method, petroleum refineries, chemical processing, water treatment, etc. Corrosion inhibitors have long been known to be the first line of protection in the oil mining, refining and chemical industries. For car radiators, one of the most important applications of inhibitors is antifreeze. There are many tools that can be used for measuring the efficacy of the inhibitor in the laboratory and in the field, such as the gravimetric method, electrochemical methods and the resistance probe. EIS is essentially a transfer function approach applicable to electrochemical systems, as we already know. There is an exciting potential for the methodology to measure essential properties [24]. Therefore, the dynamics of bound or mobile charges in the bulk or interfacial area of such a device is a very effective tool to examine. EIS may be a very important way of supplying knowledge on the dynamics of electrochemical processes that are expected to occur at the metal/electrolytic solution interface in order to assess the corrosion rate. The ability of the EIS to provide details on corrosion and safety mechanisms has already been shown, in particular when adsorbed film or organic coating is applied. The results of the EIS are used to analyze changes in the absence and presence of the inhibitor at the interface of the steel/solution.

5.1 Methodology Used for the Evaluation of Performance of the Corrosion Inhibitors

Nowadays, EIS is conducted on Potentiostat/Galvanostat with an optimised impedance analyzer through decades of continuous research and development. The electrochemical cell comprises of 3 configurations of electrodes. The working, counter and reference electrodes are used for base metal under examination, platinum/Stainless Steel wires, and Ag/AgCl. The reference electrode must be attached through a Luggin capillary and KCl salt bridge to the corrosion cell. A complex corrosion cell used for the application of EIS in the field of corrosion is represented in Fig. 7. The experiments are performed according to the method determined by ASTM G 106 Impedance spectra are usually collected at the potential of OCV or different applied potentials based on the necessity of the testing conditions under study [25].

5.2 EIS Spectra for the Evaluation of Performance of the Corrosion Inhibitors

Different corrosion mechanisms can have different characteristics in the EIS continuum (e.g., charge transfer control, diffusion control or type of mixture). The corrosion process of the device can be identified [26–28] by the study of the EIS data (Nyquist map, Bode plots).

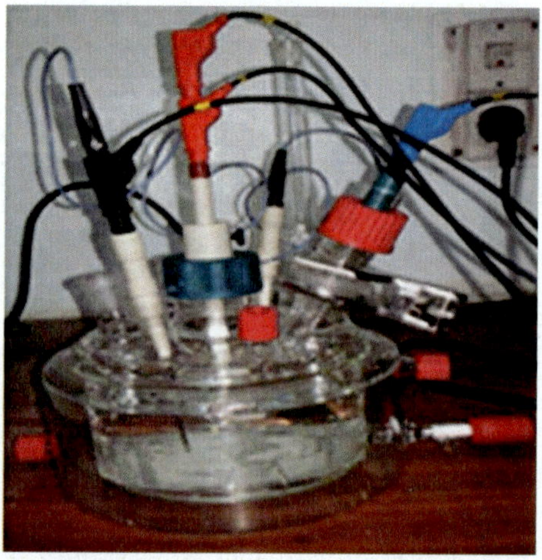

Fig. 7 Corrosion cell used in EIS for the calculation of corrosion related electrochemical parameters

As is already apparent, the Nyquist plot is a complete semi-circle, but the Nyquist plot showed a single depressed semi-circle corresponding to one or two time constants (it could be seen from the Bode plots) in addition to inhibitors to the violent media. The Nyquist plot's single capacitive loop is an example of the interface activity of the repression of the inhibitor molecules. The frequency of interfacial impedance disbursal attributable to the roughness or inhomogeneity of the MS surface is described by depression in the Nyquist plot semi-circle. The impact of a non-uniform surface (i.e. increased roughness of the surface) can easily be seen in semi-perfect circles [29].

The inhibition effect of inhibitor as visualized by Nyquist plots can be further confirmed by the corresponding Bode diagrams. The phase angle at high frequencies offers a general idea of the anti-corrosion performance of the inhibitors. The more negative the phase angle, the more capacitive the electrochemical behavior. This can be due to an improved surface coverage of the metal to be covered by the inhibitor molecules' absorbed barrier coating and, therefore, a drop in surface roughness. The involvement of the inhibitor in the hostile solution results in an improvement in absolute impedance values at low frequencies in the Bode impedance plot (impedance vs. frequency plots), showing better safety in the inhibited solution. As the suppression effect grew, the symptoms became more pronounced [30].

5.3 Evaluation of Corrosion Related Electrochemical Parameters for Examining the Performance of the Corrosion Inhibitors

In reality, EIS data is also translated as defining the electrical properties of electrochemical interfaces in terms of electrical equivalent circuits. Under the charge transfer power, a simple corrosion device can be represented by a simple equivalent circuit, ignoring the concentration olarization. By analysis of the Nyquist plot, the values of Rs, Rct and Cdl can be obtained. The parameters of the equivalent circuit, as can be seen in Fig. 4, can also be estimated by the Bode plot. The number of time constants counts the set of peaks. If the phase angle of bode plot is just one peak, the device has only one time constant. Two peaks in the Bode plot's phase angle mean that the system is defined as two time constants. The adsorbed inhibitor film on the steel surface [31] may be due to this second time constant.

Faradaic (current flow results in electrochemical reaction) and non-faradaic (current flow does not result in electrochemical reaction) components are used in the current flow due to electron transfer through the electrified interface. By overcoming an effective activation barrier, namely the polarization resistance (R*p*) and charge transfer resistance (R*ct*), the Faradaic portion emerges from the electron transfer through the interface. The non-faradaic current occurs from the charging at the interface of the double layer capacitor that requires a double layer capacitance (C*dl*). Charge transfer resistance due to slow activation reaction speeds on both the anode and the cathode is the main kinetic constraint. The charge transfer at the interface also influences the mass movement of the reactant and the product; minimal resistance is thus integrated into the diffusion.

The experimental impedance continuum for multiple reactions and interfaces can now be collected, tested and quantitatively interpreted. However, the simulation of electrode processes is the most complicated challenge in EIS for complex systems, i.e. more time constants, because different electrical circuits can be used to match the same experimental range, which is where most of the complications and errors occur.

5.4 *Interpretation of the Data Obtained from the EIS*

(a) *Solution or Electrolytic Resistance (Rs or Re)*

In an electrochemical cell, an important element relating to the total impedance of the cell is the resistance provided by the ionic solution (electrolyte). The ionic solution resistance depends on the ion form, the ionic concentration, the temperature, and the configuration of the predicted region in which the current is carried. In determining the conductivity of the electrolyte, the specification of the current flow direction and the geometry of the electrolyte solution are important parameters to consider. In the Nyquist plot, as can be seen in Fig. 5a, Rs denotes the intercept adjacent to the origin. By connecting the EIS to an equivalent electric circuit model, the electrolyte resistance can be computed. When inhibitors are incorporated to the electrolytic solution during the trial, this does not improve to any substantial degree, assuming that the Luggin probe is reproducibly located on the active electrode surface.

(b) ***Double Layer Capacitance (Cdl)***

At any interface in the polarised device, such as the interface between the electrode and the electrolyte, ion exchange membranes, etc., charging or electrical double layer isolation occurs, which is similar to a capacitor in the electrical circuit. The double layer capacitance value depends on the different parameters, such as electrode polarization, ionic concentration, temperature, ion type, oxide layer, electrode roughness, etc.

The values of *Cdl* are derived from the CPE by utilizing the following equation [24]:

$$\mathrm{C_{dl}} = Y_o(\omega_{\max})^{n-1} \tag{10}$$

where, $\omega\max = 2\pi f\max$ ($f\max$ represents the maximum frequency at which the imaginary component of the impedance has a maximum value).

A parallel drop in Cdl levels, which is also seen as the inhibitor molecules are adsorbed to the electrode surface. This may have resulted in a reduction in the local dielectric constant and/or an increase in the electrical double layer thickness. This implies that by adsorbing to the metal/solution interface, inhibitor molecules acts.

The inhibitor's adsorption on the MS surface lowers its electrical capacitance as a result of the substitution of water molecules and other ions initially adsorbed on the metal substrate. The development of a protective layer on the electrode surface can cause a drop in electrical potential with a rise in the inhibitor concentration. This observation could be inferred from the Helmholtz model [24]:

$$Cdl = \frac{\varepsilon\varepsilon_0 A}{d} \tag{11}$$

where, = ε dielectric constant of the medium, ε_0 = vacuum permittivity (8.854×10^{-14} F cm^{-1}), A = electrode surface area and = thickness of the protective layer.

In the compact layer, *Cdl* correlates positively with the local dielectric constant I and negatively with the double layer thickness (d), according to the stern model, the reduction in *Cdl* suggests a reduction in e or rise in d, which can be caused by the adsorption of inhibitor molecules at the interface of the metal/solution.

(c) ***Constant Phase Element (CPE)***

Since we already recognize, the ideal Nyquist plots are semi circles, but the form of the Nyquist plot reveals some divergence from their ideal shape experimentally. Owing to the roughness and inhomogeneity of the metal surfaces and the adsorption of the inhibitors, the divergence from the perfect semicircle, shown as a depressed semicircle in the middle, below the actual axis, is assigned to the frequency dispersion of interfacial impedance. In such cases, the constant phase element (CPE) is used instead of the pure capacitance to account for the imperfect capacitive response of the interface. The impedance of the CPE is expressed by [24]:

$$Z_{CPE} = [Y_0(j\omega)n]^{-1} \tag{12}$$

where J is the imaginary unit, and ω is the angular frequency, Y_0 is the magnitude of CPE. A match parameter without explicit physical definition is the exponential factor n. In fact, the degree of surface roughness, i.e. the lower deviation from 1, the more homogeneous surface or surface film, is also correlated with this.

(d) ***Polarization resistance (Rp)***

If the applied electrode potential differs from the equilibrium potential of the electrochemical reaction at the polarised electrode, leading to oxidation/reduction of the species at the surface of the electrode, the electrode is assumed to be polarized. The polarization resistance (R*p*) can also be calculated by fitting the equivalent electrical circuit model EIS data. As the loop of the nyquist plot increases the value of the R*p* also increases or we can say it increases as the corrosion inhibition performance of a corrosion inhibitor increases.

(e) ***Time Constant (τ)***

The time constant value (τ) is possibly high when the corrosion inhibitors are added to the aggressive solution and thus the value of τ can be used to evaluate the inhibitor adsorption process on the substrate. The value of τ is calculate utilizing the following equation [24]:

$$\tau = C_{dl} R_{ct} \tag{13}$$

(f) ***Inhibition Efficiency (%IE or %η)***

The inhibition efficiency, %IE, is obtained using equation [24]:

$$(\%\,IE) = \frac{Rct - Rct^o}{Rct} \times 100 \tag{14}$$

where, R_{ct} and $R_{ct}{}^{\circ}$ are the charge transfer resistance with and without inhibitors, respectively. It helps in prediction of the inhibiting potential of the studied compound for the corrosion of different metals in various aggressive media [32].

5.5 An Overview of the Application of EIS Data in the Evaluation of Performance of Corrosion Inhibitors

This section will review the plethora of criteria derived utilizing EIS and outline the scope of data that can be abstracted from EIS in the assessment of the efficiency of corrosion inhibitors. To this extent, the chapter will address the basic principles associated with EIS. The work of Zhang et al. is worthy of notice [33] in their recent paper in which they examined Aloe polysaccharide (AP) as an eco-friendly mild steel corrosion inhibitor in simulated acidic oilfield water, experimentally and theoretically. From the results obtained by the EIS curves (the parameters indicated in Table 2 are obtained by fitting suitable equivalent circuits). It can be inferred from Fig. 8a that, as opposed to a device without an inhibitor, the diameter of the capacitive loop increases and the inhibition intensity increases with the increased concentration of AP, which shows that the inhibitor film on the metal surface steadily becomes compact and thus enhances the safety of mild steel. In the meanwhile, both Nyquist curves show identical capacitive loops, suggesting that the corrosion process with or without AP remains the same. In addition, a slightly concaved semicircle capacitor circuit was seen in the Nyquist plots, which is due to the heterogeneity inherent in the solid electrode surface during corrosion, which is generally referred to as dispersion results. The nyquist diagrams have been provided with an effective equivalent circuit (as seen in Table 2) and the corresponding parameters are also shown in Table 2. The Rct values improved significantly with the rise in AP concentration, according to the data collected, and the limit was as high as 96.41% at 800 mg/L, suggesting that AP could effectively inhibit the action of charge transfer. However, Cdl values attributed to the declining local dielectric constant and/or the rising thickness of the electric double layer capacitor during the presence of inhibitors to the downward trend.

Three new Schiff bases based on Imidazo(1,2-a)pyridine have been investigated by Aatiaoui et al. [34], namely (E)-N-(2-phenylimidazo(1,2-a)pyridin-3-yl)-1-(1H-pyrrol-2-yl)methanimine (1), (E)-N-(2-phenylimidazo(1,2-a)pyridin-3-yl)-1-(thiophen-2-yl)methanimine (2) and (E)-1-(5-nitrothiophen-2-yl)-N-(2-phenylimidazo(1,2-a)pyridin-3-yl)methanimine (3), as corrosion inhibitor of mild steel in 1 M HCl, experimentally and theoretically. Table 2 also displays the equivalent circuit used for fitting the nyquist curves and the related R*ct*, C*dl* and IE (%) values of MS in 1 M HCl in the absence and presence of all three inhibitors. As a result, they concluded that the reaction of the MS impedance in the acid solution changed dramatically after the three ligands were applied and that the impedance increased with an increase in the concentration of each inhibitor. In fact, the decrease in C*dl* values is due to the adsorption of the imine components on the surface of the metal, resulting in the forming of a film that reduced the acid solution's corrosive effect. There is an improvement in the efficacy of corrosion inhibition due to an increase in R*ct* values for an inhibitor concentration that exceeds 97% for each inhibitor at 10^{-3} M.

The corrosion inhibition activity of 2-aminofluorene bis-Schiff bases, i.e. 2-bromoisophthaldehyde-2-aminofluorene (M1) and glutaraldehyde 2-aminofluorene (M2), in circulating cooling water was synthesised and investigated by Wei et al. [35]. In Fig. 9, the Nyquist and Bode diagrams obtained from EIS studies of carbon steel with various M1 and M2 concentrations at 25 °C are seen. The generation of the capacitive anti-arc is the consequence of the inhibitor's creation of an adsorption layer on the surface of the carbon steel and the resistance of the charge transfer. The Nyquist diagrams demonstrate that the diameters of the capacity arcs of the specimens with the inhibitors were greater in the simulated flowing cooling water than those of the non-inhibitor carbon steel, which suggests that the corrosion reaction was

Table 2 An outline of the experimental findings of EIS of some recent studies

S. no.	Name of inhibitors	Nature of metal and medium	Equivalent circuit fit	R*ct* (Ω cm^2)	C*dl* (μF cm^{-2})	R*f* (Ω cm^2)	CPE*f* (S s^n cm^{-2})	R*p* (Ω cm^2)	L/W (Ω cm^2)	R_L (Ω cm^2)	IE or η (%)	References
1	Aloe polysaccharide (800 mg/l)	Mild Steel in simulated acidic oilfield water	R(RQ)	9.2 (0*) 256.1	312.7 (0) 78.0	–	–	–	–	–	96.41	[33]
2	Bis-Schiff Bases, M_1 and M_2 (2.50 mmol/L)	Carbon Steel in circulating cooling water	R(RQ) [R([R(RQ)]Q)]	–	686 (0) 55 338 (M_1) 41 534 (M_2)	–	–	2.711 (0) 5.329 (M_1) 5.430 (M_2)	–	–	96.25 (M_1) 99.15 (M_2)	[34]
3	Three Schiff bases based on Imidazo(1,2-a)pyridine (1, 2 and 3) (10^{-3} Mol/l)	Mild Steel in 1 M HCl	R(RQ)	20.7 (0) 723.4 (1) 159.4 (2) 103.4 (3)	191.7 (0) 34.8 (1) 65.2 (2) 76.9 (3)	–	–	–	–	–	97(1) 87 (2) 80 (3)	[35]
4	Locust Bean Gum (5 mM)	Q235 steel in 0.5 M H_2SO_4 medium	R(RQ) [R(QR[RL])]	–	93.7 (0) 61.9	–	–	13.9 120.1	45.3 576.8	114.3 598.4	88.4	[36]
5	Ircinia strobilina crude extract 2.0 g/L	Mild Steel in 1 mol/L HCl	R(RQ) [R(Q[R(RQ)])]	34.6 (0) 173.0	42.7 (0) 2.8	8.31	69.4	–	–	–	80.9	[37]

(continued)

Table 2 (continued)

S. no.	Name of inhibitors	Nature of metal and medium	Equivalent circuit fit	R*ct* (Ω cm^2)	C*dl* (μF cm^{-2})	R*f* (Ω cm^2)	CPE*f* (S s^n cm^{-2})	R*p* (Ω cm^2)	L/W (Ω cm^2)	R_L (Ω cm^2)	IE or η (%)	References
6	Jujube shell extract (2 g/l)	copper in 1 M HCl	R(RQ) [R(Q[R(QR)])]	768 (0) 6247	60 (0) 19.9	769.0	89.1 (0) 64	–	–	–	89.1	[38]

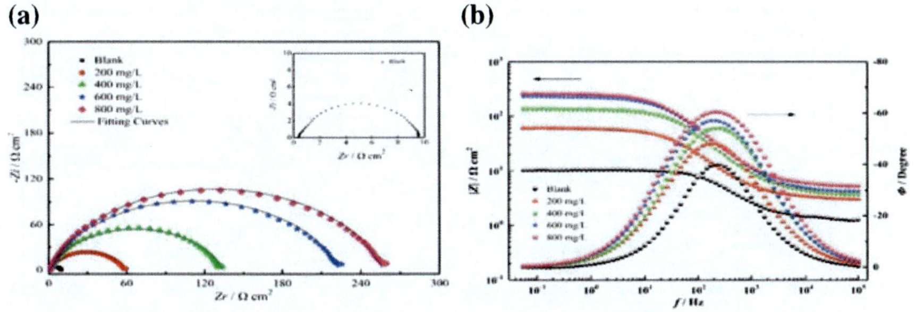

Fig. 8 **a** Nyquist plots and **b** Bode plots for mild steel in 15% HCl without and with different concentrations of AP. Republished from ref. [33] with permission from Elsevier

prevented. The corrosion inhibition performance was influenced by the concentration of the corrosion inhibitor. In simulated circulating cooling water, the concentration of the inhibitor can influence the rate of forming of the adsorption film layer; thus, with increasing concentration, the resulting capacitive loop and radius rise. The development of the time constant at the intermediate frequency might well be observed from the angular phase. In the Bode plots, one time constant for the curve without an inhibitor was reported, according to Fig. 9c–f. The angle phase plot, however, demonstrates the creation of two time constants when simulated flowing cooling water was applied to the inhibitor. This is because of the corrosion inhibitor's film resistance (RF) developed on the surface of the carbon steel. The phase angles for uninhibited and inhibited M1 solutions were 27.37° at 1.212 Hz, 24.20° at 0.026 Hz, 29.75° at 0.038 Hz, 30.04° at 0.121 Hz, 24.05° at 0.032 Hz, and 51.08° at 0.010 Hz, respectively, in Fig. 9d. The phase angles for uninhibited and inhibited M2 solutions were 27.37° at 1.212 Hz, 27.22° at 0.056 Hz, 20.97° at 0.178 Hz, 23.78° at 0.026 Hz, 45.47° at 0.010 Hz and 34.55° at 0.010 Hz, respectively, in Fig. 9f. It should be remembered that while no inhibitor was present in the simulated circulating cooling water, the values of RF and CPEF were roughly equal to 0. The Rp value is usually caused by the passage of electrons between the metal and the inhibitor, so the Rp value is correlated with the efficacy of corrosion inhibition. The data analysis indicates that with increasing inhibitor concentration, the value of Rp improved. This is because increasing inhibitor concentration can improve the inhibitor's coverage on the surface of the metal. However, as the inhibitor concentration rises, the CPE value decreases, meaning that the Cdl thickness increased and the dielectric constant decreased, meaning that the inhibitor was effectively adsorbed on the surface of the carbon steel and strengthened the electrical double-layer properties. The IE (%) value of M2 is greater than that of M1; therefore, the anti-corrosive effect of M2 on carbon steel is better than that of M1.

Guo et al. [36] in one of their study investigated Locust Bean Gum (LBG) as a green and novel corrosion inhibitor for Q235 steel in 0.5 M H_2SO_4 medium. According to the findings of EIS results, it was concluded that the capacitive loop's radius increases with the increase in the concentration of the LBG, without altering

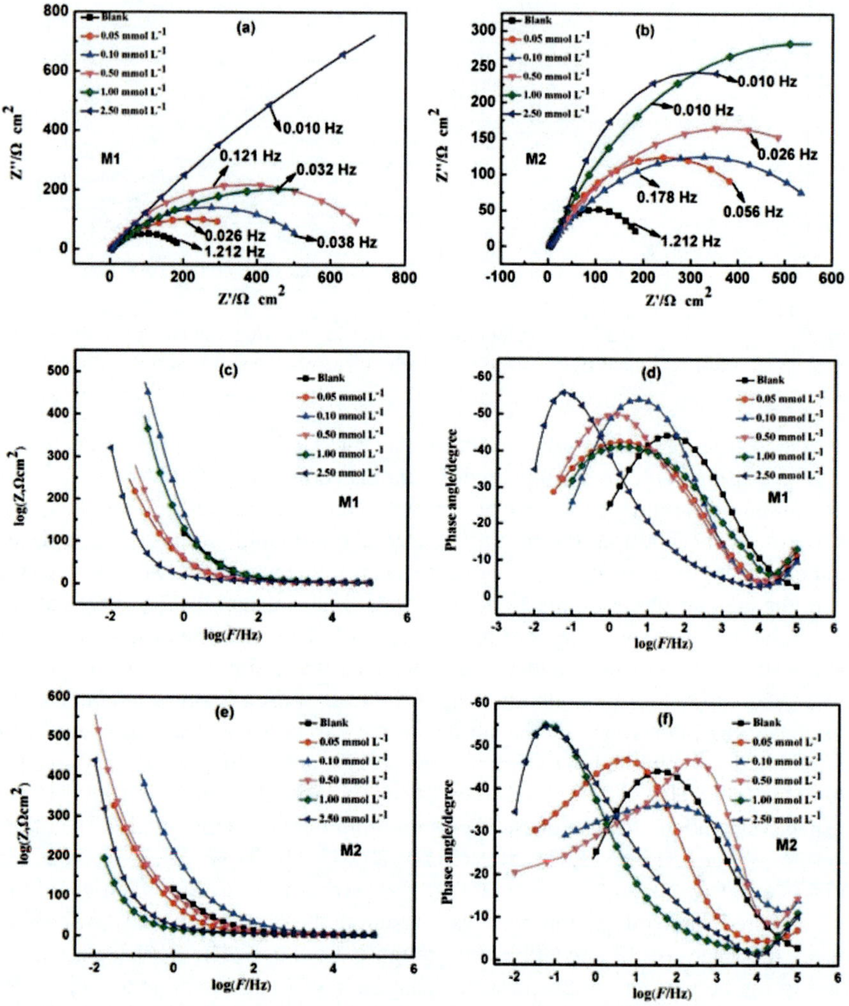

Fig. 9 Nyquist (**a** and **b**), Bode (**c** and **e**) and phase angle diagrams (**d** and **f**) for carbon steel in simulated circulating cooling water in the absence and presence of different concentrations of M1 and M2. Republished from ref. [35] with permission from RSC

the shape of the curve, as depicted in Fig. 10. The obtained plots were fitted with the suitable equivalent circuit and the parameters obtained are reported in the Table 2. It was discovered that the C*dl* value in the blank solution was 97.3 μF cm^{-2}. As the concentration of LBG increased, the C*dl* value represented a downward trend. The value of C*dl* fell dramatically to 61.9 μF cm^{-2} when the LBG concentration was 5 mM, and C*dl* showed a downward trend as the LBG concentration rose. The value of polarization resistance (R*p*) was observed to be 13.9 Ω cm^2 at 298 K in the absence of LBG. The R*p* improved dramatically with the rise in LBG concentrations to 120.1

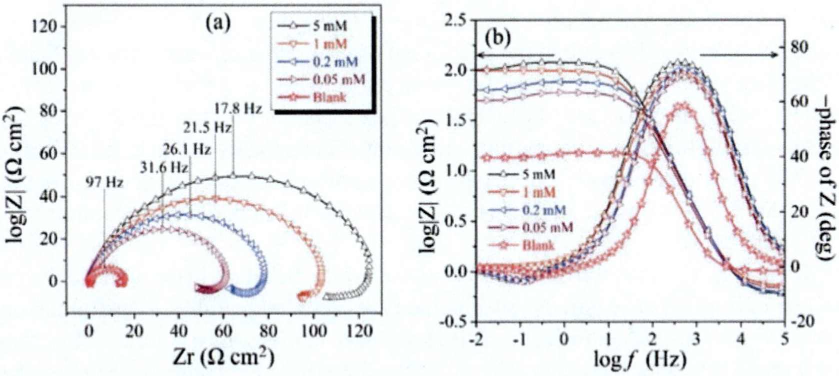

Fig. 10 The plots of the **a** Nyquist and **b** Bode of Q235 steel in 0.5 mol/L H2SO4with and without diverse concentrations of LBG at 298 K. Republished from ref. [36] with permission from Elsevier

Ω cm^2 while the LBG concentration was 5 mM. The performance of corrosion inhibition reached 88.4%. The LBG then demonstrated outstanding anti-corrosion performance.

Fernandes et al. [37] analyzed *Ircinia strobilina* crude extract as a MS corrosion inhibitor in an acid medium in one of their recent research. The Nyquist interpretation of all EIS data gathered in the presence and absence of the inhibitor at different concentrations was adapted by using Randles equivalent circuit model. To provide a more reliable and accurate fit for the capacitive loops obtained from the Nyquist plots, the double layer capacitance (C*dl*) was substituted by CPE. The CPE is often used instead of an ideal capacitor in the low frequency regions of Nyquist plots to consider depressed semi-circles or non-vertical arcs. An equivalent circuit consisting of a solution resistance (R*s*) in series with a charge transfer resistance (R*ct*) parallel to CPE had been used, as can be seen in Table 2, to fit the findings of the blank solution. For solutions with the presence of an enzyme, an equivalent circuit consisting of solution resistance (R*s*) in series with a constant phase factor (CPE*f*) that is parallel to film resistance (R*f*) had been used. The R*f* is in series with a charge transfer resistance (R*ct*) and is concurrent with a CPE. In case (a), they only have the transfer resistance of the charger in the corrosion reaction of the metal surface in contact with the acid medium and the resistance of the solution, regardless of the working electrode and the reference electrode. In case (b), with the inclusion of the inhibitor, they added a CPE and resistance owing to the formed substrate. The second capacitive loop doesn't really exist in the blank solution, as can be seen in Fig. 5, which is relevant to the adsorption of organic molecules on the surface of mild steel. The C*dl* value of the blank solution, when the inhibitor is incorporated in the medium, is much stronger than the C*dl* value. This anomaly can be due to the replacement of water molecules for inhibitor molecules, the decline in the area of the surface exposed to the corrosive medium and the rise in the thickness of the electrical double layer, and it is implied that the inhibitor molecules function by adsorption on the substrate/corrosive solution of carbon steel, demonstrating that the inhibitor molecules are adsorbed at the interface (metal/solution). The decline in C*dl* values with an increase in the concentration could be due to the adsorption of different electrode surface inhibitor molecules, resulting in a reduction in the electroactive area. It is also seen in Table 2 that R*ct* values increase with the increase in the concentration of the inhibitor, leading to increased efficiency in corrosion inhibition. This appears when more adsorbed inhibitor molecules mostly on metal surface reduces the active oxidation reaction sites. An efficiency value of 80.9% was reached at the maximum concentration of inhibitor (2.0 g/L).

Jmiai et al. [38], in their study observed the similar behavior in the nyquist plot (as shown in Fig. 11). The parameters obtained are recorded in Table 2 when fitted the Nyquist plot with an equivalent circuit. According to the analysis of this Table, they mentioned that with the concentration of the JS extract augments in the corrosive solution the values of R*ct* (788.4 $\Omega.cm^2$ for blank, 6247 Ω cm^2 for 2 g/L JS) R*f* and R*ct* (788.4 $\Omega.cm^2$ for blank, 7016 Ω cm^2 for 2 g/L JS) increase, on the other hand the values of the capacitance of CPE_1 and CPE_2 (462 $\mu F.cm^{-2}$ for blank, 64 μF cm^{-2} for 2 g/L JS) decrease respectively, which shows in parallel an increase in the inhibition efficiency (IE (%) = 89% for 2 g/L JS). This indicated that the molecules of the JS extract replace the chloride ions Cl^- on the surface of the copper electrode to form an anti-corrosion layer which prevents the corrosive molecules. To access all the frequencies of the corrosion phenomenon the Bode diagrams had been reported. These are represented in magnitude plot in Fig. 11. The amplitude curved show two very distinct regions: (i) the high frequency zone (where log |Z| the values are lower) and the low frequency zone show almost constant curves. (ii) While in the medium frequencies they reported linearity between the log |Z| and the values of log f. According to which, the concentration increased with the value of log |Z|, this growth indicates the retardation of the corrosion process and the resistivity of this region in the presence of the extract.

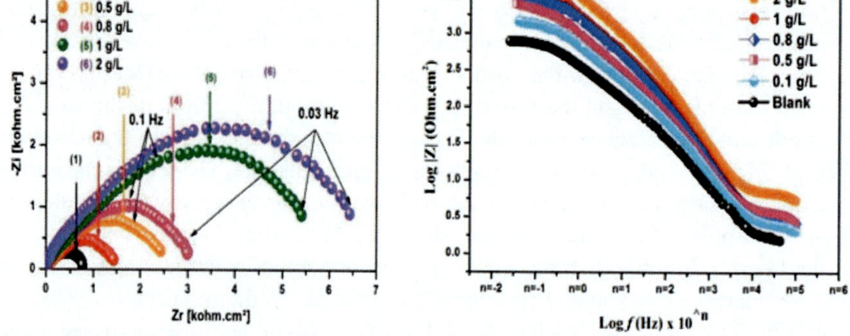

Fig. 11 **a** Nyquist, **b** Bode plots and **c** Electrical equivalent circuit model used to fit EIS data obtained for copper exposed in 1 M HCl aerated solution with and without different concentrations of JS extract at 298 K. Republished from ref. [38] with permission from Elsevier

6 Limitation and Future Challenges Related to EIS

Presently, EIS assessment can be easily monitored through the use of commercial potentiosat by decades of continuing research and development. However, for electrochemists and material scientists, the understanding and measurement of impedance data is indeed challenging and sometimes problematic. The dynamic nonlinear least square fitting of impedance data to an equivalent electric circuit model is done primarily to quantify distinguish physical quantities of interest from impedance data. However, the key problem is how to establish a physically realistic suitable model without background experience of the series of processes employed in the electrochemical system, it is the typical situation that individuals construct the equivalent circuit model as per the numbers of semicircles in the Nyquist plot of EIS data or by referring to the publications or based on past knowledge of the investigated object. Nevertheless, without a rigorous basis, this somewhat arbitrary approach becomes difficult and often inaccurate when the distinctive frequencies of several processes are identical to each other. The option of good initial estimates of the parameters in the equivalent circuits is another difficult role in fitting, which would largely decide the efficiency of the final result. Visual calculation can be used to roughly calculate the value of the resistance in the circuit. Certain parameters, such as the capacitance or exponent of the constant phase element (CPE), are, therefore, much harder to estimate. Another challenge is that much focus has been devoted predominantly to the Bode plot, there are comparatively limited experiments on graphical analysis of the Nyquist plot, and more efforts are therefore needed. There are some general problems encountered during the review of the data from the EIS. However, there are some other problems that need to be further recognized in estimating the performance of corrosion inhibitors. A significant drawback of impedance measurements is that, without prior knowledge of the values of the Tafel slopes and the Stern-Geary coefficient, the polarization resistance cannot be transformed into a corrosion rate. Secondly, the instability of the corrosion mechanism, such as relaxation of adsorbed species or pitting corrosion, can cause dispersion in the impedance values recorded at low frequency. Another major downside of global EIS measurements is its poor lateral precision, since the impedance is averaged over the entire macroscopic electrode surface. This makes it more difficult to analyse localised electrochemical corrosion reactions that average the specific electrochemical parameters of the micro-defect from those of the whole electrode surface. The key problem with EIS analysis is that it is always possible to use various equivalent circuit models to match the same EIS results, all with low chi-squared (χ^2) error values. The outcomes of resistance and capacitance are dependent on the suitable electrical model set. Therefore, the construction of an effective equivalent circuit requires deep understanding of the corrosion mechanisms of the device, the use of complementary characterization techniques including surface, optical and/or physical chemical analysis, and the assistance of models from published works.

7 Conclusion

EIS has been proved to be a very versatile, fast and informative technique for the study of several systems. In the field of corrosion inhibitors, EIS helps in predicting the mechanism of the inhibitor film formation on the metal substrate and to yield various information regarding the metal/solution interface. It helps in the estimation of several corrosion related electrochemical parameter like the resistance and capacitance of the inhibitors' film, the charge transfer resistance, and the double layer capacitance. By the aid of these parameters the inhibitor mechanism can be analyzed and the corrosion rate can also be calculated.

References

1. R.G. Kelly, J.R. Sculy, D. Shoesmith, R.G. Buchheit, Electrochemical techniques in corrosion science and engineering, 1st edn. (CRC Press, 2002)
2. D.A. Koutsouras, P. Gkoupidenis, C. Stolz, V. Subramanian, G.G. Malliaras, D.C. Martin, Chem. Electro. Chem. **4**, 2321 (2017)
3. F. Huet, A review of impedance measurements for determination of the state-of-charge or state-of-health of secondary batteries. J Power Sour **70**, 59–69 (1998)
4. N. Wagner, W. Schnurnberger, B. Muller, M. Lang, Electrochim. Acta **43**, 3785–3793 (1998)
5. X. Li, Z. Qin, H. Fu, T. Li, U. Peng, Z. Li, J.M. Rini, X. Liu, Biosens. Bioelectron. (2020). https://doi.org/10.1016/j.bios.2020.112672
6. P. Delahay, *New Instrumental Methods in Electrochemistry* (Interscience, New York, 1954)
7. M.E. Orazem, B. Tribollet, *Electrochemical Impedance Spectroscopy* (Wiley, New York, 2008)
8. J.R. Macdonald (ed.), *Impedance Spectroscopy: Emphasizing Solid Materials and Systems*, 1st edn. (Wiley, New York, 2001)
9. R. Aslam, M. Mobin, S. Zehra, I.B. Obot, E.E. Ebenso, ACS Omega **2**, 5691–5707 (2017)
10. L. Guo, I.B. Obot, X. Zheng, X. Shen, Y. Qiang, S. Kaya, C. Kaya, Appl. Surf. Sci. **406**, 301–306 (2017)
11. M. Mobin, R. Aslam, J. Aslam, Mater. Chem. Phys. **191**, 151–167 (2017)
12. J. Huang, Z. Li, B. Yann Liaw, J. Zhang, J. Power Sources **309**, 82–98 (2016)
13. C.M.A. Brett, A.M. Oliveira Brett, *Electrochemistry, Principles, Methods, and Applications* (Oxford University Press, Oxford, 1993)
14. K.S. Cole, R.H. Cole, J. Chem. Phys. **9**, 341 (1941)
15. D.W. Davidson, R.H. Cole, J. Chem. Phys. **19**, 1484 (1951)
16. D.D. Macdonald, Electrochim. Acta **51**, 1376 (2006)
17. A.J. Bard, L.R. Faulkner, *Electrochemical Methods* (Fundamentals and Applications (Wiley, New York, 2001)
18. P.B. Raja, S.B. Sethuraman, Mater. Lett. **62**, 113–116 (2008)
19. J.R. Macdonald, J. Phys. Chem. A **115**, 13370 (2011)
20. E. Warburg, Ann. Phys. Chem. **67**, 493 (1899)
21. V.S. Sastri, Corrosion Inhibitors, Principles and Application (1998)
22. M. Parveen, M. Mobin, S. Zehra, RSC Adv. **6**, 61235–61248 (2016)
23. S. Feliu Jr., Metals **10**, 775 (2020)
24. S. Zehra, M. Mobin, J. Aslam, M. Parveen, J. Adhes. Sci. Technol. **32**, 317–342 (2018)
25. M. Mobin, S. Zehra, M. Parveen, J. Mol. Liq. **216**, 598–607 (2016)
26. M. Mobin, S. Zehra, R. Aslam, RSC Adv. **6**, 5890–5902 (2016)
27. H.M.A. El-Lateef, Corros. Sci. **92**, 104–117 (2015)

28. T.H. Muster, H. Sullivan, D. Lau, D.L.J. Alexander, N. Sherman, S.J. Garcia, T.G. Harvey, T.A. Markley, A.E. Hughes, P.A. Corrigan, A.M. Glenn, P.A. White, S.G. Hardin, J. Mardel, J.M.C. Mol, Electrochim. Acta **67**, 95–103 (2012)
29. G. Gece, S. Bilgic, Corros. Sci. **52**, 3435–3443 (2010)
30. F. Zhang, J. Pan, P.M. Claesson, Electrochim. Acta **56**, 1636–1645 (2011)
31. Z. Tao, S. Zhang, W. Li, B. Hou, Corros. Sci. **51**, 2588–2595 (2009)
32. M.P. Asfia, M. Rezaei, G. Bahlakeh, J. Mol. Liq. (2020). https://doi.org/10.1016/j.molliq.2020.113679
33. W. Zhang, Y. Mac, L. Chen, L. Wang, Y.C. Wua, H.J. Li, J. Mol. Liq. **307**, 112950 (2020)
34. W. Wei, Z. Liu, C. Liang, G.C. Han, J. Hana, S. Zhang, RSC Adv. **10**, 17816 (2020)
35. A.E. Aatiaoui, M. Koudad, T. Chelfi, S. Erkac, M. Azzouzi, A. Aouniti, K. Savas, M. Kaddouri, N. Benchat, A. Oussaid, J. Mol. Struct. (2020). https://doi.org/10.1016/j.molstruc.2020.129372
36. L. Guo, R. Zhang,B. Tan, W. Li, H. Liud, S. Wua, J. Mol. Liq. **310**, 113239 (2020)
37. C.M. Fernandes, T.S.F. Fagundes, N.E. Santos, T.S.M. Rocha, R. Garrett, R.M. Borges, G. Muricy, A.L. Valverde, E.A. Ponzio, Electrochim. Acta **312**, 137–148 (2019)
38. A. Jmiai, A. Tara, S. El Issami, et al., J. Mol. Liq. (2018). https://doi.org/10.1016/j.molliq.2020.114509

Advancement in Spectroscopic and Microscopic Techniques

Applications of Liquid Cell-TEM in Corrosion Research

Recent Trends and Advancements in Spectroscopic and Microscopic Techniques

Khalid Hattar and Raymond R. Unocic

Abstract A fundamental understanding of corrosion at the nanometer scale is needed to develop physics-based predictive models of the corrosion mechanisms. Recent advancements associated with in-situ Transmission Electron Microscopy (TEM) liquid cell experiments, also known as liquid phase experiments, in the last twenty years has permitted the real time nanoscale observation of mechanisms in a range of complex materials and liquid environments. In all of these experiments, caution and care must be taken to understand the impact of the liquid cell dimension and electron beam effects on the observed reactions. Despite these cautions, over the last seven years, in-situ TEM liquid cell experiments have been expanded to experiments exploring corrosion initiation. The early studies explored simple deposited high purity metal systems exposed to deionized water, brine, or acetic acid; however, the more recent studies have shown that experiments can be performed on complex environments like site-specific regions of pipeline steel in a range of tailored corrosive environments. During this time, the number and refinement of the analytical techniques utilized during the in-situ TEM corrosion experiment has also increased, providing insight into the evolving sample thickness, composition, crystallographic orientation, bonding structure, and much more. This chapter concludes with a discussion of the potential expansion of in-situ TEM corrosion to such complex degradation phenomena, such as stress corrosion cracking and biofouling. Overall, in-situ TEM corrosion is a challenging technique with many concerns, but it has the potential to provide new fundamental insight into corrosive mechanisms that are active at the nanoscale.

K. Hattar (✉)
Center for Integrated Nanotechnologies Sandia National Laboratories, Albuquerque, NM 87185, USA
e-mail: khattar@sandia.gov

R. R. Unocic
Center for Nanophase Materials Sciences, Oak Ridge National Laboratory, Oak Ridge, TN 37831, USA
e-mail: unocicrr@ornl.gov

I. Toor (ed.), *Recent Developments in Analytical Techniques for Corrosion Research*,
https://doi.org/10.1007/978-3-030-89101-5_6

1 Introduction to TEM and the Associated Limitations

As has been detailed in other chapters in this book, it has been known for decades that corrosion in metals is a complex phenomenon that can be influenced by the local microstructural chemistry [1], mechanical loading [2], electric field [3], magnetic field [4], and even radiation environment [5]. Although corrosion is known to have deleterious impacts on structures as large and complicated as airplanes [6], bridges [7], industrial economy [8], and cultural monuments [9], the origins of corrosion typically occur at the nanometer scale where chemical degradation intersects mechanical failure. As such, a true physics-based understanding of corrosion mechanisms requires experimentally validated observation of the nanometer evolution of materials when exposed to the often-complex environments associated with corrosion.

Since its inception in 1933 [10], Transmission Electron Microscopy (TEM) and the sister technique Scanning Transmission Electron Microscopy (STEM), which when considered together will be abbreviate as S/TEM, have emerged as the go-to technique to characterize inorganic materials with nanoscale resolution. To fully comprehend the potential impact of electron microscopy over classical metallography utilizing visible light, we need only compare the theoretical resolution, δ, as determined by Rayleigh criterion:

$$\delta = \frac{0.61\lambda}{\mu \sin \beta} \tag{1}$$

where λ is the wavelength of the electron in nm, μ is the refractive index, and β is the semi-angle of collection. Assuming that the numerical aperture $(\mu \sin \beta) = 1$, and incorporating the non-relativistic de Broglie equation:

$$\lambda = \frac{1.22}{E^{1/2}} \tag{2}$$

where E is the energy of the electrons in eV. This results in a simplistic relationship between resolution and electron energy:

$$\delta = \frac{0.61 * 1.22}{E^{1/2}} = \frac{0.74}{E^{1/2}} \tag{3}$$

As a result, the theoretical resolution limit of an optical microscope is ~200 nm, a 200 kV TEM is 1.7 pm, and a 1 MV TEM is 0.74 pm. As with optical microscopy, the obtainable resolution of TEM is often significantly lower than the theoretical limit due to aberrations in the electromagnetic lens. Advancements in electron optics design and control have greatly advanced the obtainable resolution of S/TEM resulting in the development of Aberration Corrected S/TEM (AC-S/TEM) [11, 12]. These drastic improvements in resolution from optical microscopy to S/TEM and the subsequent jump from S/TEM to AC-S/TEM can be seen in Fig. 1 [13].

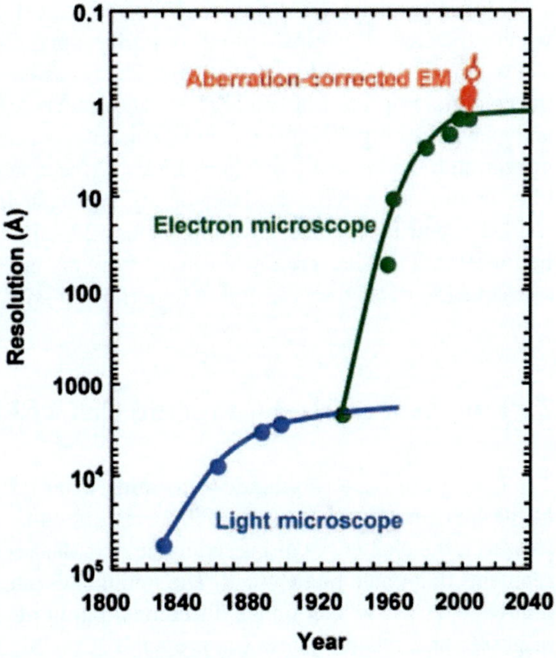

Fig. 1 Plot of the resolution possible in state-of-the-art microscopes as a function of year. Reprinted with permission from [13]. Copyright 2011, Springer Nature

This drastic improvement in resolution provided by S/TEM comes with many costs that need to be considered from the onset of any study considering the use of S/TEM for materials research. These concerns can be delineated into three categories: thin sample effect, electron beam effect, and vacuum effect. The image formed in S/TEM micrographs is a result of the complex interaction of the electron beam (both particle and wave in nature) as it passes through the chemical and crystallographic nature of the sample. Although essential for the interpretation of in-situ S/TEM studies of corrosion discussed in this chapter, the Block Wave interaction associated with S/TEM image formation is outside the scope of this book and the reader is referred to the following seminal text on the field [14–17]. The penetration depth that the elastically scattered electrons need to transmit through the sample is an inherent limit of all S/TEM techniques and is a function of the samples density and average atomic mass, as well as the accelerating voltage used in S/TEM experiments. The latter of which is usually kept in the 80–300 keV range to limit the radiation (knock-on) damage created by the electrons themselves. The expected impact of knock-on damage from electrons can be predicted and are well known for most metals [16, 18, 19]. In addition to knock-on damage that can have extensive impact in low atomic number metals or alloys containing light atomic number additions, the electron beam can also induce radiolysis and many other interesting and sometimes unexpected structural evolution [19]. Understanding the impact of the electron beam and the environments needed for S/TEM observation is essential for any experiment attempting to produce a corrosion study within S/TEM.

A subset of the S/TEM field is dedicated to real time observations of materials when subjected to external stimuli or during operation, which are often referred to as in-situ or in-operando, respectively. There has been a growing effort to increase the potential impact of in-situ S/TEM. The readers are pointed to a range of reviews and viewpoints highlighting recent advancements achieved in in-situ TEM capabilities through the inclusion of Micro-Electro Mechanical System (MEMS), microfluidic, small scale mechanics, laser optics, Artificial Intelligence/Machine Learning (AI/ML), and many other evolving fields [20–25]. Within in-situ S/TEM, in-situ liquid cell S/TEM has greatly advanced from the microfluidic development and has subsequently opened up the ability to perform in-situ S/TEM corrosion studies.

2 Emergence of In-Situ Liquid Cell TEM

The concept of exploring liquid-containing materials in a TEM is not new, as one of the first images taken in the TEM was the wing surface of the house fly [10]. However, the challenges discussed in the previous section severely limited the environments that could be explored. The initial environmental TEM experiments were limited to TEMs with complex differential pumping dedicated to a limited number of gasses and relatively low gas pressures [26–28]. These instruments were aptly named Environmental Transmission Electron Microscopes (ETEMs). In addition to a range of studies into gas environments, the development of such instruments could be utilized for studying some liquid surface interaction when great care is taken to minimize the volume and vapor pressure of the liquid. One such version of an ETEM, which can be seen in Fig. 2, was designed to explore catalytic reactions in liquid environments and was termed Wet-ETEM [29]. Despite the beauty and significant impact of these complex instruments in a range of fields from catalytic reactions to battery technology, the only corrosion related studies explored the impact of hydrogen embrittlement [30–32]. As such, ETEM development will not be a focus for the remainder of this chapter.

A true exploration of liquid environments was much more elusive. The first liquid environment that could be explored was done utilizing liquids with negligible vapor pressure in what is often now termed an open liquid cell (i.e. liquid metals [33] and ionic liquids [34]). The advantage of such an approach is that it did not require a dedicated multimillion-dollar TEM or even a dedicated TEM holder. As such, an open cell design can, with clever design, be implemented at any institution that has an operational TEM. An example of such an open cell design utilizing an ionic liquid (IL) for the study of a lithiation process can be seen in Fig. 3a [34]. In this open cell study, Fig. 3b–s, the complex twisting evolution a single SnO_2 nanowire can be observed with nanometer resolution. Despite the impact of these early open liquid cell experiments, the vapor pressure limitation severely hindered the range and scope of materials and phenomena that could be evaluated in the TEM without concerns of contaminating and subsequently damaging the instrument. Outside of studying the kinetics of liquid metal embrittlement [35], the open cell geometry has not been

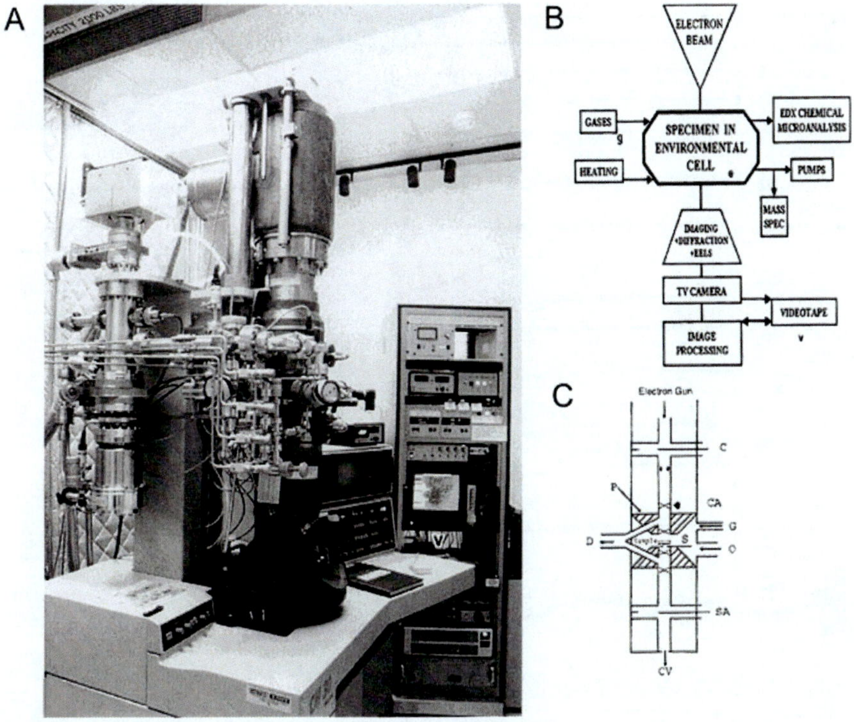

Fig. 2 **a** An image of the complex wet-ETEM developed to study catalyst in-operando. **b** Schematic of the wet-TEM systems. **c** Illustration of the differential pumping scheme utilized to provide a 'wet' environment for exploration [29], reproduced with permission

utilized for in-situ TEM corrosion experiments and will not be discussed in greater detail in this chapter.

The emergence of in-situ TEM liquid cells has really occurred with the development and commercialization of dedicated liquid cell TEM holders. The development of these holders was highly dependent on the rapid refinement of the microfabrication capabilities necessary to produce well controlled and reliable microfluidic vessels and channels with dimensions on the order of 100 nm or less. The reader is referred to textbooks on microfabrication [36] and microfluidics [37] for a deeper understanding of the multi-level processing steps needed for the development of closed and flowing cell designs that will be the core technology that makes in-situ TEM corrosion experiments possible. The maturation of this technology and subsequent near simultaneous commercialization by two vendors Hummingbird Scientific Inc. and Protochips, Inc. in the late 2000s can be directly correlated with the exponential growth of in-situ liquid cell TEM experiments, seen in Fig. 4a. Although the microfabrication approach requires the utilization of dedicated to in-situ TEM liquid cell holders, the continued refinement of the original vendors, entrance of new vendors, and the creative application by researchers has continued to broaden

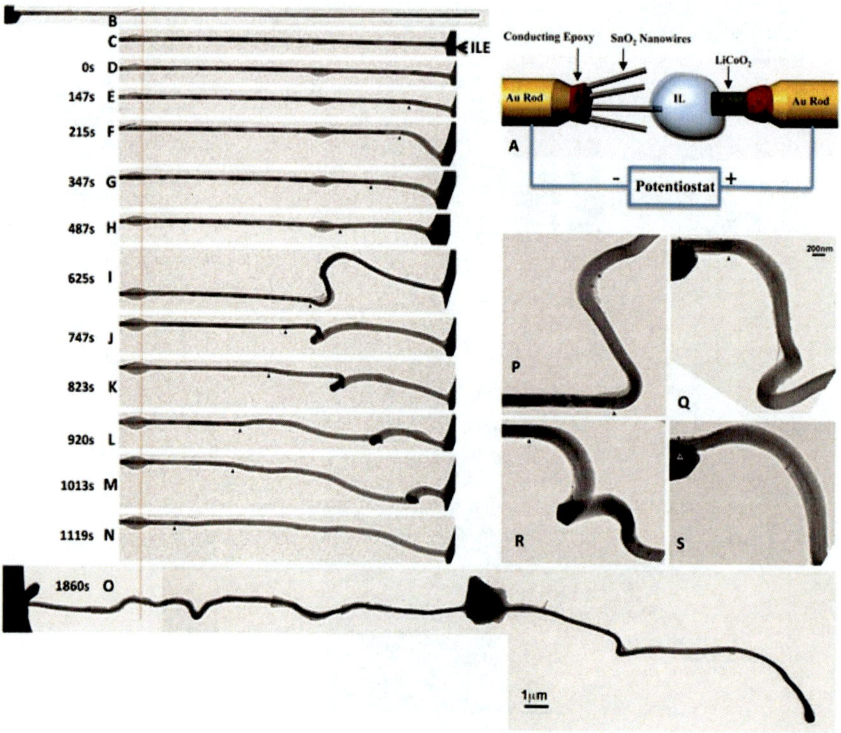

Fig. 3 Time-lapse structure evolution of a SnO_2 nanowire anode during charging at −3.5 V against a $LiCoO_2$ cathode. The single-crystal nanowire was elongated 60% and the diameter increased 45% (resulting in a 240% volume expansion) after charging for 1860s. **a** A schematic of an open cell TEM design utilizing ionic liquid (IL) to study the lithiation process during in-situ TEM observation. **b**–**c** The initially straight nanowire became significantly twisted and bent after charging (**d**–**s**). The chemical reaction front progressed along the nanowire's longitudinal direction, with the front clearly visible, as pointed out by arrowheads in **e** to **s**. The red line in **b** to **o** marks a reference point to track the change of the nanowire length. **p** to **s** are sequential high-magnification images showing the progressive migration of the reaction front, swelling, and the twisted morphology of the nanowire after the reaction front passed by. From [34]. Reprinted with permission from AAAS

the in-situ TEM liquid cells experiments that can be explored. This has resulted in over 300 publications in 2020 utilizing such technology. These studies are largely dominated by research exploring the lithiation processes core for battery technology [38] and catalytic reactions [39] of great importance to both the energy and chemical sectors. Detailed reviews of the range and complexity of modern in-situ TEM liquid cell experiments can be found in the following references [39–42].

A small subset of the in-situ TEM liquid cell studies to date explore the degradation or corrosion of the TEM samples in various liquid environments. Even an extremely generous characterization of in-situ liquid cell TEM experiments that investigate degradation or corrosion of materials resulted in less than 10% of the overall published results utilizing in-situ liquid cell being labeled as corrosion studies,

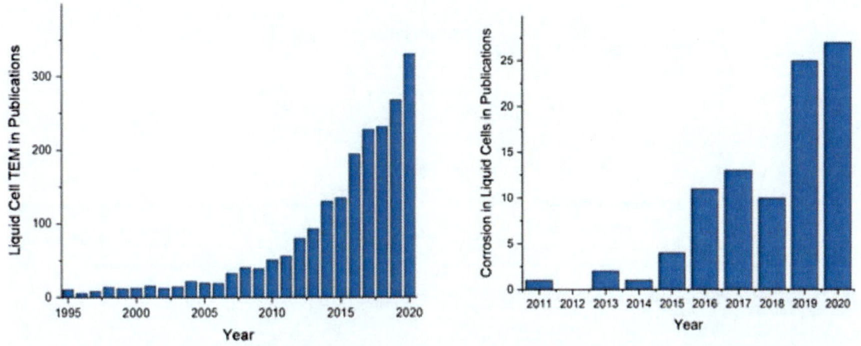

Fig. 4 Plot of publications as a function of year for **a** in-situ TEM liquid cell and **b** in-situ TEM corrosion

as can be seen in Fig. 4b. The relatively small publications count highlights both the limited effort in such complex studies relative to battery or catalytic studies and the initial wariness of accepting such a novel nanoscale approach by the larger corrosion field. The rapid growth of the associated publication is a sign of optimism as the in-situ TEM corrosion studies gain increased acceptance and support. The broad definition of in-situ TEM corrosion publications utilized to produce Fig. 4b resulted in the inclusion of early studies exploring the degradation of lithiated structures in an electrochemistry cell [43], the growth and degradation of lead [44] or lithium dendrites [45–47], and the degradation of Al-Cu alloy at elevated temperatures [48]. These more general degradation studies will not be the focus of this chapter. The first study solely dedicated to in-situ TEM corrosion was published by See Wee Chee et al. in 2014 [49]. The remainder of this chapter will focus on the challenges and opportunities associated with these complex experiments, the rapid development of in-situ TEM liquid cell experiments for corrosion studies over the last seven years, and the future potential for new in-situ TEM corrosion studies.

3 Limitations and Opportunities of In-Situ Liquid Cell TEM

The general challenges and opportunities of liquid cell experiments are well detailed by Frances Ross [40]. However, due to the vast complexity and potential for in-situ TEM corrosion studies [50], it is important to highlight several of the potential applications and concerns. As can be seen in Fig. 5, the enclosed nature of the liquid cell TEM holder provides simultaneous protection of the TEM vacuum from the corrosive liquids and the experimental set-up from the vacuum. With clever design, any sample of interest that can be placed or flowed into the microfluidic region under the electron beam can be imaged and characterized. The subsequent evolution of these samples can be studied as a function of flow rate, flow dynamics, surface

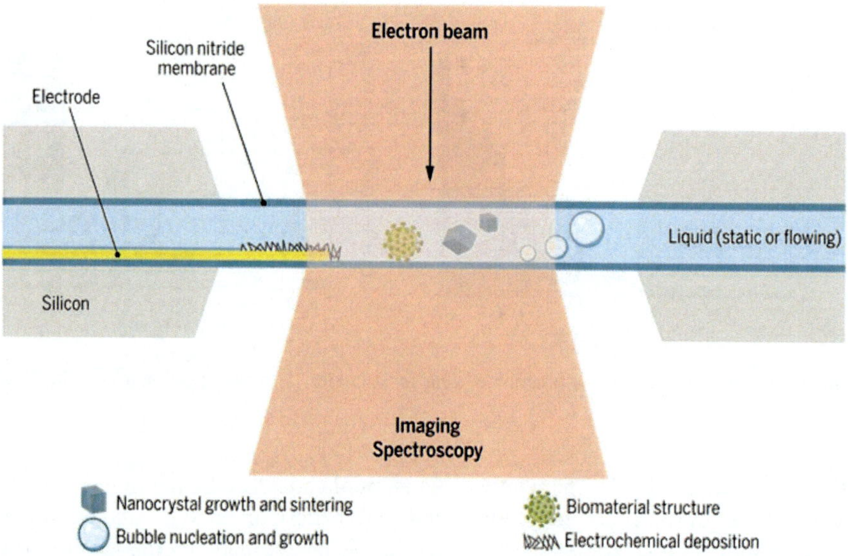

Fig. 5 A schematic showing the range of complex interactions that are possible in liquid cell TEM experiments from [40]. Reprinted with permission from AAAS

chemistry, changes in the flow chemistry, electrical biasing, heating, and many more variables. The most common variable studied in most in-situ TEM microfluidic studies is electron beam effects [51–54], which can be used to study a range of phenomena from the formation of Au nanoparticles [53] through the simultaneous formation of complex rare earth, metals, oxides, and hydrides [54] to the evolutions of dynamic two phase solutions [52]. As long as a researcher is willing to confine their imagination to a ~ 100 nm-thick liquid-tight cell and considers the impact of near relativistic electron (2.0845×10^8 m/s at 200 kV or 69.5% the speed of light) bombardment, a range of amazing experiments are possible!

The range of potential experiments should be balanced with an equal or larger range of concerns and trade-offs that should be considered for any in-situ liquid cell TEM experiment. These trade-offs are reviewed by Niels de Jonge and Frances Ross [41] and include window choice, fluid choice, sample selection (size and geometry), as well as electron beam induced effects. The window choice is a trade-off between imaging resolution and stability. The two most utilized choices are thin amorphous films (e.g. silicon nitride) or 2D materials (e.g. graphene or hexagonal boron nitride). Uniform amorphous films have the advantage of not contributing significant diffraction contrast, making the interpretation of the interactions in the cell easier. A range of window thicknesses and dimensions are available or can be produced. The general trade-off is that thicker films over smaller window dimensions are more stable and subject to less bowing resulting from the pressure differential. However, the smaller window in plan-view dimensions minimizes the field of view. In addition, thicker films result in greater electron scattering, which subsequently degrades imaging resolution. The ultimate choice for high resolution in-situ TEM liquid cell experiment is the utilization of 2D material cells. A detailed review of this approach and the associated benefits and risks were detailed by Jungjae Park et al. [55].

In a similar manner, the experimentalist should be cautious in the choice of fluid. As seen during the emergence of in-situ corrosion and all other subset of the in-situ liquid cell fields, the first choice of fluid and sample are the simplest, most benign, model systems. In all in-situ TEM liquid cell experiments, fluids of interest must be compatible with all components of the TEM holder, tubing, pumps, etc. The vendors are aware of this concern and have generally selected as inert components as possible (SiN windows, Si, Ti-alloy, PEET, etc.), but the impact of corrosion, leeching, and surface interaction of these components with the fluid should be considered. A very important and often overlooked aspect of this concern is the gap dimension in the cell and wetting nature of the cell impact on the fluid dynamic nature. If the fluid gap is too small or the wettability too poor, the already low Reynolds number commonly present in liquid cell experiments and resulting in laminar flow will transition to no fluid motion. The Reynolds number can be calculated:

$$\mathrm{Re} = \frac{uL}{\vartheta} = \frac{\rho u L}{\mu} \tag{5}$$

where u is the flow speed, L is a characteristic linear dimension (the gap dimension), ϑ is the kinematic viscosity of the fluid, ρ is the density of the fluid, and μ is the dynamic viscosity of the fluid. The Reynolds number of the proposed experiment should be considered during the design of experiment. If the surface wettability is inadequate for the proposed experiments a range of tricks can be utilized. Our favorite being pre-treating the windows with bovine serum albumin (cow's blood).

In a similar manner, the choice of sample and associated geometry should be considered in detail prior to the experiment. A plethora of techniques can be utilized in the production of TEM samples. These range from simple mortar and pestle and drop-casting through cryo-microtome and the now ubiquitous focused ion beam (FIB). The simplest, often utilized, but still elegant sample preparation approach for in-situ S/TEM corrosion studies is direct deposition of the metal onto the cell window. This has the benefit of permitting great control of the sample thickness and exposure region, but the downside of limiting the study to the microstructures possible and the type of materials that can be directly deposited via deposition. As a result, many of the early studies focused on exploring deposited nanocrystalline metals. Each technique despite its complexity has the potential to alter the samples chemistry, topology, and or internal microstructure, something that should be considered in the TEM sample preparation decisions. For in-situ liquid cell experiments, it is also important to decide if the sample should be adhered to the window and if so how. The most common approach to adhere a sample not directly deposited is to tack down the corners with a Pt-weld in the FIB. However, much care must be taken not to damage the window with the FIB process or contaminate the sample with Pt, C, or Ga. Another consideration that should be evaluated is what surfaces of the sample is exposed to the corrosive fluid. Depending on the real-world application that is being emulated, the research may consider protecting the other surfaces with an amorphous and inert coating layer to still permit good quality S/TEM images, while providing a protective coating. There is no wrong way to perform sample preparation, but each approach has associated limitations and concerns that should be considered before any work is initiated.

As with all in-situ S/TEM experiments, the influence of the electron beam should be considered. This is to a greater order during in-situ TEM fluidic experiments where radiolysis and other charge effects can greatly alter the chemical reactions observed. Two approaches to elucidate the impact of the electron beam effect should be considered in all in-situ TEM corrosion studies. The first is excluding the electron beam effect. This would entail performing detailed ex-situ experiments on a relevant scale sample with post-mortem characterization to fully understand reaction kinetics and resulting structural and chemical changes through electron beam-off experiments in similar, if not identical, experimental conditions to those utilized in the in-situ TEM corrosion study. This approach will help to delineate the impacts from thin film geometry, laminar flow conditions, exposed surfaces, and direct electron beam condition. The second approach that should be utilized whenever possible is the detailed characterization of the reaction that occur. The possibility of achieving this has greatly improved since the initial in-situ TEM corrosion studies through the incorporation of analytical microscopy and analytical chemistry techniques. The better characterization of the reaction by-products that can be identified, the greater insight the experimentalist can gain into the reaction dynamics. The impact of all these effects should be considered not only in the experimental design, but just as importantly in the selection of imaging conditions and data interpretation. It is very easy to fall prey to the over-interpretation of artifacts that are unavoidable and potentially impactful during any in-situ TEM corrosion experiment.

4 In-Situ TEM Corrosion Studies

With all of the potential benefits and limitations associated with in-situ TEM corrosion studies identified, we can now explore the advancements made over the last seven years since the first dedicated in-situ TEM corrosion experiment [49] was performed. Two recent reviews by Ali Kosari et al. [56] as well as Zhengwei Song and Zhi-Hui Xu [57] have detailed the history, challenges, recent achievements, and potential future of in-situ TEM corrosion field. This section will begin by highlighting some of the earliest in-situ TEM corrosion studies. These were often done on simple model systems with minimal characterization associate with them. Despite these limitations, a great deal of fundamental understanding of films degradation could be determined. Next, examples of experiments performed on much more complex alloy systems with increasing complex characterization will be highlighted. Finally, this section will end with the effort to draw out the most amount of analytical insight from the sample before, during, and after the in-situ TEM corrosion study. Overall, this section will highlight the rapid growth in the field and the range of potential studies now available.

4.1 In-Situ TEM Corrosion Studies in Model Systems

All of the initial in-situ TEM corrosion experiments utilized simple sample chemistry, microstructure, and geometries, as well as simple corrosive medium and imaging conditions. For example, work by See Wee Chee et al. explored the degradation of deposited Al thin films when exposed to a stagnant solution of brine (0.01 M NaCl) in a closed cell geometry [49]. Despite these limitations, it is clear in Fig. 6 that direct observation on corrosion mechanisms are possible during bright field (BF) TEM imaging with surprisingly good image quality and resolution. Their initial observation suggested that the degradation of the Al film was dominated by blistering and rupture.

This initial study proposed a blistering mechanism might be active during the initial pitting of films. Given that this is not the classical pitting mechanisms, it quickly raises the question if the same observation would be made in other high-purity deposited metal thin films that are often nanocrystalline. A follow-on experiment in nanocrystalline copper is compared in Fig. 7 [58]. The copper films appear to show the formation of nanoscale pits that link together over time to form larger pit structure, Fig. 7a–c. This early study also shows that not only can the initial corrosion mechanisms be observed, but the linkage of various corroded regions and the subsequent failure of the thin films can be studied, as is Fig. 7d–i. The study also explored the difference in closed cell vs. open cell geometries in the nanocrystalline Al films. Despite other variables being changed, it was interesting that the open cell geometry showed a directional dewetting of the Al film from the SiN window, whereas the closed cell geometry confirmed the blistering observation made in the first in-situ TEM corrosion study. This is another example that the nuance of experimental parameters can be of great importance in what active mechanisms is observed. It should also be noted that even in this very early in-situ TEM corrosion experiment, electrical biasing was being utilized following the previous electrochemical liquid cell technology development for battery technology [59, 60]. This experiment found that corrosion of the Al film occurred rapidly at the electrode when biasing was applied ultimately limiting the bias that could be utilized [58].

The next complex experimental conditions were investigated with the erosion of deposited, high-purity, nanocrystalline Fe exposed to weak acetic acid (vinegar!) [61]. Classically, corrosion is thought of as a slow process; however, this study demonstrated that under certain circumstances the process of forming a single pit with a diameter less than 100 nm can happen in less than 1 s. It is also interesting to note that in this nanocrystalline geometry, there did not appear to be a direct correlation between the grain contrast in bright-field and the selective corrosion. This suggests, though not definitive, that grain orientation may not play a significant role in the observed pitting (Fig. 8).

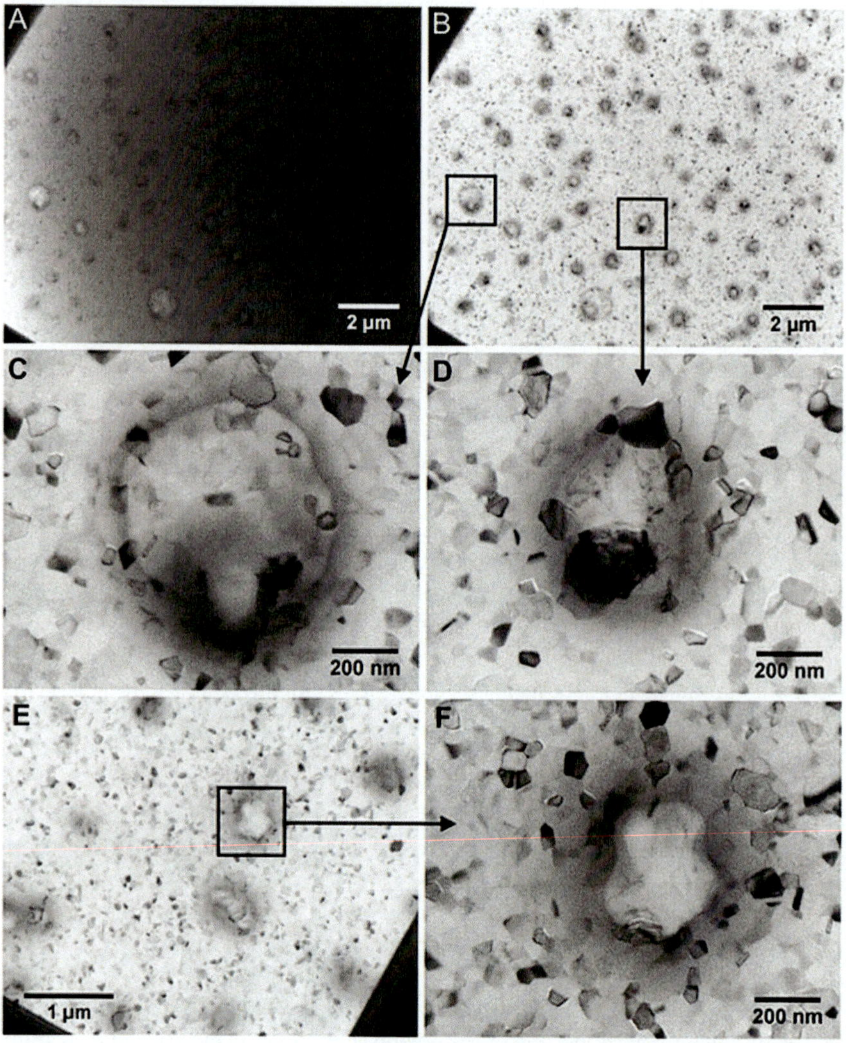

Fig. 6 In-situ TEM bright-field images of Al film blister formation during exposure to 0.01 M NaCl for 15 h. **a** Bright-field image of the cell filled with liquid before de-wetting. **c**, **d**, & **f** Higher magnification images of selected blisters. **d** A potential ruptured blister. Reproduced from [49] with permission from The Royal Society of Chemistry

In the same study [61], it was found that films annealed to create a bimodal grain size distribution prior to the same in-situ TEM corrosion experiment had a very bimodal response. The small grains pitted following the same mechanism seen in the as-deposited nanocrystalline films. In contrast, the large grains appeared to have little to no change in size or shape, as can be seen in Fig. 9. The lack of analytical characterization in these early in-situ TEM corrosion studies raises several questions

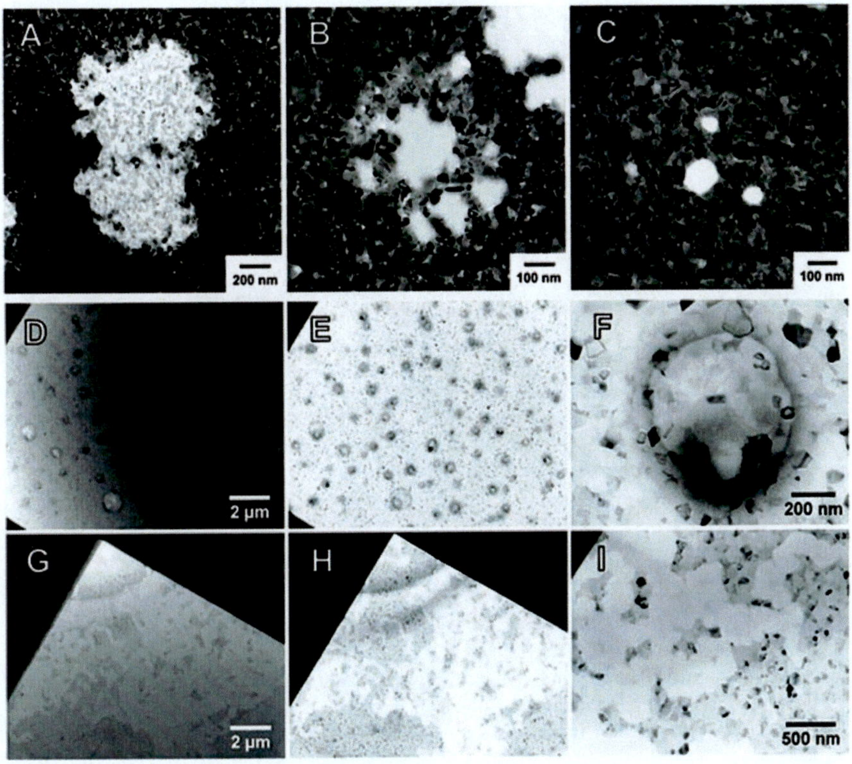

Fig. 7 **a–c** Pits at different stages of development observed in 50 nm Cu films after flow of 6 M NaCl solution for 2 h. (**d–f**) Blisters formed in a 100 nm Al film after exposure to 0.01 M NaCl solution for 15 h. (**g–i**) Meandering corrosion tracks formed in a 100 nm Al film after exposure to 1 M NaCl solution for 4 h. Images (**a–d** & **g**) are obtained with liquid in the cell and (**e**, **f**, **h** & **i**) were taken after liquid had been de-wetted with ethanol–water mixtures. Reproduced from Ref. [58] with permission from the Royal Society of Chemistry

about the results. Some of the questions left unanswered in observations like those seen in Fig. 9 are:

1. Is the preferential dissolution of the nanocrystalline grains due to the large grains decreased grain boundary area, as the mind would first assume?
2. Are the larger grains being affected at the same rate, but is a hillock with increased thickness thus delaying the full consumption?
3. Are the larger grains oxides, carbides, or any other phase that might be less prone to dissolution?
4. Is the large grain of a specific orientation that has a slower reaction rate?

In this published study, these questions could not be answered; however, all of these questions and more can be answered with modern analytical S/TEM techniques. Later sections in this chapter will detail the use and benefits of modern analytical S/TEM techniques as applied in more recent in-situ S/TEM corrosion experiments.

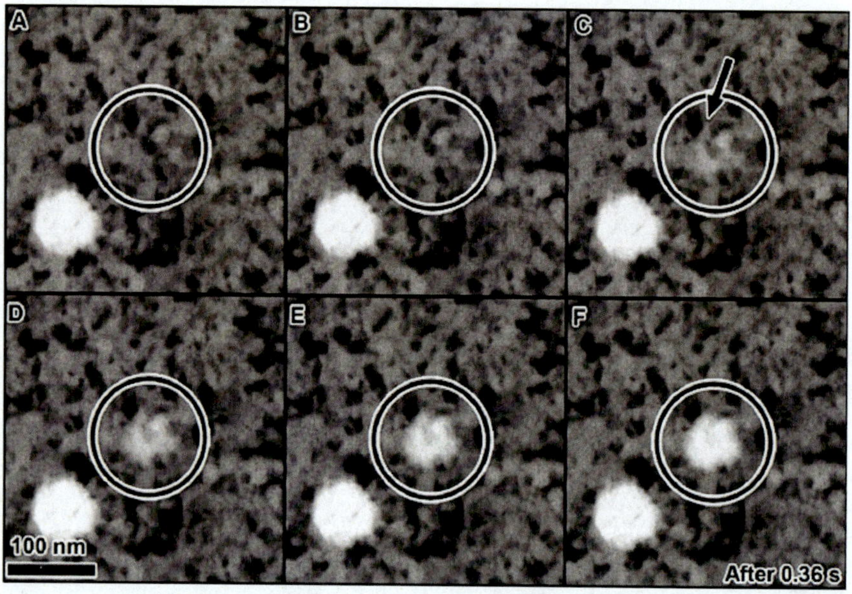

Fig. 8 **a** Time lapse series of micrographs detailing one corrosion event. Each image represent one frame from a video recorded at 13.75 frames per second. The evidence of dissolution can be seen by the change of contrast beginning in **b** that gradually spread through grains surrounding the sample. The dark grain in the center, indicated by an arrow in **c**, appears to be more resistant to corrosion. Copyright 2021 Wiley. Used with permission from [61]

Similar follow-on corrosion experiments in deposited Fe films have shown that the influence of electron irradiation can have a significant impact on the oxidation and corrosion rate of the film [62]. The appealing aspect of running these experiments in-situ is that the kinetics of the evolution can be captured and analyzed. Over the last decade, the time and effort required to track the evolution of the microstructure between frames has decreased exponentially. In this study, image processing was utilized to threshold and track the evolutions of the nanoscale pits providing great insight into every frame of the video sequence. This experimental insight provides delineation between the rapid but nearly continuous pitting rate seen in the sample exposed to high electron flux (Fig. 10i), which is significantly different from the jerky erosion rate in the sample exposed to a lower electron flux (Fig. 10j). Taken in total, these early studies into in-situ TEM corrosion of deposited films demonstrate that the approach is feasible, but great care needs to be taken in the design and control of experimental parameters ranging from film pre-treatments through fluid flow conditions to electron flux.

These in-situ S/TEM corrosion experiments in model systems are not limited to deposited thin films. Initial experiments also explored the stability of metallic nanoparticles in aqueous solutions. The use of in-situ TEM liquid cell experiments to investigate the wet etching of nanoparticles are reviewed by Mei Sun et al. [63]. Yingying Jiang et al. explored the stability of Pd nanocubes exposed to $FeCl_3$ aqueous

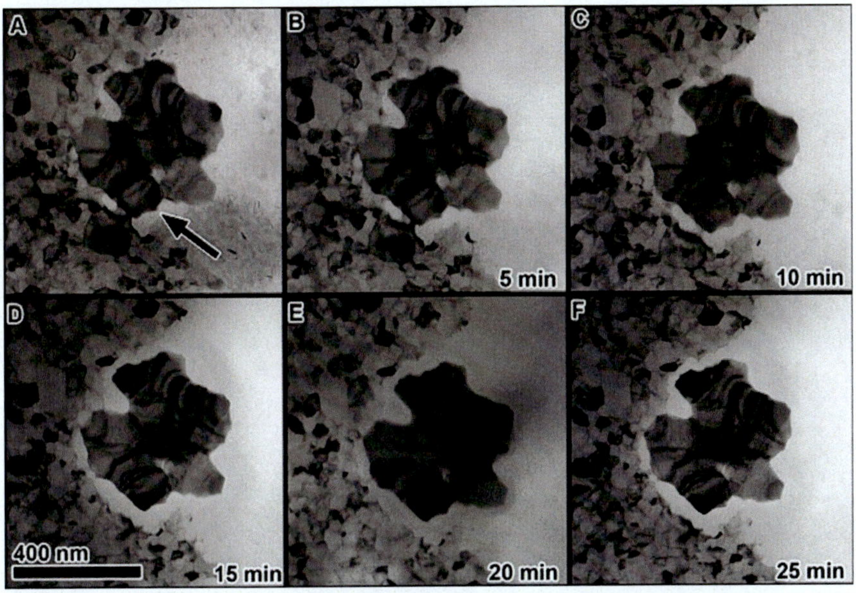

Fig. 9 Dissolution of grains surrounding an abnormally large grain. **a** Shows the initial observation of the area, and (**b–f**) demonstrate how the material surrounding the abnormal grain is dissolved leaving it isolated. A small protrusion from the abnormal grain is highlighted by an arrow in A. Copyright 2021 Wiley. Used with permission from [61]

solutions (0.228 mol/L) [64]. By performing this study in-situ direct observation of individual and assemblies of nanocubes is possible providing statistically relevant kinetic information into the stability of these nanostructures, as can be seen in Fig. 11. It is clear from this analysis that the rate of particle erosion is not constantly increasing as the particle reaches radii less ~3 nm and that the erosion rate is dependent on the particle size. Surprisingly, the largest particles do not appear to disappear last in this data set. In-situ TEM experiments also permit the direct observation of subtle nanoscale effects, such as the softening of the cube edges occurring prior to substantial shrinkage of the nanoparticles area. Since most nanostructures are far-from-equilibrium, it is extremely important to understand and experimentally validate the environmental stability of the structures in relevant environments.

If one is clever in your material selection and experimental approach, one can even explore classical metallurgical phenomena, such as the Kirkendall effect. This was done by See Wee Chee et al. who utilized an in-situ TEM liquid cell in an unconventional manner. They drop-casted wet samples, but dried the samples and operated the cell at elevated temperature to explore the impact or nanoparticle dimensions, temperature, and other experimental parameters on the Kirkendall effect during galvanic replacement reactions [65]. This was done with Ag nanocubes that became hollow via void formation inside the nanocubes, driven by galvanic replacement with Au. In a similar fashion, Hao Shan et al. showed that the corners of palladium-platinum

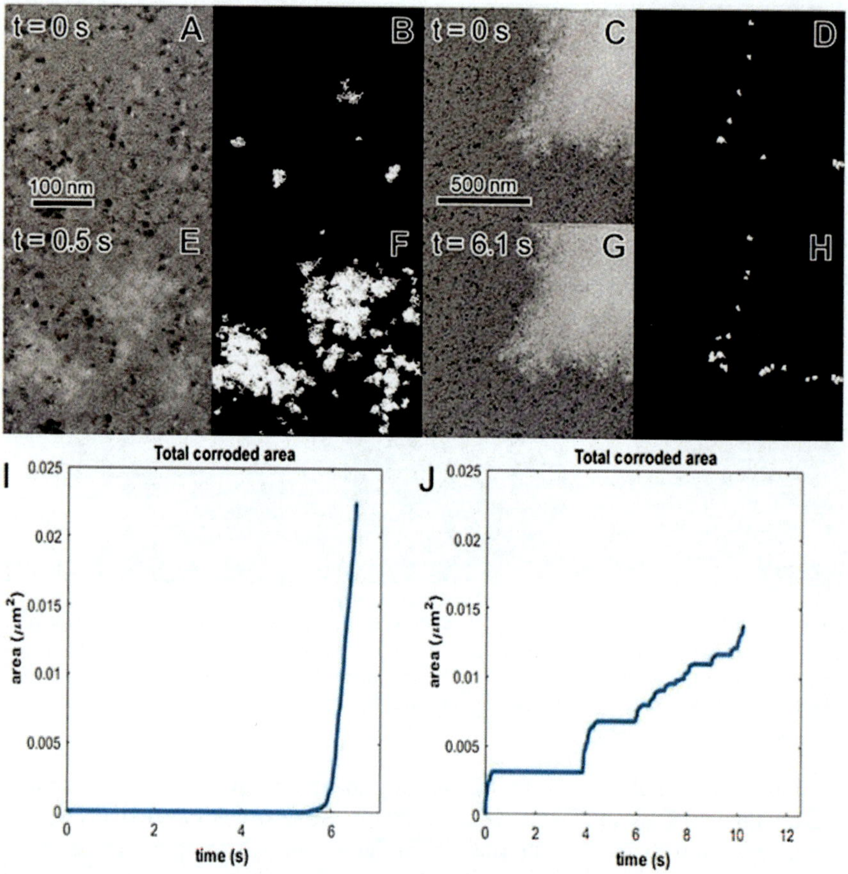

Fig. 10 Bright field TEM images captured from video showing the initiation and expansion of the oxidation front for a higher dose rate event **a**, **c** and lower dose rate event **e**, **g**. Associated binarized image differences showing the newly oxidized area over the 0.5 s time span **b**, **d** and the 6.1 s time span for the low dose rate **f**, **h**. Total corroded area as a function of time for the higher dose rate event **i** and lower dose rate event **j** which shows arresting and reinitiation behavior that is not seen in the higher dose rate event. Reprinted from [62]. Copyright 2021, with permission from Elsevier

core–shell nanocubes corrosion process started at the cube corners when exposed to 0.1 M NaBr [66]. The following erosion of the core portion of the nanoparticle then proceeds in an asymmetrical manner, which can be quantified as a function of etching area, etching rate, curvature of the etching face, and even the etching rate as a function of the curvature. All of this nanoscale data is made possible due to the in-situ nature of the experiment and can be used to directly elucidate that active kinetics operating in the system. If tailored properly, nanoparticles can be ideal TEM samples to explore the kinetics of a range of materials systems and environments (Fig. 12).

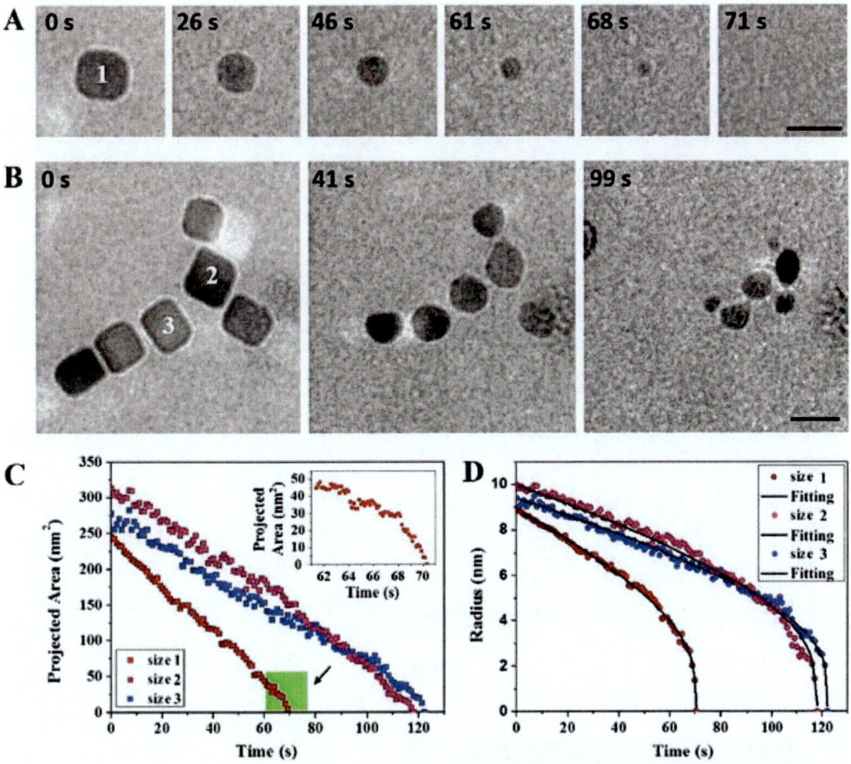

Fig. 11 TEM images of in-situ dissolution process for **a** monodispersed and **b** aggregated Pd nanocrystals. The electron dose used here is 136 electrons/Å^2 s. All scale bars are 20 nm. **c** Projected area and **d** radius variations of the three marked Pd nanocrystals in **a** and **b** with reaction time Reprinted from [64]. Copyright 2021, with permission from Elsevier

Although the majority of the model systems explored and highlighted in this chapter were high purity metals, this is not the only model material systems explored. For example, Donovan N. Leonard and Roland Hellmann explored the impact on ledge and crack evolution in a mineral Wollastonite, $CaSiO_3$, when exposed to pure deionized water [67]. Essentially, once an experimental workflow has been developed, the range of TEM samples that can be explored is immense, as we hope you will realize in the next section exploring commercially relevant alloys.

4.2 In-Situ TEM Corrosion Studies in More Complex Alloys

The previous section highlighted experiments investigating fundamental corrosion mechanisms in model materials systems. The fact that the majority of in-situ S/TEM corrosion studies are in such model systems to date should probably be attributed to

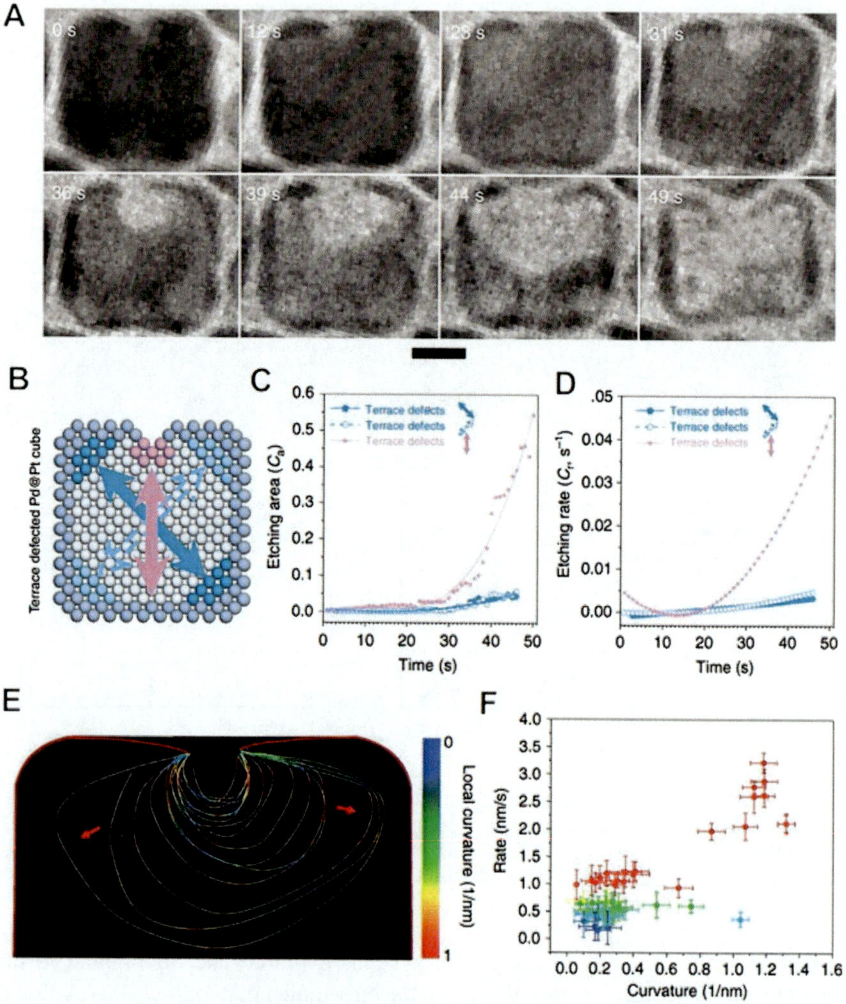

Fig. 12 Etching process of terrace defected cube. **a** Time sequential TEM micrographs showing the etching process of internal Pd atoms in a single terrace defected Pd@Pt cube. Scale bar is 5 nm. **b** The illustration of atomic structures demonstrates the calculation of C_a. **c** Scatter diagrams and fitting curves of C_a. **d**) Corresponding rates of etching areas in **c**, along each direction in the cube. **e** Time-domain contour plots of etched hole along the direction of terrace defects. Contour lines are spaced in time by 2 s. Color of curves shows the local curvature. **f** Relationship of the etching rate and local curvature. The measuring dimensions (lengths) have the error of ± 1 pixel in the image, the error bars are then calculated following error analysis during the derivation of curvature and etching rate [66]. Use permitted under the Creative Commons License CC-BY 4.0

the novelty of the field, as the model systems are far from the only samples that can be explored. Ali Kosari et al. nicely demonstrated in Fig. 13 how thoughtful S/TEM sample preparation could permit one to go from a real-work material of interest, such as an airplane, all the way down to a 100 nm or less thick sample necessary for in-situ S/TEM corrosion studies [68]. It is important to consider in the experimental design,

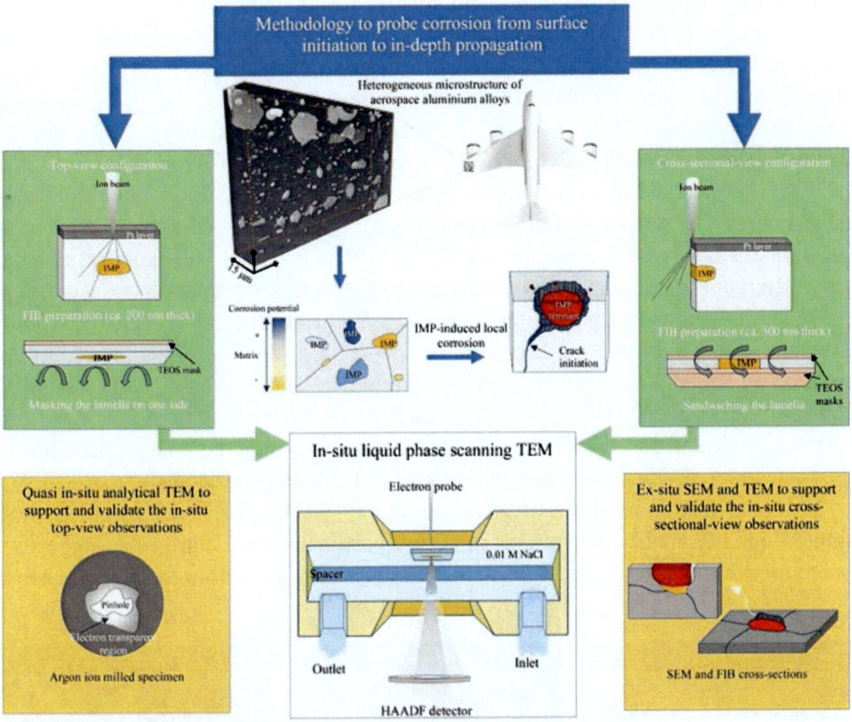

Fig. 13 Methodology for examining IMPs-induced corrosion degradation in aerospace aluminum alloys from surface initiation to in-depth propagation at the nanoscale. Three-dimensional reconstruction of aerospace AA2024-T3 microstructure which leads to complicated corrosion. IMPs acquiring different corrosion potential and electrochemical activity in aqueous solutions are potential pit initiation sites. In-situ LP-TEM are performed in HAADF STEM mode on the treated lamellae of approx. 10 μm–20 μm, representing 'real-life' corrosion. Top-view observations are provided by depositing a 25-nm-thick layer of tetraethyl orthosilicate (TEOS) using 2-keV electron-beam deposition; the corresponding schematic depicts the location of intermetallic particle (IMP) on the lamellae. For cross-sectional observations, the TEM specimen sandwiched between two 25-nm-thick layers of TEOS; note the location of intermetallic particle (IMP) on the lamellae. The prepared specimens are transferred to the home-made liquid cell using a FIB-based procedure, and each specimen is placed at the entrance side of the electron probe to mitigate the beam-broadening effect. Quasi in-situ analytical TEM experiments are conducted, in parallel, on argon ion-milled specimens (3-mm disk) to provide in-situ top-view observations with detailed structural and compositional information. Ex-situ examinations are carried out on FIB cross sections to support cross-sectional in-situ observations [68].

the variation in the local microstructure and expected attack direction of the corrosive medium, so that the S/TEM experiment can be designed to best emulate those real-world conditions. This was done during the study into the evolutions of S-phase (Al_2CuMg) of AA2024-T3, an aerospace grade Al-alloy, exposed to brine (0.01 M NaCl). The ability to do site-specific FIB lift-outs permits the study of particular interphases or grain boundaries that might be of greatest interest to enhance model refinement of these complex commercial grade alloys.

In another study, Ali Kosari et al. explored the nature of dealloying in complex intermetallic alloys present in the same aerospace alloy. By selecting a slightly different region of the alloy, they explored the stability of an individual $Al_{76}Cu_6Fe_7Mn_5Si_6$ particle exposed to a similar brine solution. Here, they observed that the complex particle undergoes dealloying with the loss of Al and Mn, and some Fe. The chemical evolution of this particle would not be possible without intermediate Energy-Dispersive X-ray Spectroscopy (EDS) maps obtained and present at the bottom of Fig. 14. The porous particle that remains would be expected to have very different structural stability than the structure initially present.

Switching from aerospace to petrochemical applications, Steven Hayden, Claire Chisholm et al. explored the evolution of low-carbon pipeline steels to a complex reagent of aqueous electrolyte (pH 6.1) containing CO_2 (*aq.*, 6 μM), O_2 (*aq.*, 281 μM), and Na_2SO_4 (*aq.*, 2.78 μM). Careful effort to protect the top and bottom surfaces promoted the corrosion to occur from the top flank of the sample. In addition, the foresight to perform thorough analytical TEM characterization prior to the in-situ STEM experiment provided a wealth of structural and compositional insight into the sample. Finally, meticulous efforts to not only measure the electron flux exposure, but to minimize the electron exposure during the experiment limit the electron beam

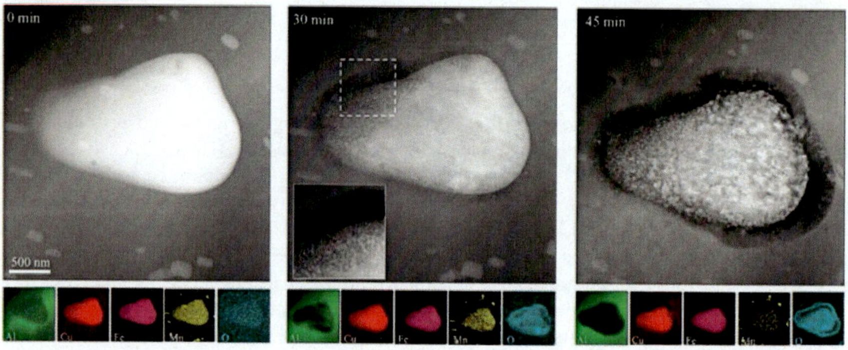

Fig. 14 Quasi in-situ STEM/EDS analysis of local corrosion induced by $Al_7Cu_2Fe(Mn)$ phase. HAADF-STEM images and corresponding EDS maps at different exposure times, revealing the dissolution of Al and Mn, and to a minor extent Fe from the particle. Due to Cu and Fe aggregation, the particle shows a globular morphology with time. The inset at 30 min is a closer look at the white rectangular region, revealing the dealloyed region of the particle. The specimen is an argon ion-milled 3-mm disk exposed repeatedly to 0.01 M NaCl [69]. Use permitted under the Creative Commons License CC-BY 4.0

impact on the observations. The combination of these efforts resulted in a rigorous study of the evolution of a single triple junction where localized corrosion initiated between a cementite and two ferrite grains, as can be seen in Fig. 15. It is worth noting the low electron flux resulted in poor signal-to-noise during the experiment (Fig. 15c–f). Despite this being only one of a near infinite number of triple junctions between various phases in the miles and miles of low-carbon pipeline steel used around the world, this observation provides clear insight that re-initiating a galvanic

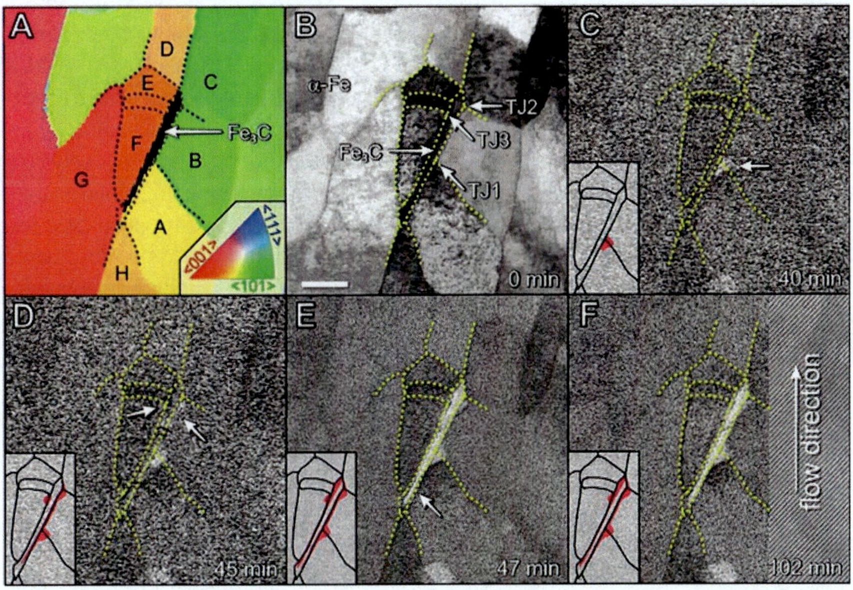

Fig. 15 Aqueous corrosion of 1018 steel visualized by in-situ STEM. Scale bar indicates 200 nm. The PED-enhacned automated crystallographic orientation map (ACOM) and bright-field STEM micrographs depict the region where localized corrosion was observed. Liquid exposure times are provided in each panel. Regions of increasing intensity resulted from increased transmission of the beam (**c–f**) and correspond to localized corrosion events. Grain boundaries are delineated in yellow to facilitate comparison between panels, and insets show a schematic of the grain boundaries with regions experiencing localized corrosion highlighted in pink. **a** ACOM indicating the various crystallographic orientations in the body-centered cubic (α-ferrite) reference plane. Black pixels denote non-BCC phases (i.e., cementite). **b** BF-STEM micrograph showing the microstructure of this region in vacuum prior to exposure to the aqueous electrolyte; the cementite grain inclusion was completely bordered by a ferrite matrix. TJ1, TJ2, and TJ3 are indicated with white arrows. (**c–f**) BF-STEM micrographs depicting this region during exposure to flowing aqueous solution. **c** Localized corrosion at the inclusion was first observed at the triple junction (white arrow) between Fe_3C, αA, and αB (**TJ1**). Total electron flux 0.905 $e^-/Å^2$*s. **d** At 45 min, localized corrosion was observed to have claimed the interface between the Fe_3C grain and grains αA/αB/αC, as well as the triple junctions (white arrows) formed by Fe_3C/αB/αC (**TJ2**) and Fe_3C/αE/αF (**TJ3**). Total electron flux 0.915 $e^-/Å^2$*s. **e** At 47 min, the localized corrosion area stopped evolving. Total electron flux 1.23 $e^-/Å^2$*s. **f** Final in-situ micrograph acquired after 102 min of solution exposure. Total electron flux 5.15 $e^-/Å^2$*s [70]. Use permitted under the Creative Commons License CC-BY 4.0

corrosion mechanism is possible in similar environments and should be included in the refinement of corrosion models of low-carbon steel.

The previous two aerospace and pipeline steel examples show a clear impact of in-situ S/TEM corrosion to provide fundamental mechanistic insight into complex commercial alloys. As a result of these fundamental studies, hypotheses can be developed, tested, and validated to provide more corrosion-based chemistries and microstructures for such applications. It should be reinforced that these nanoscale observations alone should not result in the change of any commercial alloy composition or processing. The example from Sibylle Schilling et al., seen in Fig. 16, comparing the drastically different electrochemical corrosion response between in-situ, ex-situ microcell, and bulk response in 304 stainless steel (SS) exposed 0.001 M H_2SO_4 [71]. In this study, the difference between ex-situ microcell and in-situ experiments shows that even with similar geometries the response of a material during TEM observation may vary significantly from the same configuration when not exposed to the electron beam and other artifacts created by the TEM environment. This example reminds us that in-situ S/TEM experiments should be considered a peep-hole into the active nanoscale mechanisms. Any hypothesis developed based on these observations will need to be delineated from experimental artifacts, weighted against other active mechanisms, and scaled up to account for all the multiscale effect governing the ultimate properties and performance of the commercial alloy.

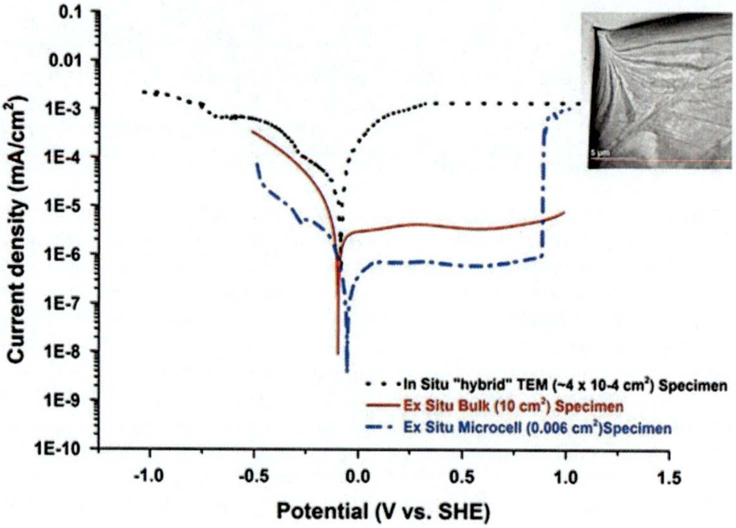

Fig. 16 Potentiodynamic polarization curve of Type 304 SS in 0.001 M H_2SO_4. The black curve was measured using the in-situ electrochemical cell with a 500 nm gap, blue curve was measured using the microcell, and the red curve was measured using the conventional bulk sample technique. Insert: BF-TEM image shows the area imaged during the in-situ electrochemical test [71], reproduced with permission

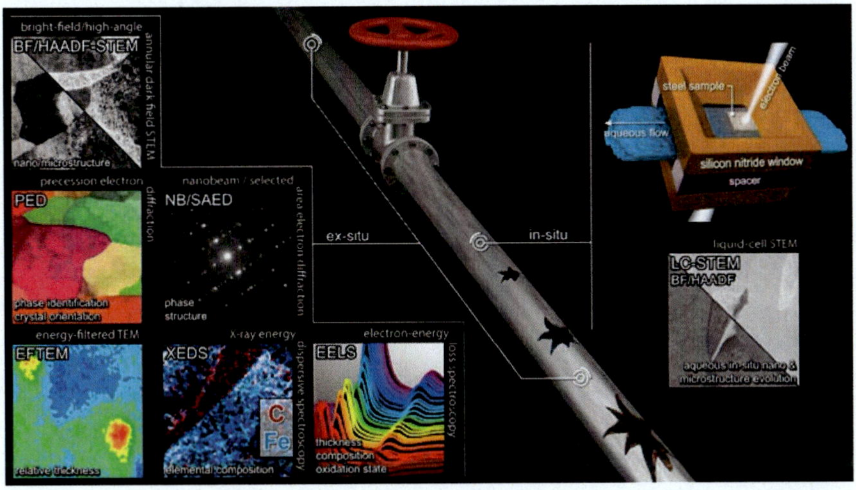

Fig. 17 Experimental workflow for examining aqueous corrosion initiation in 1018 steel in real time at the nanoscale. The initial sample was characterized using a suite of electron microscopy techniques, whose names, abbreviations, primary data types, and representative data sets are provided. The fully characterized sample was placed in a liquid cell for in-situ real-time observation under aqueous flow. The corroded sample was then characterized using the same suite of ex-situ electron microscopy techniques to aid in the interpretation of the in-situ observations [70].

4.3 Coupling Analytical Microscopy with Liquid Cell TEM

To fully elucidate the influence of various experimental factors on the observed nanoscale evolution, a detailed understanding of these parameters is needed. Fortunately, a range of analytical S/TEM permits the characterization of these parameters. Figure 17 highlights the range of imaging, diffraction and spectroscopy techniques possible in a real-world pipeline steel alloy. The vast majority of the experiments shown above were conducted in standard bright field (BF) imaging at zero-degree tilt. Despite the tilt limitations due to the size of the in-situ TEM liquid cell holders, it should be realized that significantly higher quality images providing more information can be obtained if proper imaging conditions can be determined and obtained. Some of the possible imaging conditions/techniques to consider utilizing would include BF-STEM, Weak-Beak Dark Field (WBDF), High Angle Annular Dark Field (HAADF), and Energy Filtered TEM (EFTEM). Each of these techniques has the benefit of minimizing the contrast from specific selected defects, ideally artifacts, while accentuating the contrast of importance. In a similar fashion the information on phase, grain orientation, grain boundary character can be collected in real time through an in-situ Select Area Electron Diffraction (SAED) or Nanobeam Electron Diffraction (NBED) experiment or mapped a key steps in the corrosion process through Precession Electron Diffraction enhanced Automated Crystal Orientation Mapping (ACOM). The ACOM technique permits mapping of the phases and grain orientations present, as well as all but one of the grain boundary characteristics. Finally, EDS and Electron Energy Loss Spectroscopy (EELS) are two commonly employed analytical TEM techniques that can provide the local chemistry and the local bonding structure and thickness, respectively. The utilization of these advanced, but well-tested, S/TEM characterization techniques provides a direct and rapid way to understand at a deeper level the governing mechanisms active during in-situ S/TEM corrosion experiments. The ability to rapidly switch between these various imaging, diffraction, and spectroscopy conditions/techniques is not outside the realm of possibilities in the not too distant future [24].

5 Future Directions Liquid Cell-TEM for Corrosion Studies

In the near future, it is also expected that the recent advancements of combining in-situ TEM liquid cell experiments for corrosion studies will continue to explore an increasing number of material systems in an increasing number of abrasive environments utilizing an increasing number of advanced characterization techniques. It will not be long before publication will utilize emerging techniques like 4D STEM to image the local strain during liquid cell corrosion [72]. Similar studies in the electrochemistry field have already been produced utilizing nanodiffraction to study in-situ lithiation and delithiation of MoS_2 Nanosheets [73]. In the future it might also be possible to couple electron tomography or post-mortem FIB-based approaches to elucidate the 3D nanoscale structure, as has been done for in-situ TEM gaseous oxidation experiments [74]. Work is also underway to utilize controlled cooling of the sample to temporarily alter the fluid content and subsequently improve the image resolution. This has been successfully shown for O_2 and H_2S corrosion of the AA2024-T3 alloy and cold-rolled HCT980X steel lamellae [75]. It is also expected that in the near future work will be incorporated to characterize the ejected fluid using advanced analytical techniques. These advancements in characterization applied to in-situ liquid cell experiments, as well as the continued effort to incorporate increasingly complex material systems, are important and will both highlight the usefulness and increase the trust of corrosion scientists in the potential of liquid cell TEM experiments for corrosion studies.

It is also expected that the complexity of the in-situ experiments will continue to grow to gain better understanding of the factor governing the fundamental corrosion mechanisms. A straight forward, but unexplored, field would be the combination of in-situ ion irradiation TEM [76] with in-situ corrosion TEM studies to understand the complex interaction of radiation damage that has been shown to enhance and impede corrosion [5, 77]. Two further out in-situ TEM corrosion sub-fields of immense importance that are outside the current reach of in-situ liquid cell experiments are Stress Corrosion Cracking (SCC) and biofouling. SCC is a complex degradation of metals in which both mechanical loading and corrosive environment is present in just the right ratio. Although the field has been extensively studied for over 50 years [78], there are still many unanswered questions of the active mechanisms occurring at the nanoscale and the best way to predict the evolution of SCC [79]. A recent study was able to explore the inherent strain impact on the corrosion kinetics in palladium-platinum core–shell octahedra nanoparticles [80]. Although this noble metal nanoparticle is an extreme model system, it clearly highlights the potential of elucidating strain's role in corrosion at the nanometer scale. One possible solution on the horizon to further elucidate SCC would be the development of an in-situ TEM technique to explore in-situ TEM SCC. In-situ TEM deformation studies have a rich history as one of the oldest forms of in-situ TEM studies and have been explored to a limited extent with gas environments [30] and during liquid metal embrittlement [33]. However, to the best of our knowledge, no experiments have been performed to explore the impact of aqueous or other classical corrosive environments during in-situ TEM deformations studies. This is due to the difficulty of creating a liquid cell that permits simultaneous mechanical deformation. Recent efforts underway led by Katherine Jungjohann at Sandia National Laboratories is developing a MEMS-based platform to permit simultaneous thermal, mechanical, and corrosive environment exposure all during in-situ S/TEM observation, as can be seen in Fig. 18 [30, 81]. If successful, this MEMS-based approach would permit real time nanoscale evolution of SCC initiation as a function of loading conditions, corrosive environment, and even temperature, which would greatly advance our fundamental understanding of the complex evolution that occurs during SCC.

In a similar fashion, biofouling of metals is a complex interplay between biochemistry and corrosion and has been shown to have a significant deleterious impact on marine transport and related fields [83]. Previous work has shown that everything

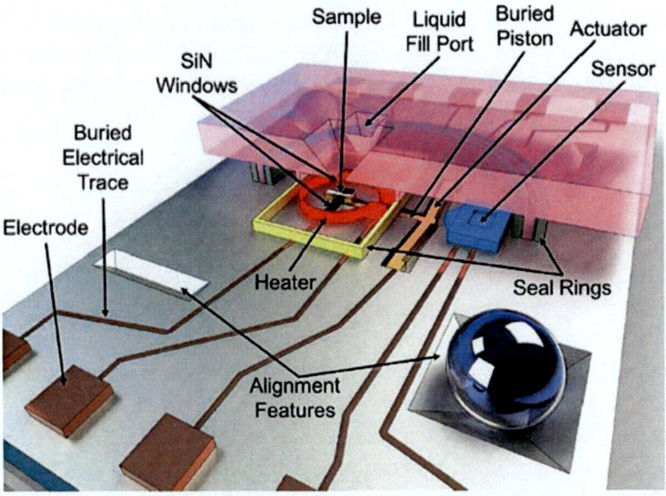

Fig. 18 Schematic of an in-situ TEM MEMS-based heated stress corrosion cracking platform designed and developed at Sandia National Laboratories [82] reproduced with permission

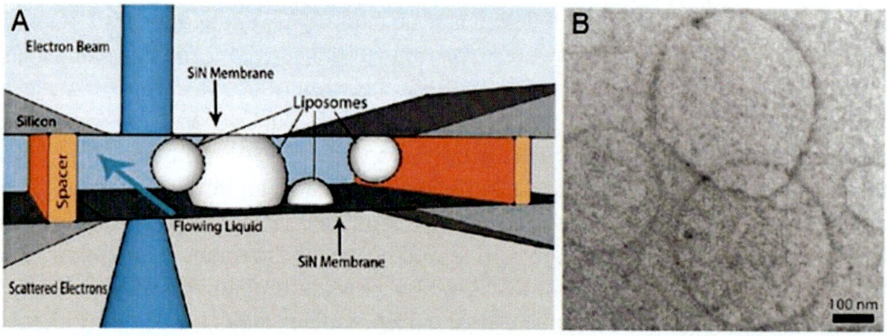

Fig. 19 **a** schematic of liposomes during an in-situ TEM liquid cell experiment. **b** image of liposome present during in-situ TEM liquid cell experiments Reprinted with permission from [84]. Copyright 2021 American chemical society

from biochemicals, such as bovine serum albumin (a portion of cow's blood) to full liposomes can be imaged during in-situ liquid cell-TEM experiment, as seen in Fig. 19 [84]. It is not a far leap to consider an electron-transparent metal film being tacked down in this cell during the introduction of any biochemical of interest that can fit within the microfluidic channel of the liquid cell. Such an experiment would provide great insight into the nanoscale interactions of biofouling, despite the concerns and challenges.

Once one has a good fundamental understanding of the physical limitation of electron beam interaction with mater, the ability for in-situ S/TEM corrosion study is limited by the creativity and perseverance of the scientist exploring the field. Despite this simplistic and optimistic view, the challenges are immense and we think Ernst Ruska's quote regarding the invention of the TEM from his Nobel lecture still holds, "We should not, therefore, blame those scientists today who did not believe in electron microscopy at its beginning. It is a miracle that by now the difficulties have been solved to an extent that so many scientific disciplines today can reap its benefits."

6 Conclusion

The juxtaposing recent advancements in microfluidic technology and S/TEM techniques has resulted in the emergence of in-situ S/TEM corrosion studies over the last decade in liquid cell environments. The first studies in this field utilized simple bright-field TEM imaging on simplistic and classical corrosion examples, such as high purity iron exposed to acetic acid. More recent studies have utilized increasing complex S/TEM imaging characterization tools on an increasing variety of industrial relevant alloys. The field will always be limited by the inherent interaction of electrons, as it transmits through all the materials in the full liquid cell sandwich. This will result in compromising decisions between sample thickness, cell size, radiation damage, resolution, etc. in the planning and implementation of all in-situ S/TEM corrosion studies. Despite these limitations, recently published advancements in the field of in-situ corrosion provides promise for the further study of chemical degradation of metal alloys in more and more complex environments. Overall, in-situ S/TEM corrosion is a rapidly emerging field aimed at elucidating materials degradation with nanometer resolution in real time.

Acknowledgements The authors would like to thank Drs. See Wee Chee, Katherine Jungjohann, Eric Lang, and Kinga Unocic, as well as Mr. Daniel Ranke and Ms. Kaitlyn Moo for their helpful discussion. This research was conducted in part at the Center for Nanophase Materials Sciences, which is a DOE Office of Science User Facility. This work was performed, in part, at the Center for Integrated Nanotechnologies, an Office of Science User Facility operated for the U.S. Department of Energy (DOE) Office of Science. Sandia National Laboratories is a multimission laboratory managed and operated by National Technology & Engineering Solutions of Sandia, LLC, a wholly owned subsidiary of Honeywell International, Inc., for the U.S. DOE's National Nuclear Security Administration under contract DE-NA-0003525. The views expressed in the article do not necessarily represent the views of the U.S. DOE or the United States Government.

Disclaimer The full description of the procedures used in this book includes the identification of certain commercial products and their suppliers. The inclusion of such information should in no way be construed as indicating that such products or suppliers are endorsed by the authors or affiliated institutions or the U.S. Government or are recommended by authors or affiliated institutions or the U.S. Government or that they are necessarily the best materials, instruments, software or suppliers for the purposes described.

References

1. Comstock, R.J., G. Schoenberger, and G.P. Sabol. *Influence of processing variables and alloy chemistry on the corrosion behavior of ZIRLO nuclear fuel cladding.* in *Zirconium in the Nuclear Industry: Eleventh International Symposium.* 1996. ASTM International.
2. J. Pickens, J. Gordon, J. Green, The effect of loading mode on the stress-corrosion cracking of aluminum alloy 5083. Metall. Trans. A **14**(4), 925–930 (1983)
3. ZHANG, X., et al., *Effect of Direct Current Electric Field on Corrosion Mechanism of Zn Exposed to Simulated Industrial Environment.* Journal of Chinese Society for Corrosion and protection, 2017. **37**(5): p. 451–459.
4. K. Shinohara, R. Aogaki, Magnetic field effect on copper corrosion in nitric acid. Electrochemistry **67**(2), 126–131 (1999)
5. Schmidt, F., et al., *Effects of Radiation-Induced Defects on Corrosion.* Annual Review of Materials Research, 2021. **51**.
6. S. Benavides, Corrosion in the aerospace industry, in *Corrosion Control in the Aerospace Industry.* (Elsevier, 2009), pp. 1–14
7. M. Yunovich, N.G. Thompson, Corrosion of highway bridges: Economic impact and control methodologies. Concr. Int. **25**(1), 52–57 (2003)
8. G.H. Koch et al., Cost of corrosion in the United States, in *Handbook of environmental degradation of materials.* (Elsevier, 2005), pp. 3–24
9. F. Di Turo et al., Impacts of air pollution on cultural heritage corrosion at European level: What has been achieved and what are the future scenarios. Environ. Pollut. **218**, 586–594 (2016)
10. E. Ruska, The development of the electron microscope and of electron microscopy. Rev. Mod. Phys. **59**(3), 627 (1987)

11. S.J. Pennycook et al., Aberration-corrected scanning transmission electron microscopy: from atomic imaging and analysis to solving energy problems. Philosophical Transactions of the Royal Society A: Mathematical, Physical and Engineering Sciences **2009**(367), 3709–3733 (1903)
12. Kisielowski, C., et al., *Detection of single atoms and buried defects in three dimensions by aberration-corrected electron microscope with 0.5-Å information limit.* Microscopy and Microanalysis, 2008. **14**(5): p. 469.
13. S.J. Pennycook et al., Materials advances through aberration-corrected electron microscopy. MRS Bull. **31**(1), 36–43 (2006)
14. Fultz, B. and J.M. Howe, *Transmission electron microscopy and diffractometry of materials.* 2012: Springer Science & Business Media.
15. Hayat, M.E., *Basic techniques for transmission electron microscopy.* 2012: Elsevier.
16. D.B. Williams, C.B. Carter, The transmission electron microscope, in *Transmission electron microscopy.* (Springer, 1996), pp. 3–17
17. De Graef, M., *Introduction to conventional transmission electron microscopy.* 2003: Cambridge university press.
18. R. Egerton, Control of radiation damage in the TEM. Ultramicroscopy **127**, 100–108 (2013)
19. R. Egerton, P. Li, M. Malac, Radiation damage in the TEM and SEM. Micron **35**(6), 399–409 (2004)
20. Spurgeon, S.R., et al., *Towards data-driven next-generation transmission electron microscopy.* Nature materials, 2020: p. 1–6.
21. I.M. Robertson et al., Visualizing the behavior of dislocations—seeing is believing. MRS Bull. **33**(2), 122–131 (2008)
22. M.L. Taheri et al., Current status and future directions for in situ transmission electron microscopy. Ultramicroscopy **170**, 86–95 (2016)
23. Q. Yu, M. Legros, A. Minor, In situ TEM nanomechanics. MRS Bull. **40**(1), 62–70 (2015)
24. Hattar, K. and K.L. Jungjohann, *Possibility of an integrated transmission electron microscope: enabling complex in-situ experiments.* Journal of Materials Science, 2021: p. 1–12.
25. I.M. Robertson et al., Towards an integrated materials characterization toolbox. J. Mater. Res. **26**(11), 1341–1383 (2011)
26. Kishita, K., et al., *Development of an analytical environmental TEM system and its application.* journal of Electron Microscopy, 2009. **58**(6): p. 331–339.
27. T. Lee et al., An environmental cell transmission electron microscope. Rev. Sci. Instrum. **62**(6), 1438–1444 (1991)
28. P.L. Gai, R. Sharma, F.M. Ross, Environmental (S) TEM studies of gas–liquid–solid interactions under reaction conditions. MRS Bull. **33**(2), 107–114 (2008)
29. P.L. Gai, Development of wet environmental TEM (Wet-ETEM) for in situ studies of liquid-catalyst reactions on the nanoscale. Microsc. Microanal. **8**(1), 21 (2002)
30. G. Bond, I. Robertson, H. Birnbaum, On the mechanisms of hydrogen embrittlement of Ni3Al alloys. Acta Metall. **37**(5), 1407–1413 (1989)
31. I.M. Robertson et al., Hydrogen embrittlement understood. Metall. and Mater. Trans. B. **46**(3), 1085–1103 (2015)
32. D. Shih, I. Robertson, H. Birnbaum, Hydrogen embrittlement of α titanium: in situ TEM studies. Acta Metall. **36**(1), 111–124 (1988)
33. Hugo, R. and R. Hoagland, *In-situ TEM observation of aluminum embrittlement by liquid gallium.* Scripta materialia, 1998. **38**(3).
34. J.Y. Huang et al., In situ observation of the electrochemical lithiation of a single SnO2 nanowire electrode. Science **330**(6010), 1515–1520 (2010)
35. R. Hugo, R. Hoagland, The kinetics of gallium penetration into aluminum grain boundaries—in situ TEM observations and atomistic models. Acta Mater. **48**(8), 1949–1957 (2000)
36. Madou, M.J., *Fundamentals of microfabrication: the science of miniaturization.* 2018: CRC press.
37. Tabeling, P., *Introduction to microfluidics.* 2005: OUP Oxford.

38. A. Kushima et al., Liquid cell transmission electron microscopy observation of lithium metal growth and dissolution: Root growth, dead lithium and lithium flotsams. Nano Energy **32**, 271–279 (2017)
39. H.-G. Liao, H. Zheng, Liquid cell transmission electron microscopy. Annu. Rev. Phys. Chem. **67**, 719–747 (2016)
40. Ross, F.M., *Opportunities and challenges in liquid cell electron microscopy.* Science, 2015. **350**(6267).
41. N. De Jonge, F.M. Ross, Electron microscopy of specimens in liquid. Nat. Nanotechnol. **6**(11), 695–704 (2011)
42. A.V. Bharda, H.S. Jung, Liquid electron microscopy: then, now and future. Applied Microscopy **49**(1), 1–6 (2019)
43. X.H. Liu, J.Y. Huang, In situ TEM electrochemistry of anode materials in lithium ion batteries. Energy Environ. Sci. **4**(10), 3844–3860 (2011)
44. M. Sun et al., Structural and morphological evolution of lead dendrites during electrochemical migration. Sci. Rep. **3**(1), 1–6 (2013)
45. R.L. Sacci et al., Nanoscale imaging of fundamental Li battery chemistry: solid-electrolyte interphase formation and preferential growth of lithium metal nanoclusters. Nano Lett. **15**(3), 2011–2018 (2015)
46. A.J. Leenheer et al., Lithium electrodeposition dynamics in aprotic electrolyte observed in situ via transmission electron microscopy. ACS Nano **9**(4), 4379–4389 (2015)
47. B.L. Mehdi et al., Observation and quantification of nanoscale processes in lithium batteries by operando electrochemical (S) TEM. Nano Lett. **15**(3), 2168–2173 (2015)
48. P. Palanisamy, J.M. Howe, In situ observation of Cu segregation and phase nucleation at a solid–liquid interface in an Al alloy. Acta Mater. **61**(12), 4339–4346 (2013)
49. S.W. Chee et al., Metastable structures in Al thin films before the onset of corrosion pitting as observed using liquid cell transmission electron microscopy. Microsc. Microanal. **20**(2), 462–468 (2014)
50. T. Xu, L. Sun, Investigation on material behavior in liquid by in situ TEM. Superlattices Microstruct. **99**, 24–34 (2016)
51. K. Jungjohann et al., In situ liquid cell electron microscopy of the solution growth of Au–Pd core–shell nanostructures. Nano Lett. **13**(6), 2964–2970 (2013)
52. J.M. Grogan et al., Bubble and pattern formation in liquid induced by an electron beam. Nano Lett. **14**(1), 359–364 (2014)
53. J.E. Evans et al., Controlled growth of nanoparticles from solution with in situ liquid transmission electron microscopy. Nano Lett. **11**(7), 2809–2813 (2011)
54. C.A. Taylor et al., Synthesis of complex rare earth nanostructures using in situ liquid cell transmission electron microscopy. Nanoscale Advances **1**(6), 2229–2239 (2019)
55. J. Park et al., Graphene liquid cell electron microscopy: progress, applications, and perspectives. ACS Nano **15**(1), 288–308 (2021)
56. A. Kosari et al., Application of in situ liquid cell transmission Electron microscopy in corrosion studies: a critical review of challenges and achievements. Corrosion **76**(1), 4–17 (2020)
57. Z. Song, Z.-H. Xie, A literature review of in situ transmission electron microscopy technique in corrosion studies. Micron **112**, 69–83 (2018)
58. S.W. Chee et al., Studying localized corrosion using liquid cell transmission electron microscopy. Chem. Commun. **51**(1), 168–171 (2015)
59. A.J. Leenheer et al., A sealed liquid cell for in situ transmission electron microscopy of controlled electrochemical processes. J. Microelectromech. Syst. **24**(4), 1061–1068 (2015)
60. C. Han, M.T. Islam, C. Ni, In Situ TEM of Electrochemical Incidents: Effects of Biasing and Electron Beam on Electrochemistry. ACS Omega **6**(10), 6537–6546 (2021)
61. D. Gross et al., In situ TEM observations of corrosion in nanocrystalline Fe thin films. Processing, Properties, and Design of Advanced Ceramics and Composites II **261**, 329 (2017)
62. Key, J.W., et al., *Investigating local oxidation processes in Fe thin films in a water vapor environment by in situ liquid cell TEM.* Ultramicroscopy, 2020. **209**: p. 112842.

63. Sun, M., J. Tian, and Q. Chen, *The studies on wet chemical etching via in situ liquid cell TEM.* Ultramicroscopy, 2021: p. 113271.
64. Y. Jiang et al., Probing the oxidative etching induced dissolution of palladium nanocrystals in solution by liquid cell transmission electron microscopy. Micron **97**, 22–28 (2017)
65. S.W. Chee et al., Direct observation of the nanoscale Kirkendall effect during galvanic replacement reactions. Nat. Commun. **8**(1), 1–8 (2017)
66. H. Shan et al., Nanoscale kinetics of asymmetrical corrosion in core-shell nanoparticles. Nat. Commun. **9**(1), 1–9 (2018)
67. D.N. Leonard, R. Hellmann, Exploring dynamic surface processes during silicate mineral (wollastonite) dissolution with liquid cell TEM. J. Microsc. **265**(3), 358–371 (2017)
68. Kosari, A., et al., *In-situ nanoscopic observations of dealloying-driven local corrosion from surface initiation to in-depth propagation.* Corrosion Science, 2020. **177**: p. 108912.
69. Kosari, A., et al., *Dealloying-driven local corrosion by intermetallic constituent particles and dispersoids in aerospace aluminium alloys.* Corrosion Science, 2020. **177**: p. 108947.
70. Hayden, S.C., et al., *Localized corrosion of low-carbon steel at the nanoscale.* npj Materials Degradation, 2019. **3**(1): p. 1–9.
71. S. Schilling et al., Practical aspects of electrochemical corrosion measurements during in situ analytical transmission electron microscopy (TEM) of austenitic stainless steel in aqueous media. Microsc. Microanal. **23**(4), 741–750 (2017)
72. C. Ophus, Four-dimensional scanning transmission electron microscopy (4D-STEM): From scanning nanodiffraction to ptychography and beyond. Microsc. Microanal. **25**(3), 563–582 (2019)
73. Z. Zeng et al., In situ study of lithiation and delithiation of MoS2 nanosheets using electrochemical liquid cell transmission electron microscopy. Nano Lett. **15**(8), 5214–5220 (2015)
74. W. Harlow, M.L. Taheri, Toward 3D imaging of corrosion at the nanoscale: Cross-sectional analysis of in-situ oxidized TEM samples. Micron **120**, 91–95 (2019)
75. M. Ahmadi et al., Locally Condensed Water as a Solution for In Situ Wet Corrosion Electron Microscopy. Microsc. Microanal. **26**(2), 211–219 (2020)
76. J. Hinks, A review of transmission electron microscopes with in situ ion irradiation. Nucl. Instrum. Methods Phys. Res., Sect. B **267**(23–24), 3652–3662 (2009)
77. G. Marsh et al., The influence of radiation on the corrosion of stainless steel. Corros. Sci. **26**(11), 971–982 (1986)
78. M.O. Speidel, Stress corrosion cracking of aluminum alloys. Metall. Trans. A **6**(4), 631 (1975)
79. S. Persaud, J. Smith, R. Newman, Nanoscale precursor sites and their importance in the prediction of stress corrosion cracking failure. Corrosion **75**(3), 228–239 (2019)
80. F. Shi et al., Strain-Induced Corrosion Kinetics at Nanoscale Are Revealed in Liquid: Enabling Control of Corrosion Dynamics of Electrocatalysis. Chem **6**(9), 2257–2271 (2020)
81. Jungjohann, K.L., et al., *Active mechanical-environmental-thermal MEMS device for nanoscale characterization.* 2020, Google Patents.
82. Chisholm, C., et al., *Mechanical-Environmental-Thermal MEMS Platform for In-Situ Nanoscale Materials Characterization.*
83. M.E. Callow, J.A. Callow, Marine biofouling: a sticky problem. Biologist **49**(1), 1–5 (2002)
84. S.M. Hoppe et al., In-situ transmission electron microscopy of liposomes in an aqueous environment. Langmuir **29**(32), 9958–9961 (2013)

Applications of Electrochemical Scanning Tunneling Microscopy in Corrosion Research

Muhammad Baseer Haider, Shankar Kunwar, and Khan Alam

Abstract Scanning tunneling microscopy is a technique that is used to image atomic resolution conducting surfaces, find the density of states of the sample surfaces and is used for atomic manipulation of species on the sample surfaces. Electrochemical scanning tunneling microscopy is a specialized scanning tunneling microscopy where the tip and the sample are in in an electrolyte environment. Electrochemical scanning tunneling does not only have the tunneling current but also Faraday current and capacitative current. Special care is given to keep the faradic current and capacitative current at the lowest so mostly tunneling current is achieved to image the surface. This is achieved by coating the scanning wire with insulating material and keep the tip of the wire exposed so the tunneling current is through the tip of the wire only. Electrochemical scanning tunneling microscopy is based on the continuous oxidation and reduction of the electrolyte between the tip and the sample which creates the tunneling current. Corrosion is a natural process of converting pure metal to a more stable, compound state thus releasing energy in the process. The underlying reason for corrosion is the thermodynamic instability of chemically active sites against water, air, or other chemicals. Since the corrosion reaction is mostly electrochemical, corrosion behavior of metals can be evaluated with electrochemical techniques. Application of Electrochemical scanning tunneling microscopy in the field of corrosion has elevated the research to another dimension and has advanced the understanding of the corrosion nucleation sites, pitting initiation, crevices, surface disintegration, passive coating formation, and helped in developing corrosion inhibitors. In this chapter, different examples of electrochemical scanning tunneling microscopy and spectroscopy study of surfaces like nickel, copper and gold in different chemical environments are given to understand the importance of this technique in study corrosions in metals.

M. B. Haider (✉) · S. Kunwar · K. Alam
Physics Department, King Fahd Univesity of Petroleum and Minerals, Dhahran 31261, Saudi Arabia
e-mail: mhaider@kfupm.edu.sa

I. Toor (ed.), *Recent Developments in Analytical Techniques for Corrosion Research*,
https://doi.org/10.1007/978-3-030-89101-5_7

1 Scanning Tunneling Microscopy

1.1 History of Scanning Tunneling Microscopy

In a real world, a ball thrown at the brick wall (strong barrier) is bound to bounce back. However, in quantum world there is a chance that the subatomic particle can tunnel through the strong barrier. Such a strange spooky phenomenon of the Quantum world, where a sub-atomic particle disappears from one side of the potential barrier and re-appears on the other side of the barrier without impacting the barrier and without manifesting itself in the barrier, is called quantum tunneling. The best example of quantum tunneling could be the electron tunneling through vacuum between two conducting surfaces [1–3]. In 1982, two scientists, Gerd Binning and Heinrich Rohrer, working at IBM Zurich, applied quantum tunneling and invented a machine which they called as Scanning Tunneling Microscope (STM).

1.2 Working Mechanism of STM

The STM consists of a very sharp probe (tip), usually made of metals like Platinum-iridium alloy or tungsten, attached to a piezoelectric transducer. The piezoelectric material gently brings the tip extremely close to the surface (within a few nanometers). Such a condition is maintained by the piezoelectric transducer driven by small voltage and the feedback loop. The schematic of STM is shown in Fig. 1a.

When an atomically sharp tip is brought closer to the sample surface as shown in the magnified image of the tip-sample in Fig. 1a, electrons tunnel through the vacuum between the tip and the sample. Normally the rate of electron tunneling from the tip to the sample and the sample to tip is almost identical, and hence there is no net tunneling current. To get net tunneling current a bias voltage (V) is applied between the sample and the tip. The tunnel current depends on many factors such as density of states (ρ_s) of the sample, density of states of the tip (ρ_t), and the tunneling matrix (M) which is the manifestation of the media between the tip and the sample. The tunnel current (I_T) is the cumulative result of tunneling electrons of energy ε in the range of E_f and $E_f + eV$, which can be mathematically expressed as:

$$I_T = \frac{4\pi e}{\hbar} \int_{E_f}^{E_f+eV} \rho_s(E_f - eV + \varepsilon)\rho_t(E_f + \varepsilon)|M|^2 d\varepsilon$$

where, E_f is Fermi energy of the sample, e is electronic charge and $\hbar$ is the reduced plank constant. The current thus tunneled is extremely small of the order of nanoamperes (nA) and is being amplified billions of times to make it recordable. Most of the time tips are chosen to have uniform density of states or s-wave symmetry. In such cases, the tunneling current is only dependent on the sample density of states and the

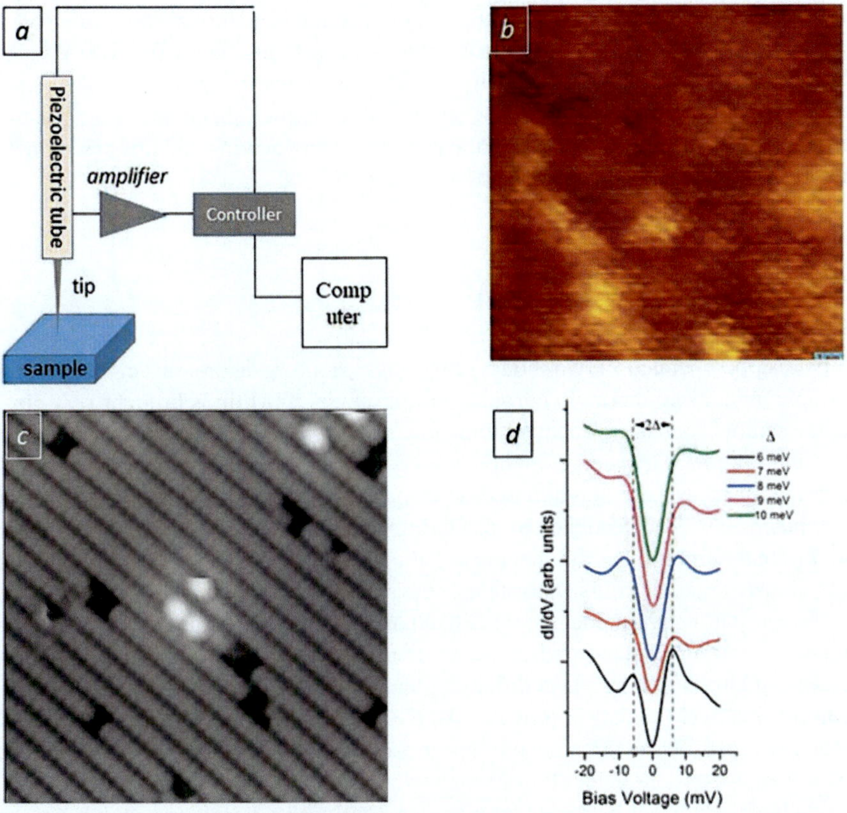

Fig. 1 **a** Schematic diagram of Scanning Tunneling Microscopy. **b** Real topographic STM image of 12.8 nm × 12.8 nm surface of $FeSe_{0.5}Te_{0.5}$ [4]. **c** Atomic resolution STM image of H-terminated Si(001) surface, where 2 × 1 dimer atomic rows are clearly visible with three bright protrusions due to Si dangling bonds [courtesy of Muhammad Baseer Haider]. **d** Scanning tunneling spectra of a superconducting $Pr_{0.88}$ $LaCe_{0.12}CuO_{4-\delta}$ (PLCCO) sample [5]

applied voltage. At low bias conditions, the tunneling current depends exponentially on the tip sample separation d and can be expressed by the following equation: [1]

$$I_T = eV\rho_s e^{-2\kappa d}$$

Here, constant $\kappa = \frac{\sqrt{2m\phi}}{\hbar}$ $and\ \phi = tip - sample\ work\ function.$

STM is commonly used for surface imaging of conducting surfaces with sub-angstrom image resolution. A typical topographic image of $FeSe_{0.5}Te_{0.5}$ and H-terminated Si(001) surfaces with visible atomic resolution are shown in Fig. 1b and c, respectively [4]. Figure 1c clearly shows 2 × 1 dimer formation on H-terminated Si(001) surface with three bright protrusions due to Si dangling bonds that were not capped by hydrogen atoms [image courtesy of Muhammad Baseer Haider]. STM is also used for tunneling spectroscopy in the conductance mode where (dI_T/dV) is measured as a function of applied bias between the sample and the tip, which is used to measure the sample surface density of states, as shown in Fig. 1c [5]. STM can be used for atomic manipulation on the surface of a sample. The atomic manipulation through STM is quite important for the nano-scale engineering and it is based on the interaction between the atoms of tips and samples.

1.3 Scanning Modes of STM

STM can be operated in two modes, constant current mode or constant height (tip-sample separation) mode. In constant current mode, STM tip is brought very close to the sample so that the quantum tunneling of electrons takes place between the tip and the sample at a specific applied bias. A specific tunnel current value is set so that the tip adjusts the height based on the set value of the tunneling current using computer controlled feedback loop. During scanning, the tip moves up and down at different height to get the specific value of the current. The height profile on the two-dimensional matrix is equivalent to the surface electronic image.

In constant height mode, the tip is fixed at a specific height from the sample and tunneling current is measured. If the tip is moved over a two-dimensional sample matrix and measures current at different points from a fixed height, the profile of the current size over the matrix provides the electronic surface image. Constant height mode is mostly used for tunneling spectroscopy in which the density of states is measured at constant height by varying the bias voltage [4].

Although STM has been evolved significantly since its discovery its fundamentals remains the same, i.e. the current between the tip and sample and its variation with the gap between them. If we specialize the current that would lead to the specialized Scanning Microscope. For example: if the current is sensitive to the electron spin, such STM is called the spin polarized STM (SP-STM), which is one of the most promising field of future technologies such as Spintronics. SP-STM is used to image the magnetic samples and can be used to differentiate between the surface atoms or domains with different spin orientations. A magnetically polarized tip is used for spin polarized STM. Another specialized STM is Electro-Chemical Scanning Tunneling Microscope (EC-STM), in which the current between the tip and the sample is ionic in nature in an electrolyte atmosphere.

1.4 Electrochemical Scanning Tunneling Microscopy (ECSTM)

Allen J. Bard, in 1986, used Scanning Probe into electrolyte environment and demonstrated current at large tip-sample gap, which was inconsistent with electron tunneling. Instead of tunneling current, he observed Faradic current. It was the first attempt to use STM in electrolyte environment and it was called as Electrochemical Scanning Tunneling Microscope (ECSTM) [6]. However, in 1989, Bard successfully develop the current between tip and the sample, which is equivalent to the tunneling current [7]. Since then it has been widely used to different disciplines such as chemistry, biology and material sciences, to measure the local electrochemical behavior of different kind of samples.

As in normal STM, an ECSTM consists of a sharp tip probe and a sample, but both the probe and the sample are dipped into an electrolyte solution. ECSTM probe is a metallic wire coated with an insulating material with a sharp nano or micro metallic part exposed at the tip. Making ECSTM tip is quite tedious and is the most challenging part of the microscopy. In ECSTM, there are several components of tip current: (i) tunneling current, (ii) Faraday current, (iii) capacitative current.

Tunneling current is the current between the sample and the tip. Faraday current is due to the chemical reaction between the tip and the electrolyte in which sample and tip are immersed. Capacitative current is due to the charging discharging of electrochemical double layer at the tip electrolyte interface.

1.5 ECSTM Working Mechanism

In ECSTM the care is taken to minimize the Faraday current and capacitative current compared to the tunneling current otherwise no meaningful image or spectrum can be acquired. To minimize the unwanted Faraday current and capacitative current ECSTM tip are usually made of platinum iridium alloy and are coated with electrochemical insulators like Apiezon wax, melt glass, copolymers or electrophoretic paint. [8–17] Apex of the tip is left uncoated so that tunneling can take place but the area of the apex is small enough to have tunneling current orders of magnitude larger than the Faraday current and capacitative current. Tunneling current is usually in 1–10 nA whereas Faraday current and capacitative current are in the range of pA.

The ECSTM tip thus prepared is brought very close to the sample in an electrochemical environment. Two extra electrodes, the counter or axillary electrode and a reference electrode, ionize the electrolyte. Generally, the tip is virtually grounded and a bi-potentiostat is used to control the potential of two electrodes with respect to the reference electrode. The current between the sample and the tip is amplified by a preamplifier and then fed to the computerized control unit. The schematic of the ECSTM setup is shown in Fig. 2 [18].

Due to the proximity of the tip and the sample, tunneling current depends on the height between them and hence ECSTM works in a similar manner as the normal STM. The working mechanism of ECSTM is based on the continuous oxidation and reduction of the electrolyte between the tip and the sample. This process creates a current equivalent to the tunneling current (i_T) which is given by:

$$i_T = 4nFCDa$$

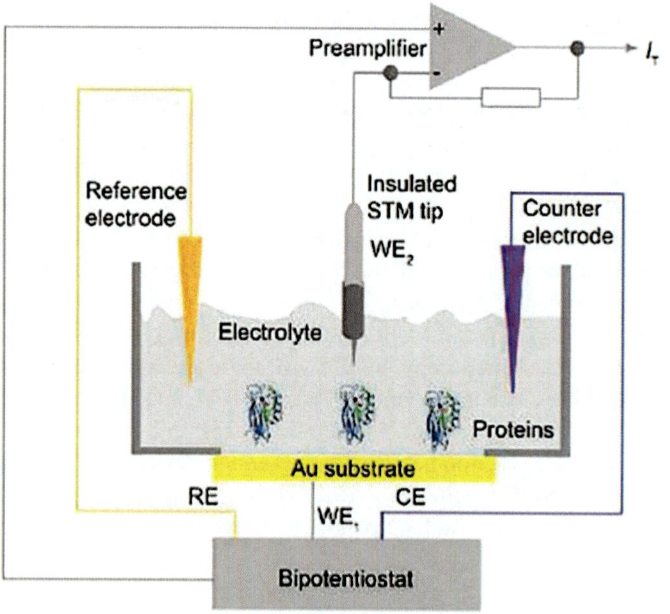

Fig. 2 Schematic of EC-STM operating mechanism [18]

where, n = number of electrons transferred at the tip, F = Faraday constant, C = concentration of oxidized species in the solution, D = diffusion coefficient and a = radius of the tip.

When an oxidized molecule reaches the electro-negative tip, it gets reduced. This reduced molecule reaches to the sample, which is very close to the tip, leaves the electron at the positively biased sample and gets reduced again. Process continues and a steady current i_T flows within the hemisphere from the surface to the tip. The current i_T varies with the gap between the tip and the sample. If the gap is large, i_T is steady because steady number of ions reach the tip and gets reduced. However, as the gap between the sample and the tip reduces significantly the current would be dependent on the sample. For an insulating sample, the tunneling current would decrease exponentially with the decreasing gap as depicted in Fig. 3. However, for a conducting sample, the tunneling current would increase exponentially with the decrease in the gap as shown in Fig. 3. In insulating samples as the tip sample distance decrease, fewer and fewer ions get oxidized and reduced between the sample and the tip. However, in the case of conducting sample as the gap decreases the oxidation and reduction process becomes faster and hence results in increased tunneling current.

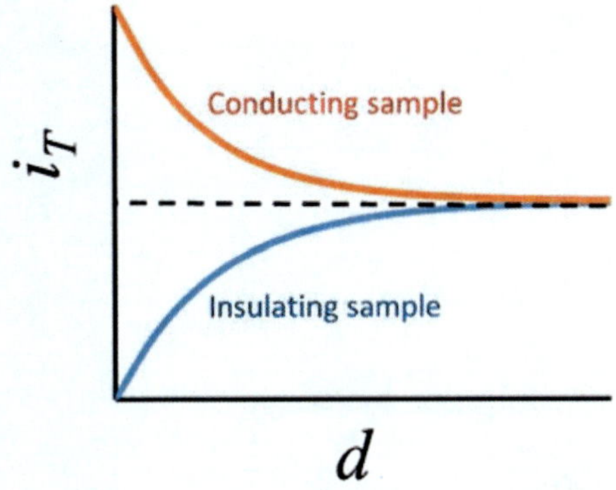

Fig. 3 Schematic of the tunneling current variation with tip-sample gap in an insulating sample and in a conducting sample

1.6 Application of ECSTM to Corrosion

Corrosion is a natural process of converting pure metal to a more stable, compound state thus releasing energy in the process. The underlying reason for corrosion is the thermodynamic instability of chemically active sites against water, air, or other chemicals. The chemistry of corrosion is quite complex that is still not fully understood, but it is widely regarded that it happens due to electrochemical processes. Metal corrosion occurs in two different ways: first, general corrosion that engulfs the entire exposed surface to the electrolyte solution and second, localized corrosion that nucleates at specific sites such as pits and crevices. Nucleation sites of corrosion are found to be chemically active and are generally considered to be the pits, scuffs, or other impurity sites at surfaces and interfaces. Corrosion reactions happen due to oxidation and reduction; the oxidation is known as anodic reaction and the reduction is called cathodic reaction. Metal acts like an anode and change its valence to a positive value in corrosion reaction as it transfers electrons to electrolytic environment. Metallic surfaces are often protected from oxidation by applying passive layers, but sometimes, harsh conditions and chemical activity produce localized instabilities in the passive layers that prompts pitting corrosion as depicted in Fig. 4a [19].

Corrosion rate, which is the conversion of the metal to other compound, requires continuous monitoring of a site for a period of time as shown in Fig. 4b–e [19]. These lap time images show the accumulation of chlorella vulgaris on the carbon steel surface, which is directly proportional to the site-specific corrosion and pit formation as shown in Fig. 4a [19].

Using STM, Blades et al. captured the initial stage of oxidation on two metallic surfaces Ni(100) and Ni-8 wt % Cr(100). The metals were exposed to oxygen for a period of time at 500 °C temperature and STM images were recorded step by step as shown for Ni(100) and Ni−8 wt % Cr(100) in Fig. 5a–h, respectively. Impact of oxygen exposure on the metals is obvious. The Ni(100) surface reconstructs and change into faceted structure due to dissociative chemisorption of oxygen initially and making the surface inert for further oxidation. This behavior, however, is in contrast with the Ni-Cr alloy, where oxidation triggers the formation of islands and nanoscale

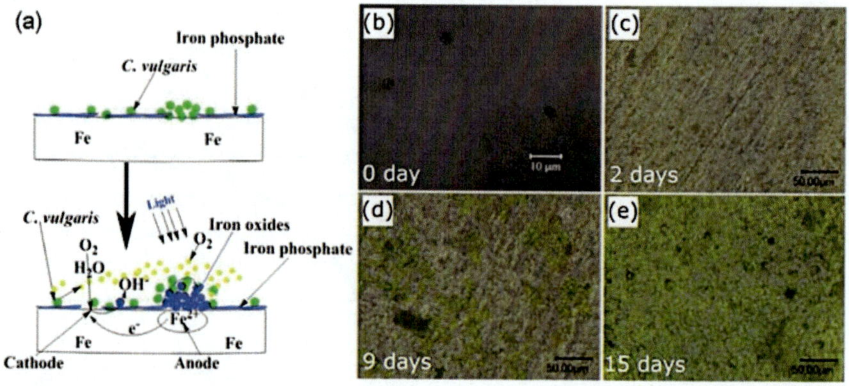

Fig. 4 **a** Formation mechanism of pitting corrosion at initial stage and reaction pathways. **b** Morphology of chlorella vulgaris and Q235 carbon steel surface corrosion after **c** 2 days, **d** 9 days, and **e** 15 days [19]

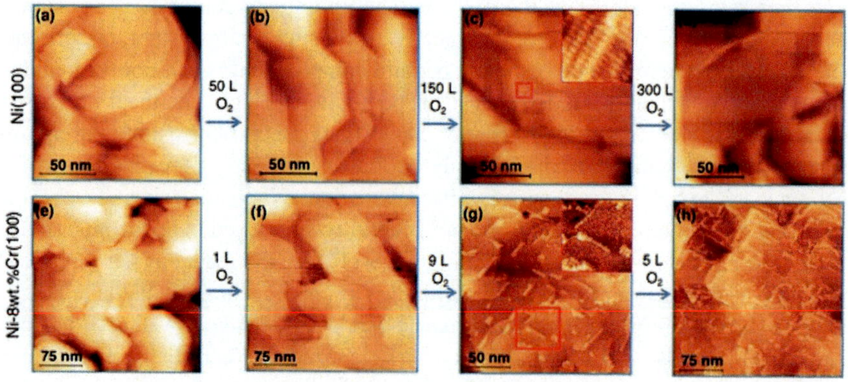

Fig. 5 STM images show oxidation for Ni(100) (**a–d**), and Ni − 8 wt % Cr(100) (**e–h**). The oxidation steps are performed at 500 °C and are shown in the figure. The insets in (**c**) and (**g**) taken from the red squared section in the corresponding images. Vbias = 2.0 V and It = 0.1 nA for all images in this series, except parts (**a**), (**e**), and (**f**) which were taken at Ubias = 0.15 V and It = 0.5 nA [20]

clusters. The islands form on top of the terraces and long chain-like structures form along the edges of the terraces. With more oxidation, these structures evolved into well-ordered reconstruction that decorates the entire surfaces [20].

Since the corrosion reaction is mostly electrochemical, corrosion behavior of metals can be evaluated with electrochemical techniques. Application of ECSTM in the field of corrosion has elevated the research to another dimension and has advanced the understanding of the corrosion nucleation sites, pitting initiation, crevices, surface disintegration, passive coating formation, and helped in developing corrosion inhibitors.

Traditional electrochemical experiment techniques measure average response over a wide surface of electrode or interface of electrode and electrolyte, which often lead to misinterpretation of data. ECSTM images chemically active surfaces via tunneling current between the surface and its tip and hence the surface and tip remains in the tunneling contact. ECSTM complements SECM, which is another widely used technique for studying corrosion. SECM relies on Faradic current and the vertical distance between the surface and tip remains from 5 to 30 nm, out of the tunneling current range [21].

In ECSTM a biased voltage is applied between surface and its tip, the tip can have positive or negative voltage with respect to the surface thus enabling the electron to tunnel from surface to the tip or from the tip to the sample, respectively. In the case of metallic surface the electron tunnel from the Fermi sea of the tip to the Fermi sea of surface whereas in the case of semiconducting surfaces the electrons tunnel from the Fermi sea of the tip to Highest Occupied Molecular Orbital (HOMO) or Lowest Uocoppied Molecular Orobital (LUMO) of the semiconducting surface. ECSTM can't be used for insulating surfaces due to the absence of tunneling current. A unique advantage of the ECSTM technique over other high resolution techniques such as Transmission Electron Microscopy (TEM) is that ECSTM measures electron densities at local sites on surfaces that allow the correlation between electrochemical intensity with visual inspection at atomic scale. It happens when microscopic (atomistic) reactivity at solid–liquid interfaces is accompanied with physical changes at the interface such as deposition, adsorption or dissolution reactions [21]. With many obvious advantages of ECSTM over other techniques, a few limitations need to be addressed for future experiments. It is difficult to adjust tip size to tune Faradic current at a fixed potential. During electrochemical process tip interferes with the working electrode. Molecules adheres to tip and drops at random places over the surfaces thus creating unwanted effects. Noise and drifts are other issues that hinder high-resolution images, and these issues are solved by employing mechanical dampers, acoustic box, and thermal isolation. The drifts are also corrected by using advanced drift corrections algorithms such as choosing a reference point on an image the system will autocorrect its drift every time.

An insight of grain-dependent corrosion properties is very useful for engineering material growth for a specific environment. For such studies, both lateral and vertical resolution of ECSTM are very useful to observe corrosion locally at nanometer scale. It provides important information about microstructure of the oxide layer and anodic dissolution at nucleation stage. Martinez-Lombardia et al. successfully applied ECSTM to study in-situ grain-dependent corrosion properties cryogenically rolled microcrystalline copper in a HCl electrolyte [22]. Their ECSTM cell was mounted on top of the copper electrode with a small area exposed to the electrolyte. Platinum wires were used for reference electrode and counter electrode, and tungsten tip was insulated with Apiezon wax except the tip of the tip for ECSTM. They observed that the thickness of the air-formed oxide layer depends on the orientation of the copper grain and noticed that anodic dissolution of the electrode in HCl was also grain-dependent. Shown in Fig. 6 are the ECSTM topographic images of the copper electrode. Figure 6a is recorded when the native oxide layer fully covers the

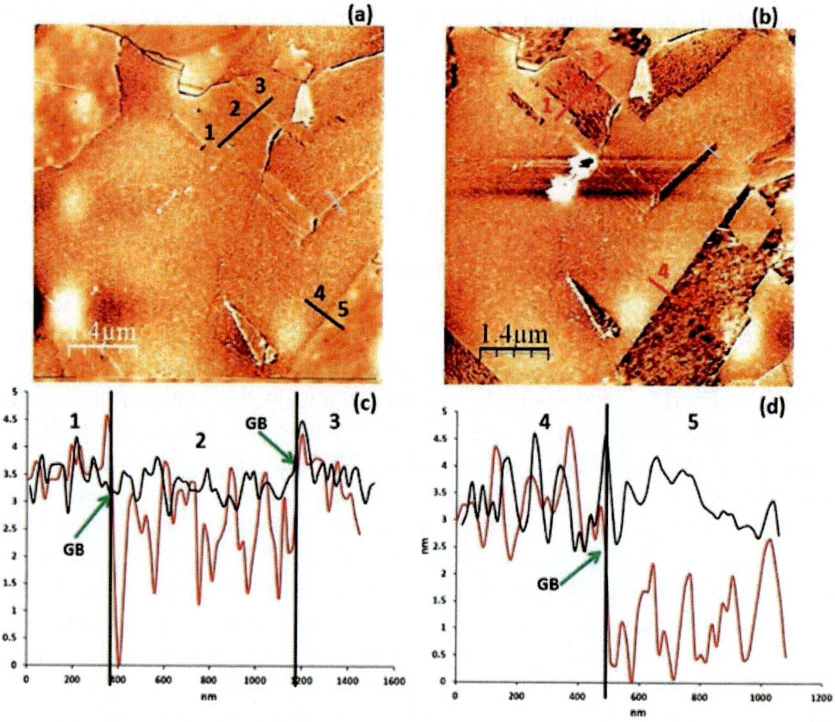

Fig. 6 **a** ECSTM image of polycrystalline copper covered by the air-formed oxide. **b** Metallic surface of the same area after cathodic reduction of the oxide film. **c** Comparison of line profile across grains 1-2-3, and **d** grains 4-5 before and after the cathodic reaction [22]

surface and Fig. 6b is recorded after the full reduction of the oxide layer. There is a stark difference between the morphology of the two images as one can notice in the line profiles. In Fig. 6c, the line profile is across three grains labeled 1, 2, and 3. In electrolyte, the native removes from the all three grains, but grain 2 is significantly lower in height as compared to grain 1 and 3. Similarly, in Fig. 6d, grain 5 reduces more than grain 4. This indicate that the oxide layer on grain 2 and 5 was much thicker than on the other neighboring grains. After the reduction the topography of the two grains 2 and 5 gets rougher than the other grains which could be due to less homogeneous redistribution of the reduced copper atoms [22]. The orientation of these grains is closer to the ⟨111⟩ than to ⟨001⟩ or to ⟨110⟩. After cathodic reduction, the sample is then set for anodic dissolution forced by sweeping the potential in the positive direction until a desired anodic current density is achieved. Anodic potentiodynamic polarization induces copper dissociation. Anodic dissolution cycles reveal that the grain with possibly <111> orientation are dissolving faster while grains with other orientation remained almost unaffected. Furthermore, it was noticed that grain boundaries are more prone to the corrosive attack [22].

Estimating pits on a metallic surface caused by corrosion gives important information about the reactivity of a metal with a specific electrolyte. ECSTM is very useful technique for carrying out such experiments. Gold (Au) is an important noble metal for applications in medicine, plating industry, and high performance electronics and resistant to corrosion under many conditions, but corrodes under appropriate chemical or electrochemical conditions in solutions containing Cl^-, Br^-, I^-, CN^-, and $CS(NH_2)_2$ [23]. Zamborini et al. studied pits formation in passive layer of *n*-alkanethiol self-assembled monolayers (SAMs) on Au surface in aqueous Br^- solutions. To understand which factor of SAMs provide better passivation layer, the experiment included four different types of molecules: $HS(CH_2)_{15}CH_3$, $HS(CH_2)_{11}CH_3$, $HS(CH_2)_{10}COOH$, and $HS(CH_2)_{11}OH$. These molecules were selected due to their complementary terminal functional groups and end groups. The as grown SAMs on Au substrate were crystalline, but after the sample exposure to Br^- solutions they became disordered and oriented parallel to the surface. Experiments reveals that for SAMs with the same terminal functional group, the corrosion resistance is proportional to passive SAM layer thickness. The SAMs with different terminal functional groups and the end groups show more corrosion resistance. The corrosion resistance order follow OH > COOH > CH_3. Figure 7 show a 3 μm × 3 μm STM images of the naked Au as well as SAMs passivated Au samples and the corresponding samples for each condition are grouped in rows from left to right. The first image in each row shows sample before corrosion; the second image shows after corrosion at 800 mV for 2 min, and the third image is obtained after corrosion at 1.0 V for 1 min. The first and last image in each row are recorded on same sample but different regions, where the second image in each row recorded on a separate sample that is processed in the same manner as the first sample[23]. Not all SAMs passivated Au sample are susceptible to pitting corrosion, the hydrophilic SAM passivated surfaces corrode uniformly terrace by terrace, whereas the methyl-terminated, hydrophobic SAM passivated surfaces show pitting corrosion [23]. For example, naked Au, $C_{11}OH$ SAM passivated Au dissolve in stepwise manner, whereas $C_{11}CH_3$, $C_{10}COOH$, and $C_{15}OH$ leads to pits corrosion. The ECSTM images also reveals that the pits depend on the corrosion voltage. For example, the naked Au shows pits in Fig. 7b which is absent in Fig. 7c; similarly, the $C_{11}CH_3$ passivated sample shows well defined pits in Fig. 7e, but that sort of pits are absent in Fig. 7f although the surface is rougher [23].

ECSTM is a surface sensitive technique and has unparalleled advantage over other instruments in directly observing surface reconstruction or surface structural transformation in the vicinity of oxidation or reduction reactions as function of an applied voltage. Correlation surface reconstructions with redox reactions would be useful for product identification. Figure 8 shows ECSTM images of a sequential transformation of a polycrystalline copper Cu(pc) surface from grainy unreconstructed surface to a Cu(pc)-[Cu(511)] reconstructed surface [24]. Image sizes and steps are shown in the figure. The Cu(pc) image shown in Fig. 8a is recorded after dipping the sample in KOH electrolyte. The grains are visible, but atomic resolution was unobtainable on the surface. After 30 min at − 0.9 V in 0.1 M KOH solution in the vicinity of CO_2 reduction reactions, the polycrystalline Cu surface has transformed into single

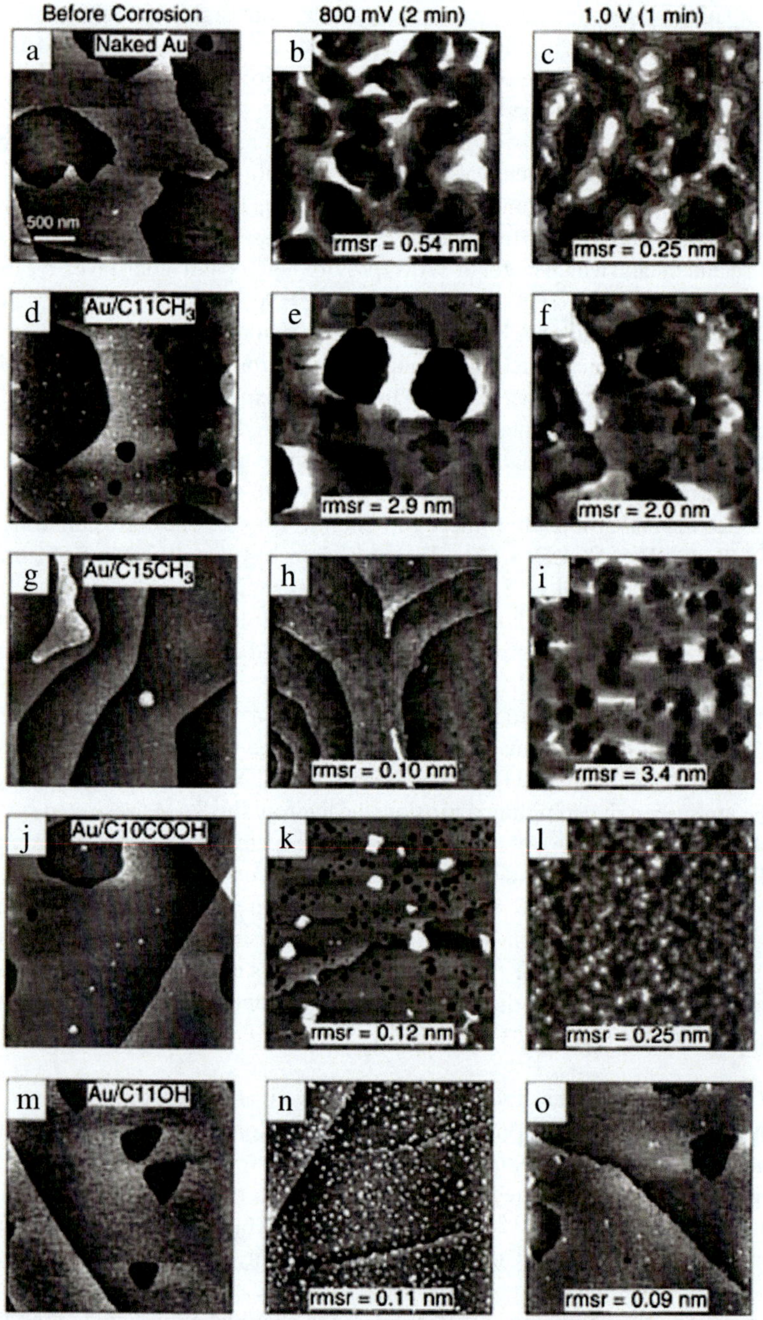

Fig. 7 3 μm × 3 μm STM images of **a–c** Naked Au, **d–f** Au/$C_{11}CH_3$, **g–i** Au/$C_{15}CH_3$, (j–l) Au/$C_{10}COOH$, and (m–o) Au/$C_{11}OH$. The first image in each row is obtained before corrosion; the second image is obtained after corrosion at 800 mV for 2 min, and the third image is obtained after corrosion at 1.0 V for 1 min [23]

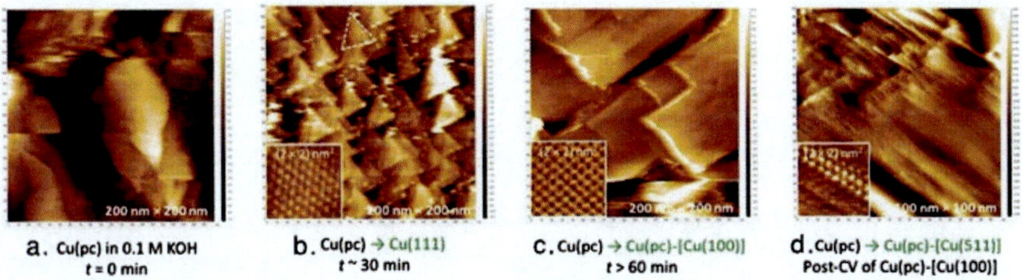

Fig. 8 **a** ECSTM image of polycrystalline copper Cu(pc surface at V(SHE) = −0.90 V in 0.1 M KOH in the vicinity of CO_2 reduction reactions. **b** Cu(pc) reconstruct into (111) oriented surface. **c** Reconstruction transforms into Cu(pc)-[Cu(100)] surface. **d** Further potential cycling of C between −0.90 V and 0.10 V in 0.1 M KOH gives rise to a stepped surface. V_{bias} = 250 mV, I_t = 2 nA for wide-scale images; 5 nA for the insets [24]

crystal Cu(111) surface, which is revealed by ECSTM image in Fig. 8b. The entire image is decorated with a distinctive array of triangular islands with hexagonal rotational symmetry and the inset shows atomically resolved image of a terrace on the triangular features that confirms a close-packed (111) structure. After an additional 60 min, the surface reconstruction evolved from triangular shape to a rectangular shaped reconstruction. The ECSTM image shown in Fig. 8c shows wide rectangular terraces and atomically resolved image of the terrace reveals (1 × 1) reconstruction represent (100) surface of Cu, which is succinctly denoted as Cu(100). Application of steady-state cyclic voltammetry to the Cu(pc)-[Cu(100)] electrode transform the rectangular surface to stepped surface (Fig. 8d). High-resolution ECSTM image shows a 3-atom wide (100) terraces bound by a (111)-oriented monoatomic step. The step reconstruction is recognized as Cu(S)-[3(100) × (111)] or shortly Cu(511) [24].

Electrochemical Scanning Tunneling Spectroscopy (ECSTS) is an extremely useful technique for probing electrical or electrochemical properties of a single molecule or a cluster of atoms. It can be used to investigate the influence of molecular structure, orientation, and environment on tunneling current, electron density, and conductance electronic junction. Ricci et al. investigated electrical properties of redox-active [Os(bipyridine)(pyridine)Cl] complex tethered to the Au surface as Au-C (AC sample) or Au-S (AS sample) bonds as shown in Fig. 9 [25]. The blue colored (noisy) curve in Fig. 9b, c represent tunneling current (I_t) versus (E_s − E) for the AC sample and AS sample, respectively. In these two images, the black colored curve represent cyclic voltammograms of the respective samples. The two curves show that peak in the I_t curve (E_{max}) is close to the equilibrium redox potential of the samples. The samples go from being reduced to fully oxidized while sweeping potential from (E_s − E) = −0.2 V to +0.2 V, and in between the tunneling current goes from 1 to 3 nA. The correlation between the two sets of results indicates that Os complex is involved in electron transport mechanism for both the AS and AC samples. The E_{max} show linear dependence on E_{bias} as shown in Fig. 9e, f, which agrees with two-steps electrochemical electron transport process that involve redox in Os complex and the negative and positive biased electrodes [25].

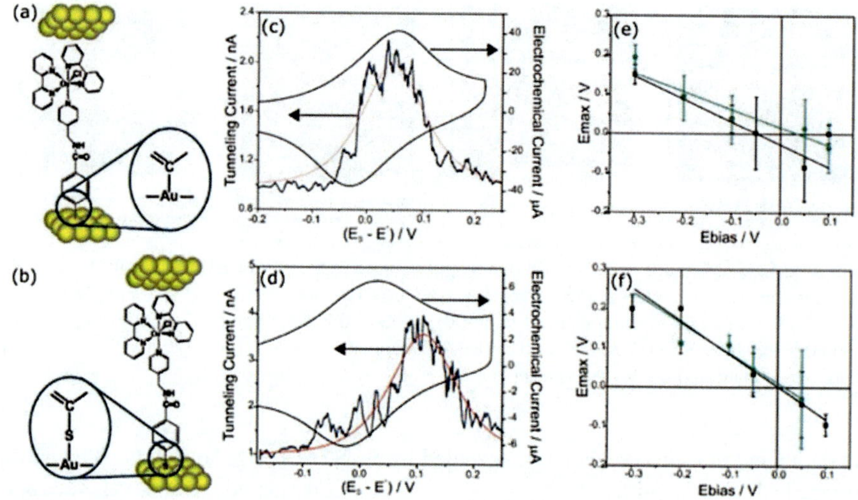

Fig. 9 Schematics of [Os(bipyridine)(pyridine)Cl] complexes bonded to Au surface by **a** C and **b** S atoms. Blue colored curve (left scale) represent I_t versus ($E_S - E$) data taken in 0.1 M $KClO_4$ 234with $V_{bias} = +0.05$ V for **c** Au(111)/PhCOOH/Os and **d** Au(111)/SPhCOOH/Os. The red lines is the best fits of I_t. The black colored smooth CV curves taken in 0.1 M NaClO4 at 500 mV s^{-1} (right scale). **e** and **f** show peak position in I_t (E_{max}) versus V_{bias} corresponding to (c) and (d). The circled and squared symbols represent anodic and cathodic scans [25]

References

1. J. Chen, in *Introduction to Scanning Tunneling Microscopy* (Oxford University Press, 1993), pp. 1–2
2. D.K. Roy, Quantum Mechanical Tunnelling. in *Tunnelling and Negative Resistance Phenomena in Semiconductors* (Science Direct, 1977), pp. 1–35
3. A.J. Bard, F.-R, Scanning electrochemical microscopy. Introduction and principles, Anal. Chem. **61**(2), 133 (1989)
4. S. Kunwar, R.M. Hamad, K.A. Ziq, Coexistence of weak and strong coupling mechanism, in an iron-based superconductor F eSe0.5T e0.5: possible signature of BCS-BEC crossover. J. Supercond. Novel Magn. **30** 3183–3188 (2017)
5. S. Kunwar, Existence of pseudogap at the transition temperature of an electron-doped copper-oxide superconductor. J. Supercond. Novel Magn. **27**, 2461–2466 (2014)
6. A. Bard, H.-Y. Liu, F.-R.F. Fan, C.W. Lin, Scanning electrochemical and tunneling ultramicro-electrode microscope for high-resolution examination of electrode surfaces in solution. J. Am. Chem. Soc. **108**(13), 3838–3839 (1986)
7. A.J. Bard, F.R. Fan, J. Kwak, O. Lev, Scanning electrochemical microscopy. Introduction and principles. Anal. Chem. **61**(2), 132–138 (1989)
8. R.M. Penner, M.J. Heben, T.L. Longin, N.S. Lewis, Fabrication and use of nanometer-sized electrodes in electrochemistry. Science **250**, 1118–1121 (1990)
9. L.A. Nagahara, T. Thundat, S.M. Lidsay, Preparation and characterization of STM tips for electrochemical studies. Rev. Sci. Instrum. **60**(10), 3128–3130 (1989)
10. R. Sonnenfeld, P.K. Hansma, Atomic-resolution microscopy in water. Science **232**, 211–213 (1986)
11. M.J. Heben, M.M. Dovek, N.S. Lewis, R.M. Penner, C.F. Quate, Preparation of STM tips for in-situ characterization of electrode surfaces. J. Microsc. **152**, 651–661 (1988)

12. R.A. Clark, A.G. Ewing, Characterization of Electrochemical Responses in Picoliter Volumes. Anal. Chem. **70**, 1119–1125 (1998)
13. R.M. Penner, M.J. Heben, N.S. Lewis, Preparation and electrochemical characterization of conical and hemispherical ultramicroelectrodes. Anal. Chem. **61**, 1630–1636 (1989)
14. Y. Shao, M.V. Mirkin, G. Fish, S. Kokotov, D. Palanker, A. Lewis, Nanometer-sized electrochemical sensors. Anal. Chem. **69**, 1627–1634 (1997)
15. B.D. Pendley, H.D. Abruna, Construction of submicrometer voltammetric electrodes, Anal. Chem. **62**, 782–784 (1990)
16. D.K.Y. Wong, L.Y.F. Xu, Voltammetric studies of carbon disk electrodes with submicrometer-sized structural diameters. Anal. Chem. **67**, 4086–4090 (1995)
17. A.G. Güell, I. Díez-Pérez, P. Gorostiza, F Sanz, Preparation of reliable probes for electrochemical tunneling spectroscopy, Anal. Chem. **76**, 5218-5222 (2004)
18. A.K. Yagati, J. Min, J.-W. Choi, Electrochemical ECSTM from thoery to future applications. InTech OPen, (2014)
19. H. Liu, D. Xu, A.Q. Dao, G. Zhang, Y. Lv, H. Liu, Study of corrosion behavior and mechanism of carbon steel in the presence of Chlorella vulgaris, Corros. Sci. **101**, 84–93 (2015)
20. H. William, Blades and Petra Reinke, from alloy to oxide: capturing the early stages of oxidation on Ni–Cr(100) alloys. ACS Appl. Mater. Interfaces **10**(49), 43219–43229 (2018)
21. T.H. Treutler, G. Wittstock, Combination of an electrochemical tunneling microscope (ECSTM) and a scanning electrochemical microscope (SECM): application for tip-induced modification of self-assembled monolayers, Electrochim. Acta **48**, 2923–2932 (2003)
22. E. Martinez-Lombardia, V. Maurice, L. Lapeire, I. De Graeve, K. Verbeken, L. Kestens, P. Marcus, H. Terryn, In situ scanning tunneling microscopy study of grain-dependent corrosion on microcrystalline copper. J. Phys. Chem. C **118**, 25421–25428 (2014)
23. . F.P. Zamborini, R.M. Crooks, Corrosion passivation of gold by n-alkanethiol self-assembled monolayers: effect of chain length and end group, Langmuir, **14**(12), 3279–3286 (1998)
24. Y.-G. Kim, J.H. Baricuatro, A. Javier, J.M. Gregoire, M.P. Soriaga, the evolution of the polycrystalline copper surface, first to Cu(111) and then to Cu(100), at a Fixed CO2RR potential: a study by operando EC-STM, Langmuir, **30**, 15053−15056 (2014)
25. A.M. Ricci, E.J. Calvo, S. Martin, R.J. Nichols, Electrochemical scanning tunneling spectroscopy of redox-active molecules bound by Au−C bonds. J. Am. Chem. Soc. **132**(8), 2494–2495 (2010)

Recent Trends in Applications of X-ray Photoelectron Spectroscopy (XPS) Technique in Coatings for Corrosion Protection

Adnan Khan, Osama Fayyaz, R. A. Shakoor, and Bilal Mansoor

Abstract Corrosion—primarily a surface phenomenon, strongly influences an asset's performance, durability, and reliability, as costs of corrosion-related repairs and refurbishment can run into millions of dollars. Notably, in the Middle East region, direct corrosion costs are estimated to be approximately 5% of the gross domestic product (GDP), a value that represents an enormous economic burden. In this context, the corrosion of metals is one of the most fundamental causes of components and equipment failure across different industrial sectors. In the oil and gas sector, extreme operating conditions and corrosive environments exacerbate the extent and seriousness of corrosion and result in a vast majority of component failures. Therefore, a comprehensive understanding of corrosion mechanisms and processes taking place on a component's surface is vital for efficient and cost-effective asset integrity management. Various advanced analytical techniques such as X-ray Photoelectron Spectroscopy (XPS), X-ray fluorescence (XRF) spectroscopy, Localized Electrochemical Impedance Spectroscopy (LEIS), Scanning vibrating electrode technique (SVET) and Scanning Ion-Selective Electrode Technique (SIET) provide useful quantitative information about corrosion. Amongst those, XPS is one of the most sophisticated technique that can be smartly utilized to study corroded surfaces, protective coatings and understanding of various corrosion mechanisms. XPS can also be effectively used to assess corrosion byproducts to understand the various surface reactions, localized surface chemistry and alteration in electrical responses under different corrosive environments. This chapter focuses on the utilization of XPS technique in coatings for materials protection. Special attention is paid to cover its role in compositional analysis of various types of coatings (metallic and polymeric)

A. Khan and O. Fayyaz—Equal Contribution.

A. Khan · O. Fayyaz · R. A. Shakoor (✉)
Center for Advanced Material (CAM), Qatar University, 2713 Doha, Qatar
e-mail: shakoor@qu.edu.qa

B. Mansoor
Mechanical Engineering Program, Texas A&M University at Qatar, Ar-Rayyan, Qatar

A. Khan · B. Mansoor
Department of Materials Science and Engineering, Texas A&M University, College Station, TX 77843, USA

I. Toor (ed.), *Recent Developments in Analytical Techniques for Corrosion Research*,
https://doi.org/10.1007/978-3-030-89101-5_8

that are actively investigated for corrosion protection. At the same time, use of XPS can also be extended to explore smart polymeric coatings confirming the self-release of various inhibitors and self-healing agents loaded in nanocarriers. Furthermore, the role of XPS in post corrosion analysis is described, which provides ample information to understand the corrosion protection mechanism. A short note on future scope and challenges faced by XPS technology in corrosion research is also provided for the interest and motivation of the readers to excel in this important area of research.

Keywords XPS · Corrosion · Composite coatings · Polymeric · Metallic

1 Introduction

The photoelectric effect discovered by Hertz and correctly explained by Einstein, for which he received the Nobel Prize in 1921, has led to the invention of many tools and techniques [1]. X-ray photoelectron spectroscopy (XPS) is one such application which is among the most frequently utilized characterization technique in the fields of chemistry, materials science, and various engineering sectors to evaluate elemental composition, structural bonding, interfaces, surface, and sub-surface chemistry [2]. According to the data obtained from Scopus, owing to its importance and effectiveness, its utilization has increased from the first paper in 1969 to more than 11,500 published research articles using XPS in 2020, last year as shown in Fig. 1a. A near exponential incremental slope within the previous half-century of XPS utilization is also observed in weighted adjacent averaging of seven data points clearly proving its importance and utilization among the research community [3].

Corrosion is a natural phenomenon of tracing back of furnished goods to its original raw state. It initiates at the product's surface, due to which the entire bulk loses its value, efficiency, and effectiveness in producing the results as per the requirement. Coatings are one of the best surface modification techniques utilized to improve the corrosion resistance of the bulk material to prolong its life and effectiveness [4].

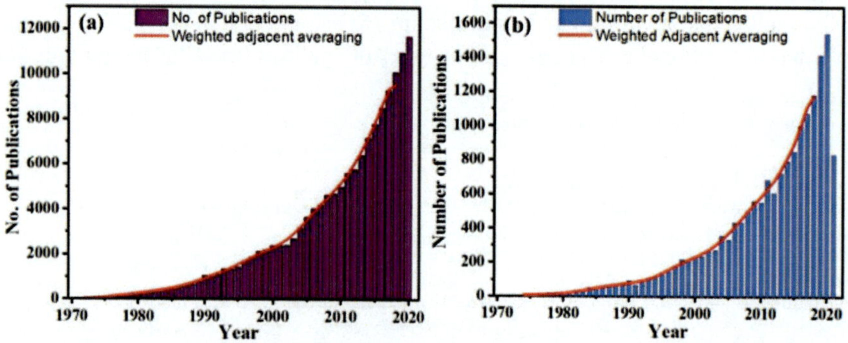

Fig. 1 Utilization of (a) XPS in material research (b) in corrosion-related studies since 1969

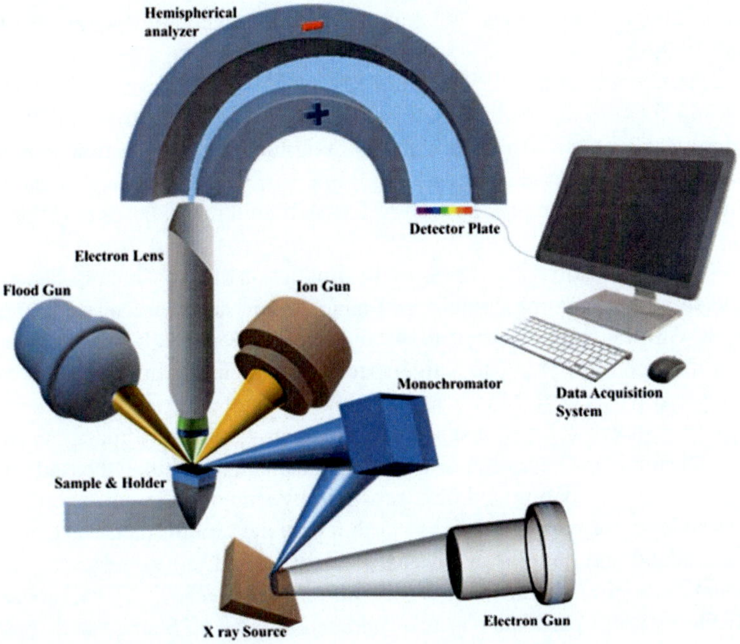

Fig. 2 Schematic diagram of the instrumentational operation of XPS for corrosion analysis

Understanding the most accurate corrosion mechanism is necessary to select the best possible prevention strategy. XPS plays an essential role in understanding the corrosion occurring at the surface with its proper mechanism to provide the best prevention methodology accurately [5]. The utilization of XPS in corrosion studies dates to 1973, and more than 1500 research articles in 2020 prove its significance, as presented in Fig. 1b. Weighted adjacent averaging also shows the appreciative incremental slope confirming its current use and future consideration in investigating corrosion and other related aspects.

XPS works on a simple principle of the photoelectric effect. When atoms or molecules absorb X-ray photons at the exposed surface, electrons are ejected with varying kinetic energy, which is being recorded over a photoelectron spectrum. The kinetic energy depends upon the binding energy of the electron and photon energy represented by the formula [6]:

$$BE = h\nu - KE \quad (1)$$

where BE refers to the binding energy of an electron, KE is the kinetic energy of the emitted electron, and $h\nu$ is the incident energy of X-ray photons on the surface which may be any corrosion resistance film. The peaks in photoelectron spectra lead to the exhaustive characterization of elements present at the surface in terms of their identity, chemical states, nature of bonding, electronic structure, and even accurate thickness of the film up to 10 nm [5]. Figure 2 shows a schematic diagram of working principle of XPS.

The effectiveness of the XPS technique is due to the assessment of the apparent relationship between the binding energies (BE) of core-level electrons and the chemical affinity of an atom. This leads to the determination of structural bonding of surface and thereby providing insight regarding surface synthesizing methodology. Researchers typically compare their BE values to literature databases to validate their results [2, 7].

XPS is utilized readily at all levels in the production of corrosion resistance films. In thick polymeric coatings, micro and nanocarriers are characterized, and their reactivity with the loaded chemical species (inhibitors and self-healing agents) is assessed through XPS [8]. The active species and their stability with the carriers are keenly studied as well. Moreover, the inhibitor-loaded carriers embedded within the organic matrices are also investigated for their stability and reactivity through XPS. Furthermore, the corrosion resistance mechanism is postulated based upon the post-analysis of the coating and bare substrate by studying the corrosion products and passive layer formation [9]. To assess thin metallic film for their corrosion resistance, elemental and compositional analyses are carried out to accurately express their chemical state and their stability in terms of corrosion [10, 11]. Moreover, reinforcement of various particulate is also validated through XPS along with their inert and unreactive nature [12]. Post corrosion analysis of substrates and various families of coatings are also carried out to understand the origin of corrosion and its protection mechanism. Figure 2 presents a schematic diagram of the working principle of a XPS instrument. It starts with an electron gun which excites the X-ray generating source. Produced X-rays are subjected to a monochromator for separating the specific X-ray for the sample under consideration. The separated X-ray falls on the sample and produces photoelectron as a result of the photoelectric effect [1]. Flood gun directs the electron flood to the sample neutralizing it from charging because of energy-intensive X-ray. Ion gun is utilized to remove the topmost layer of the area to study the sub-surface behaviors and remove contaminants from the sample surface. Photoelectrons are then made to pass through series of electron lenses into a hemispherical analyzer that separates the electrons of respective energy, which are detected at the detector plate. A data acquisition system is used to interpret the energy spectrum detected by the detector plate.

2 Analysis of Nanocomposite Coatings for Corrosion Protection

Nanocomposite coatings are emerging as a great tool in tailoring the surface properties of a large variety of materials to improve their corrosion resistance characteristics and to extend their field of applications. The corrosion resistance of nanocomposite coatings is mostly studied by electrochemical impedance spectroscopy and potentiodynamic testing. In the case of inorganic coatings, the reason for enhanced corrosion resistance is reported primarily due to reduction in the active area of the metallic matrix by the presence of inert micro/nano-sized reinforcements. In this regard, validating the presence of reinforcements in the matrix is essential particularly if it and is not detectable with other techniques. Moreover, the presence of active species released during the self-healing process of polymeric smart self-healing coatings is efficiently confirmed with XPS.

Figure 3 reveals the XPS survey spectra of various composite coating systems confirming the presence of the various elements at the surface and sub-surface levels. Furthermore, the binding energy peaks indicate the oxidation state of the elements present in the composite coating. In this way, the nature and presence of the reinforcements (inert or active) and chemistry of composite coatings can be confirmed by XPS. The survey spectra of four such systems that correspond to (a) Ni–P–Y_2O_3 [13], (b) Ni–P–ZrO_2 [14], (c) Ni–B–CeO_2 [15], and (d) Ni–W–Si_3N_4 [16] as demonstrated in Fig. 3 clearly illustrates the presence of nickel (Ni), phosphorus (P), boron (B), tungsten (W), yttrium (Y), zirconium (Zr), cerium (Ce), silicon (Si), oxygen (O) and carbon (C) as the constituents of different coating systems. The presence of carbon is necessary as it provides a reference for other elements to be plotted correctly to obtain highly magnified XPS spectra.

Highly magnified XPS spectra of various corrosion-resistant composite coatings enumerated as (a) Ni–P–TiC [17], (b) Ni–P–Y_2O_3 [13], (c) Ni–W–Si_3N_4 [16], (d) Ni–B–AlN [18], (e) Ni–P–ZrO_2 [14] and (f) Ni–P–CeO_2 [15] is presented in Fig. 4 to elucidates the presence of desired reinforcement in respective composite coating

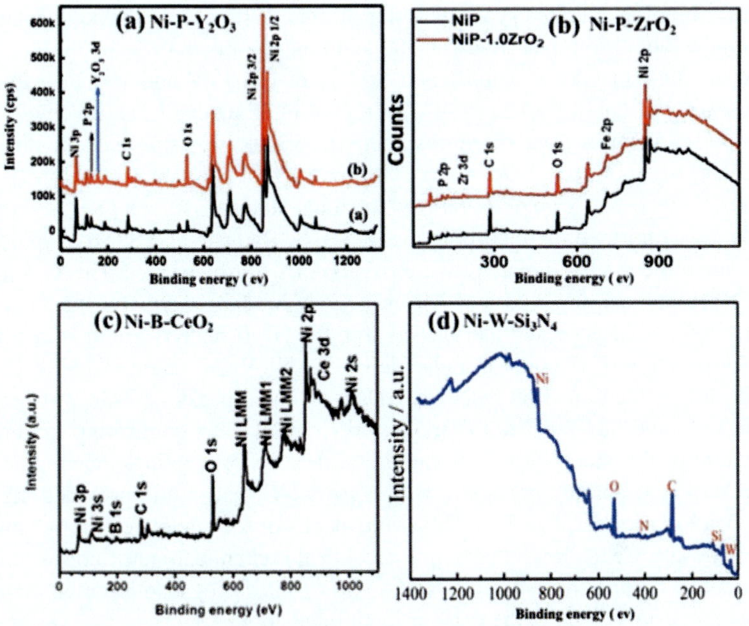

Fig. 3 XPS survey spectra of various metallic coating systems **(a)** Ni–P–Y_2O_3, **(b)** Ni–P–ZrO_2, **(c)** Ni–B–CeO_2, and **(d)** Ni–W–Si_3N_4 [13–16]

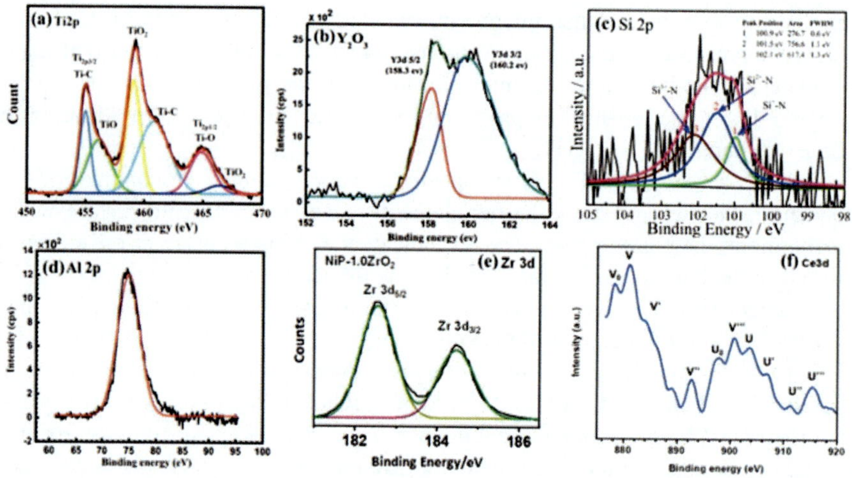

Fig. 4 Highly magnified XPS spectra confirming the presence of respective reinforcements in various coating systems developed for corrosion protection **(a)** Ti in Ni–P–TiC, **(b)** Y in Ni–P–Y_2O_3, **(c)** Si in Ni–W–Si_3N_4, **(d)** Al in Ni–B–AlN, **(e)** Zr in Ni–P–ZrO_2 and **(f)** Ce in Ni–P–CeO_2 [13–18]

systems. Osama et al. reported the results of incorporation of TiC in Ni–P–TiC composite coating. The presence of titanium was proved from high resolution XPS spectra at binding energies peaks of 454.8 eV and 461.3 eV for the confirmation of titanium carbide. Moreover, Radwan and team incorporated Y_2O_3 in Ni–P reported the deconvoluted peaks of Y3d5/2 and Y3d3/2 at 158.3 eV and 160.2 eV to prove its presence and stability in the Ni–P–Y_2O_3 coating. Baosong Li and Weiwei Zhang also reported a similar trend of improvement in corrosion resistance due to the incorporation of Si_3N_4 in the Ni-W matrix. The presence of peaks at 100.9 eV, 101.5 eV and 102.1 eV were ascribed to various chemical states of silicon (Si) as shown in Fig. 4c and confirmed the incorporation of Si_3N_4. Radwan and Shakoor reinforced aluminum nitride in Ni–B matrix to improve its corrosion resistance. The XPS spectra of Al2p presented a single peak of AlN at 74.3 eV binding energy for the Al-N bond confirming its existence and inertness within the Ni–B matrix. Sliem et al. studied the incorporation of zirconia in pulse-deposited Ni–P and made a similar conclusion regarding the corrosion resistance. Deconvoluted peaks of Zr3d was reported at 182.5 eV and 184.7 eV for Zr3d5/2 and Zr3d3/2, thus confirming the unreactive nature of the incorporated zirconium oxide in Ni–P coating. Pancrecious and co-workers incorporated nanoceria within the Ni–B matrix and provided the high magnification spectra of Ce3d with various peaks of Ce^{3+} and Ce^{4+} corresponding to $Ce3d_{5/2}$ and $Ce3d_{3/2}$. The presence of both the chemical states corresponds to the oxides of cerium in the form of CeO_2 and Ce_2O_3. They also reported increased corrosion resistance due to the reasons mentioned above.

XPS is also widely used to analyze various families of polymeric coatings synthesized for corrosion protection and to understand the prevailing corrosion mechanism in different coating systems under numerous environment [19]. The corrosion behavior of two different types of hybrid saline-epoxy coatings before and after immersion in NaCl solution was investigated. High-resolution XPS spectra of Si 2p of γ–ASP silane and BTSE doped epoxy coating before (a, c) and after (b, d) immersion in NaCl solution for 24 h is presented in Fig. 5. The Si signal detection clearly showed the incorporation of Si in the epoxy matrix. The high-resolution data fitting of Si 2p spectra of both the γ–aminopropyltrimethoxy (ASP) silane and bis-1,2-[triethoxysilyl] ethane (BTSE) doped epoxy coating can be deconvoluted into mainly two peaks associated with Si–O–Si and Si–O–C. A dominant Si–O–Si peak before immersion (Fig. 5a, c) compared to after immersion (Fig. 5b, d) reflects a completed condensation reaction of silane components on the surface of the epoxy matrix during their storage stage. This further explains that a small quantity of un-hydrolyzed Si–O–C is remaining. The depth peak analysis showed that the Si–O–Si peak in BTSE doped coatings declined, followed by an increase in the Si–O–C peak. The inset in Fig. 5d showed a peak of Si–O–Si and Si–O–C group in BTSE doped

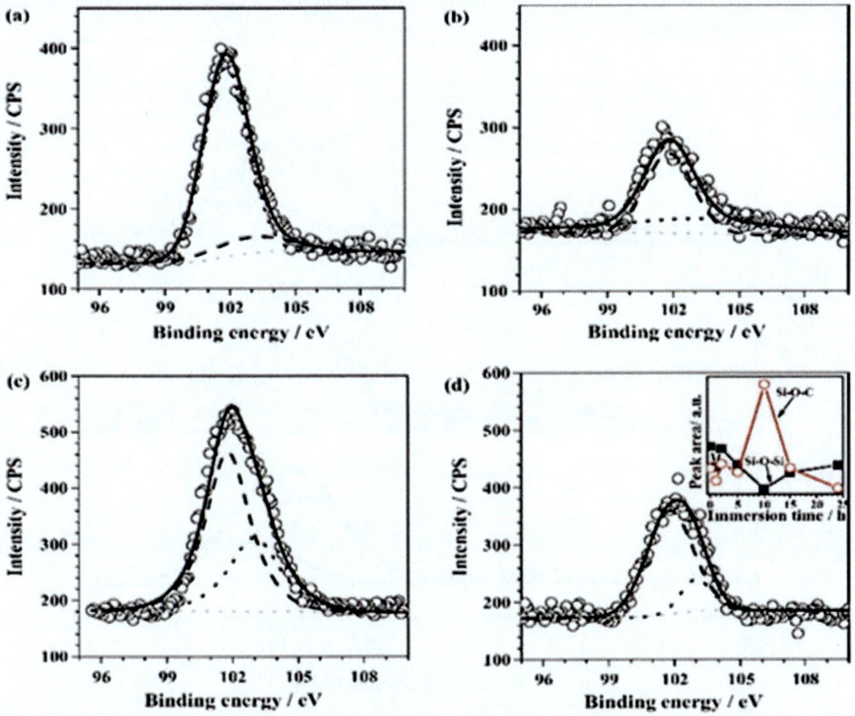

Fig. 5 High-resolution XPS spectra of Si 2p of γ–ASP silane and BTSE doped epoxy coating before (**a**, **c**) and after (**b**, **d**) immersion in NaCl solution for 24 h. The inset in "d" shows the peak area of Si–O–Si and Si–O–C group in BTSE-doped epoxy coatings [19]

coatings during the immersion time. A transition reaction takes place from Si–O–C to Si–O–Si components, which is used to investigate the self-healing property of the silane-modified polymeric coatings [20].

XPS was utilized to confirm the release of corrosion inhibitor from the nanocapsules to self-heal the defected zone of epoxy coatings [21]. Water-based epoxy coatings were applied on steel alloy having a commercial primer coat and were investigated. Polyelectrolyte (PE) nanocapsules containing (2-methylbenzothiazole (BT) and 2-mercaptobenzothiazole (MBT)) were reinforced in the topcoat. An artificial scratched coating is having three regions 1. Wet region 2. The dry region, and 3. The unscratched region was studied with the XPS spectrometer. Figure 6 presents high-resolution S 2p spectra of the wet, dry, and unscratched (top to bottom) coating regions. The unscratched spectrum showed doublet S $2p_{3/2}$ binding energy at 168.4 eV, which is associated with sulfate. A small peak observed at the binding energy of 163.8 eV representing sulfur bonded with carbon can be because of the MBT present in the coating. The sulfate traces are evidence of docusate sodium salt used in the synthesis of nanocapsules. However, in the case of the dry sample, the spectrum shows a clear signal at the bonding energy of 163.5 eV showing the leaching of MBT from the destroyed nanocapsules. In the wet sample spectrum, the signal moved to the 162.4 eV explaining the absorption of MBT on the surface of aluminum substrate.

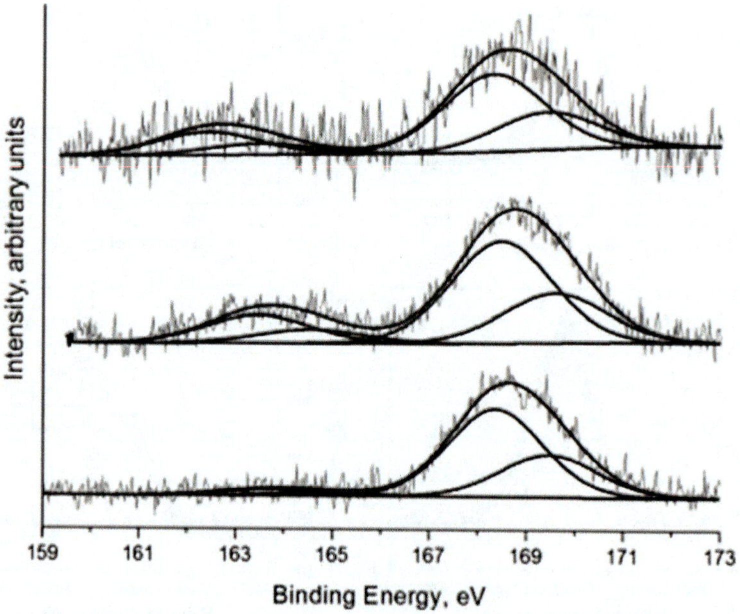

Fig. 6 The S 2p ionization region in the wet, dry, and unscratched regions (top to bottom) of epoxy coatings modified with nanocapsules [21]

3 Characterization of Nanocarriers

Binding energy is an identification of elemental data. Thus, XPS has mostly been utilized for elemental composition and its chemical state, making it helpful for investigating complex designed nanocarriers used in corrosion-resistant polymeric coatings. XPS spectra effectively provide precise information on the composition and chemical state of the individual elements present in these nanocarriers. XPS analysis has been utilized to analyze verity of novel designed nano/micro carriers [22–24] for mitigation of corrosion phenomenon in polymeric coatings. Individual element existing in the nanocarriers might be in many oxidation states, and due to the small binding energy shift, these peaks frequently overlap. Peak fitting and deconvolution are required in such cases, which require the usage of many functions such as Gaussian, Lorentzian, Gaussian–Lorentzian mixed functions, and so on.

Graphene oxide decorated with silica nanoparticles (GO–SiO_2) were synthesized by sol–gel process using two methods for corrosion protection in epoxy coating [25]. In the first method, the GO sheets were added to the silane solution and hydrolysis were carried out various times, whereas, in the other method, pre hydrolyzed GO sheets were added to the mixture of silanes. The synthesized nanohybrid carriers were characterized with the XPS technique to understand its elemental and compositional structural property. Figure 7a_1 and b_1 depicts the high-resolution survey spectrum while Figure 7a_1 and b_1 represents the Si2p spectrum of GO–SiO_2 nanohybrid carriers prepared through method 1 and 2, respectively. O 1 s, N 1 s, C 1 s, and Si 2p are the primary elements observed in both XPS spectrum the GO-SiO_2 nanocarriers. Si2p spectra of GO–SiO_2 nanohybrid carriers (by both methods) were fitted further to confirm the absorption of SiO_2 on GO sheets. The fitted Si2p spectra show Si–C and Si–O–Si bonds at 100.2 and 102.3 eV, respectively. Moreover, the higher peak of Si–O–Si than Si–C peak in each method confirms the successful SiO_2 particles incorporated on GO nanohybrid carriers. Also, it was concluded from the higher intensity of Si in the survey spectrum of method 2 compared to 1 that more SiO_2 nanoparticles covered the GO sheet in method 2.

Cobalt doped zinc oxide nanoparticles (Co: ZnO) were synthesized by the combustion method to analyze its corrosion inhibition efficiency in epoxy coatings [26]. The successful incorporation of Co into the ZnO nanoparticles and the chemical state of Co was examined using the XPS technique. Figure 8a depicts the XPS spectrum of the Co: ZnO nanoparticle. The peaks in Fig. 8a at 765–800 eV correspond to Co $2p_{3/2}$ and Co $2p_{1/2}$, confirming the incorporation of Co in ZnO nanoparticles. Figure 8b illustrated the high-resolution spectrum of the Co 2p bonding energy. The fitted curves further explain the peak of Co $2p_{3/2}$ at 781.8 eV and Co $2p_{1/2}$ at 798 eV, respectively. The peak position of Co $2p_{3/2}$ is associated with the oxidation state of Co, and hence the corresponding bonding energy (781 eV) indicates that Co is present as Co^{2+} having 2^+ oxidation state. Moreover, the energy difference of both the Co peaks is almost 15.5, which explain the presence of Co^{2+} in the structure of synthesized Co: ZnO nanoparticle [27].

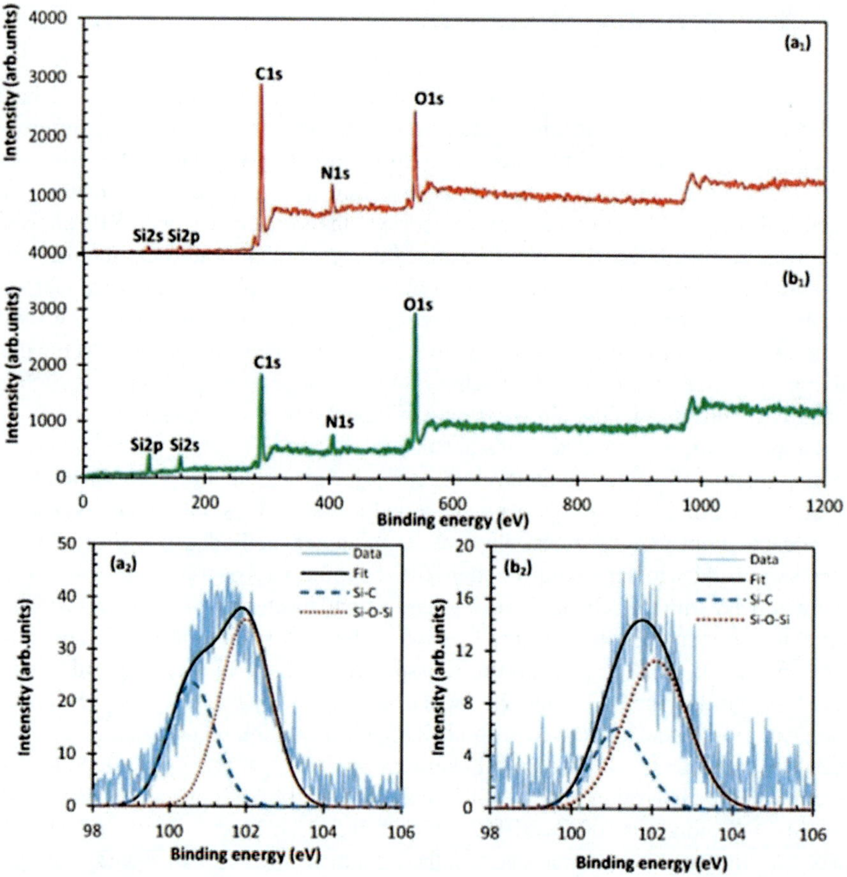

Fig. 7 XPS spectra of GO–SiO_2 nanohybrid carriers (a1, b1) is the survey spectrum of the nanohybrid carriers while (a2, b2) is the SiO_2 spectrum by method 1 and 2, respectively [25]

Graphene oxide platform decorated with zinc doped conductive polypyrrole (GO-PPy) nanoparticles has also been used to synthesize the dual functional nanocarriers with self-healing and corrosion inhibition capability [28]. XPS analysis was carried out to investigate the atomic composition and the presence of functional group of GO-PPy nanoparticles. Figure 9a represents the survey spectra of GO and GO-PPy. The broad range spectrum of GO has two dominant peaks at (287.08 eV and 532.08 eV), attributed to C 1 s and O 1 s whereas, their peaks were detected at 285.08, 400.08 and 532.08 eV, which is linked to the C 1 s, N 1 s and O 1 s of the GO-PPy nanoparticles. The additional signals in the GO-PPy spectrum confirm the successful bonding of PPy to the GO surface. Figure 9b and c depicts the high-resolution C1s spectra, which explains the chemical bonding of carbon in GO and GO-PPy. Four significant peaks in the GO sample at 283.78, 284.58, 286.68 and 288.38 eV can be noticed, which are associated with C–C, C–O, C = O, O–C = O, respectively. In contrast, six peaks are

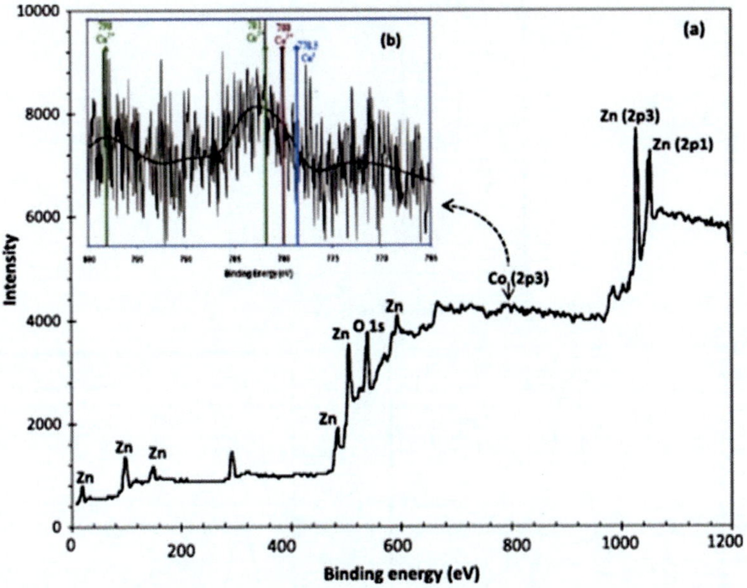

Fig. 8 **(a)** XPS survey spectrum of Co: ZnO nanoparticle **(b)** high resolution fitted spectrum of Co 2p [26]

observed in the GO-PPy sample located at 284.00, 284.80, 285.53, 286.49, 288.52 and 291.01 eV. The additions peaks of sp^2 hybridized carbon, sp^3 hybridized carbon, C–N in GO-PPy confirm the presence of PPy nanoparticles of GO. Furthermore, the XPS spectrum of O1s and N1s (Fig. 9d and e) further confirms the successful surface modification of the GO nanoplatform.

XPS analysis was also utilized to confirm the adsorption of polymeric layers on the surface of halloysite nanotubes (HNTs) in order to synthesize hybrid HNTs [29]. Hybrid HNTs were designed to be a dual inhibitor delivery system for corrosion protection in epoxy coating. The surface elemental composition of both HNTs and the newly designed hybrid HNTs was studied. Figure 10a–c depicts the major elements detected in the surface of HNTs. The high-resolution spectra of Al 2p, Si 2p and O 1 s reflects the typical aluminosilicate structure of HNTs. Contrary, the surface chemistry of hybrid HNTs (Fig. 10d and e) showed carbon and nitrogen as major bonding energy explaining the structure of deposited polyelectrolyte's structure (polyethyleneimine and polyether ether ketone). The XPS study confirms the successful synthesis of hybrid HNTs.

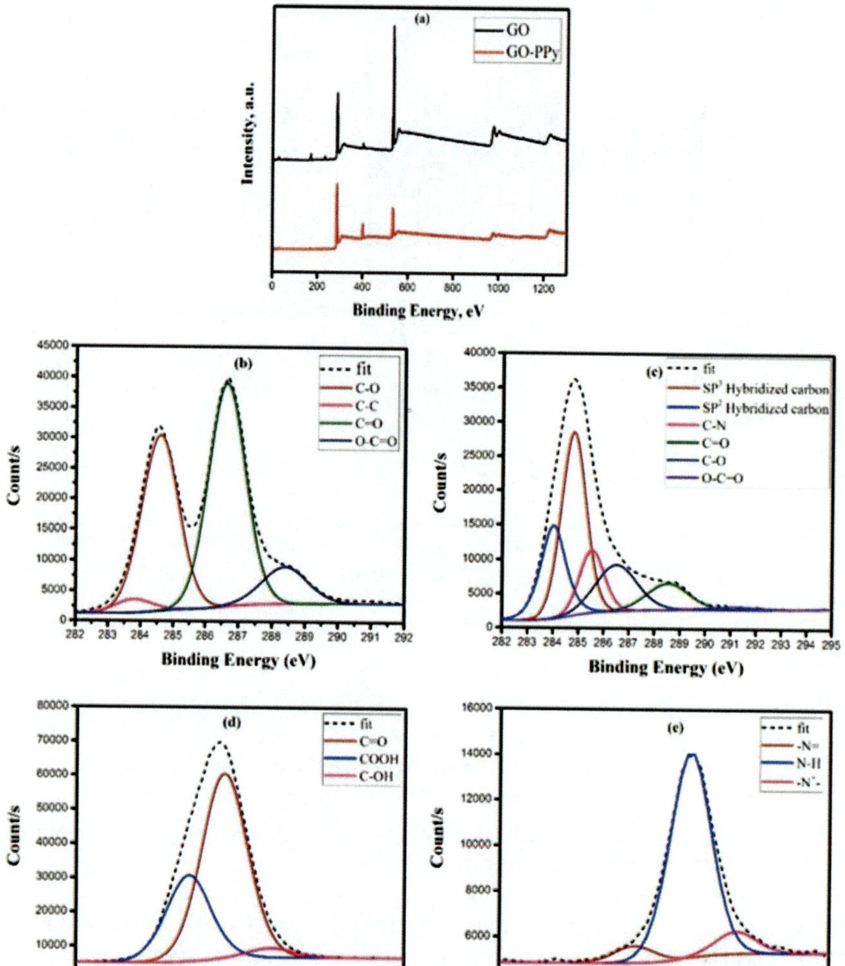

Fig. 9 **(a)** survey spectra of GO and GO-PPy, **(b)** and **(c)** high-resolution C1s spectra of GO and GO-PPy and **(d)** and **(e)** is the high-resolution O1s spectra of GO and GO-PPy [28]

4 Post Corrosion Analysis of Substrates Using XPS

XPS has been extensively utilized in post corrosion analysis to investigate the type, nature, and chemical composition of corrosion products. Industrially various substrates such as carbon steel, alloys of aluminum, and copper are rigorously tested for corrosion under different conditions to understand their corrosion behavior before recommending for a specific application. Moreover, metallic coated samples (inorganic coatings), as well as polymeric-based coated samples, are also rigorously

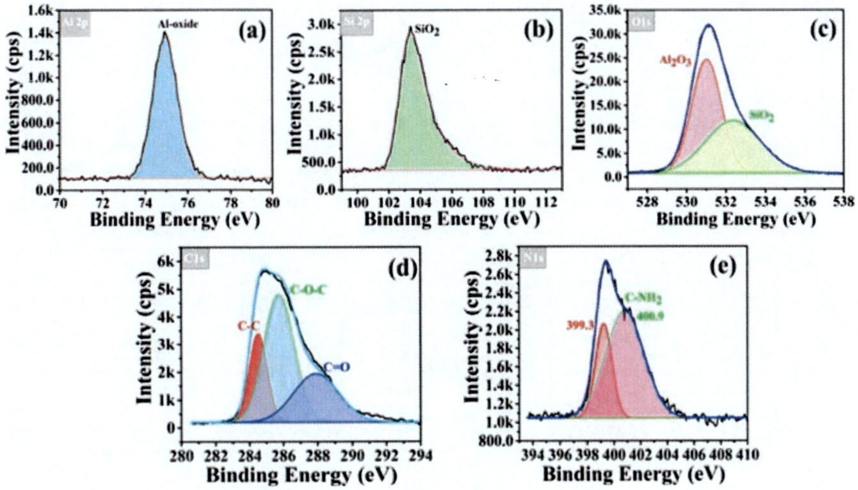

Fig. 10 XPS spectra presenting the surface elemental composition of HNTs (**a–c**) and newly designed hybrid HNTs (**d**, **e**) [29]

subjected to post corrosion analysis to investigate their corrosion behavior. The post corrosion analysis of polymeric coated samples requires the removal of coatings, and the resulted bare substrate and corrosion products are analyzed to understand their corrosion mechanism. Xiao et al. carried out post corrosion analysis on 13Cr steel with varying high temperatures, CO_2, and Cl^- content. XPS spectra in Fig. 11a, b, and c shows the peaks of O at 530.7 eV and 531.8 eV inferring the presence of O^{2-} which can be referred to as the different ionic forms of oxygen in Fe_2O_3, $FeCO_3$, and Cr_2O_3. Spectral peaks of chromium at 577.3 eV and 587.1 eV refers to the $Cr2p_{3/2}$ and $Cr2p_{1/2}$ states pointing to the formation of $Cr(OH)_3$ and Cr_2O_3. Ferrous peaks at 724.02 eV ant 710.61 eV are observed for $Fe2p_{3/2}$ and $Fe2p_{1/2}$ for the oxides formation of $FeCO_3$ and Fe_2O_3. The presence of various oxides of Chromium and Ferrous forms a passive film delaying corrosion rate and enhancing corrosion resistance of the substrate [30]. Jiang et al. further developed $Ni_{60}Nb_{40}$, $Ni_{50}Nb_{50}$, and multilayer $Ni_{50}Nb_{50}/Ni_{60}Nb_{40}$ coatings and tested them for corrosion resistance of longer-duration through XPS as shown in Fig. 11d, e, and f. They reported enhanced corrosion resistance for $Ni_{50}Nb_{50}$ in which doublet peaks of niobium are observed at 210.4 eV and 207.6 eV binding energy corresponding to the $Nb3d_{3/2}$ and $Nb3d_{5/2}$, indicating the Nb^{5+} state of Nb_2O_5 which forms the passive film over the coated surface, enhancing its corrosion resistance. Nickel peaks at 852.6 eV and 856.3 eV indicate the metallic nickel and Ni^{2+}, which leads to the formation of NiO and $Ni(OH)_2$ during the corrosion process. Deconvoluted peaks of oxygen 1S at 530.5 eV, 532.2 eV, and 533.3 eV imply the presence of O^{2-}, OH^- and oxygen of water in the $Ni_{50}Nb_{50}$ magnetic sputtered coating. As metal oxide forms the passive layer for corrosion protection higher content of O^{2-} represents efficient corrosion resistance of the coating [11]. Sun et al. investigated the corrosion product over Ni–P coatings

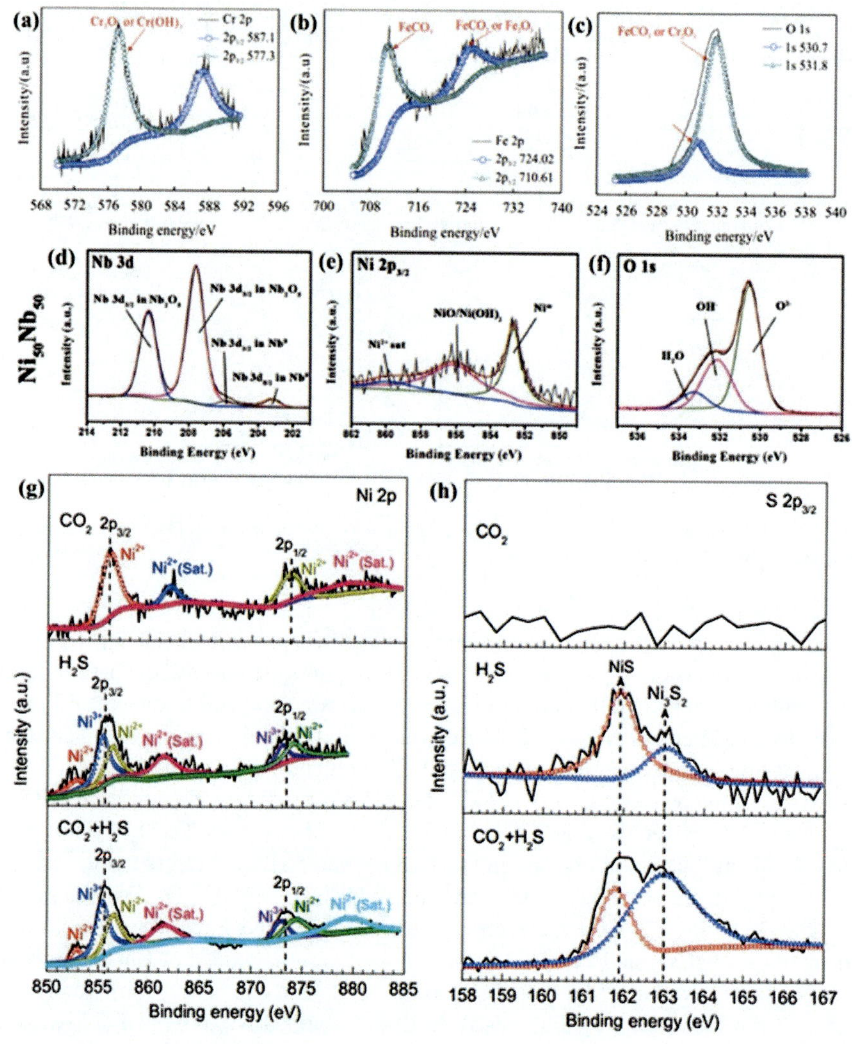

Fig. 11 Post corrosion analysis of 13Cr Steel (a, b and c), $Ni_{50}Nb_{50}$ Alloy (d, e and f) and Ni–P coatings (g and h) using XPS [11, 30]

subjected to various corrosive media such as CO_2, H_2S, and CO_2-H_2S environment to deduce the reasons for improved corrosion resistance of Ni–P coatings. XPS data generated after 168 h of testing are presented in Fig. 11g and h. No peaks of the phosphorus are evident, which implies that the products of phosphorus are already dissolved in the solution. Peaks of nickel Ni2p are clearly visible in all the corrosive media around 855 eV and 873 eV indicating the presence of $Ni2p_{3/2}$ and $Ni2p_{1/2}$ chemical states. In CO_2 assisted corrosive condition, presence of Ni^{2+} leads to the formation of NiO which forms 20% and $Ni(OH)_2$ which is contributing to rest of the

80% of corrosion products. In H_2S and CO_2-H_2S assisted corrosion environment, peaks of Ni^{2+} and Ni^{3+} are associated to oxides, hydroxides and sulfides of nickel. S2p spectra confirms the formation of NiS and Ni_3S_2 in the presence of hydrogen sulfide. In only H_2S environment, NiS contributes to the major portion of corrosion in the form of flaky products than Ni_3S_2 whereas in CO_2–H_2S environment, Ni_3S_2 forms the major portion of corrosion in the form of needle-rodlike structure along with NiS and $Ni(OH)_2$. Moreover, other nickel sulfides and $Ni(OH)_2$ forms the claylike products, which constitute to total corrosion products. Flaky product provides stability to the corrosion scale by the formation of a staggered and connected corrosion scale filled with claylike products. This solid film structure has strong adhesion impeding the mass transfer, hence increasing corrosion resistance in H_2S and CO_2–H_2S aggressive corrosion environments.

Similarly, in the case of organic coatings, the nature and chemical composition of the passive layer formed on the surface of the substrates can also be precisely analyzed using XPS technique [31]. During one study, epoxy coatings modified with microcapsules encapsulated with polydimethylsiloxane (PDMS) were developed. Later, the developed coatings were artificially scratched, subjected to corrosion process, and then investigated using XPS. Figure 12a depicts the survey spectrum of the scratched area of the coating, the elemental composition of the passive layer is identical to the standard spectra of PDMS. It can be evident from the survey that the polymer present in the scratched area is PDMS. Figure 12b shows that the Si 2p bonding energy at 101.78 eV is similar along the scratch area of the coating, which further confirms the presence of PDMS at different regions of the scratch. Moreover, the intensity of Si 2p in the intact coating scan is much lower than the scratched coating explaining the release of PDMS from the broken capsules and act as passivation in the scratched area of the coating.

In another study, the amine cured tetrafunctional epoxy coatings were applied on the steel substrate for corrosion protection [32]. XPS analysis was carried out

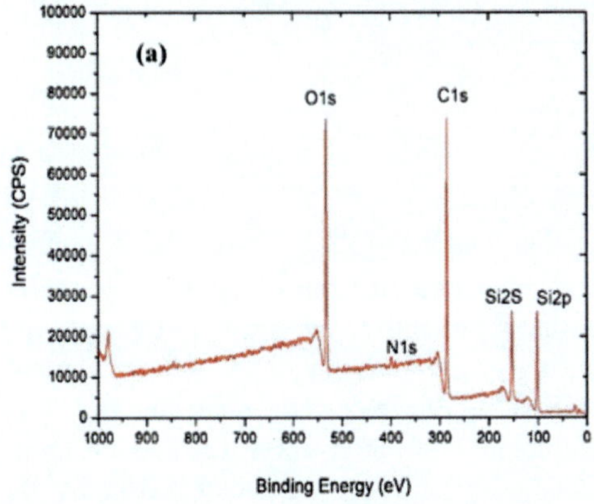

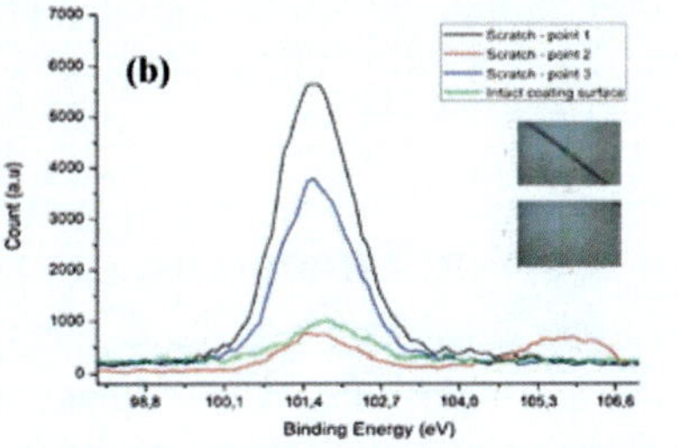

Element/ Transition	Atom %			
	Scratch - point 1	Scratch - point 2	Scratch - point 3	Intact coating surface
C1s	53.43	51.35	63.70	80.06
O1s	21.84	22.29	19.60	12.61
Si2p	21.66	26.35	12.47	2.63
N1s	3.07		4.23	2.89

Fig. 12 **(a)** XPS survey spectrum of the defected area of the coated sample and **(b)** High resolution of S2p at a different point along with the scratch and intact coating [31]

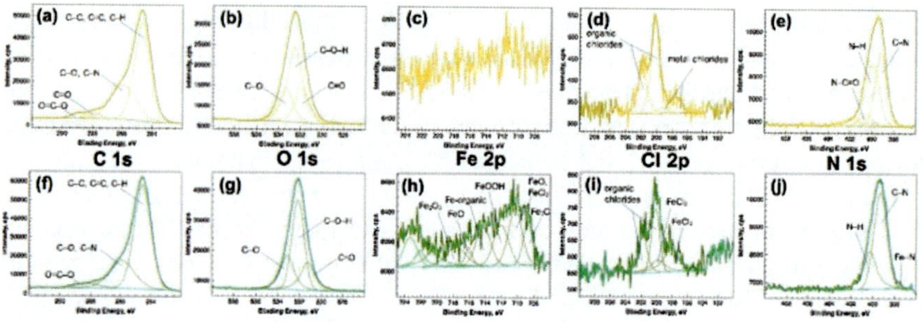

Fig. 13 XPS high resolution spectra of C 1 s, O 1 s, Fe 2p, Cl 2p, and N 1 s, for intact Coating (**a**)–(**e**) before and (**f**)–(**j**) after 30 days of immersion [32]

to investigate the changes in the chemical composition that occurred in the intact coating after exposing it to 3.5 wt% NaCl solution for 30 days. XPS high resolution spectra of C 1 s, O 1 s, Fe 2p, Cl 2p, and N 1 s, for intact coating before and after 30 days of immersion are shown in Fig. 13. Carbon, oxygen, and nitrogen are the main elements detected before immersion, accompanied by small traces of chlorine. The deconvoluted Cl 2p further explained that these traces are because of organic and metal chlorides. The high-resolution spectra of C1s (Fig. 13a) before immersion shows (C–C, C = C, and C–H) at a binding energy of 284.6 eV, (C–O and C–N) at a binding energy of 285.9 eV, (C = O) at 287.9 eV and (O = C–O) at 288. Moreover, the O 1 s (Fig. 13b) spectrum shows three characteristic peaks at 531.8, 532.5, and 533.2 eV representing C = O, C–O–H, C–OH, and C–O, respectively. Moreover, the traces of nitrogen in C1s are also confirmed from the high resolution of the N 1 s spectrum (Fig. 13e). After 30 days of the immersion (Fig. 13f–j), the traces of Fe 2p are detected, representing the initiation of corrosion activity because of the degradation of coating thickness accompanied by the increase of porosity. The shifts and growth in the intensity of Cl 2p peaks explain the presence of iron oxides. The conversion of the carbonyl group to hydroxyl through hydrolysis can be noticed in the C 1 s, N 1 s, and O1s spectrum after the immersion.

5 Current Limitations

XPS is a highly effective tool that can be smartly used to understand the corrosion (nature, type, and mechanism) and the role of protective coatings to mitigate corrosion. It is one of the techniques that confirms the presence of elements, their chemical state, and the passivation of substrates and coatings developed to understand and mitigate corrosion. It can also measure the thickness of the layers if it is under 10 nm. However, various issues hinder its applicability on the commercial and industrial scale. The sample size is restricted to an area of less than 1 cm^2 and height of less than a few mm. Cutting the sample for the required dimensions further adds up contamination along with the removal of corrosion product, especially in terms of the formation of concentration cell in pitting corrosion which propagates along with the depth of the sample. Reproducibility is another challenge in terms of corrosion studies, especially when it comes to analyzing the corrosion products. Samples also undergo high vacuum conditions and argon bombardment, which may affect the observation in terms of hydrogen-assisted corrosion. Apart from technological challenges, some other factors such as inappropriate sample preparation, storage, mounting, and handling can also limit the frequent and effective use of the XPS. In terms of instrumentation, inadequate setup and verification of instruments require regular calibration. Reliability of the samples during preparation, storage, mounting, handling even during measurements requires utmost care. Analysts and operators with limited knowledge and understanding regarding the instrumentation and analysis of various kinds of samples may produce erroneous results. Recording and reporting is another aspect that requires enormous attention in XPS as this is a comparatively newer technique. For instance, plotting the data according to international conventions should be strictly followed. Presenting only the fitted data rather than experimental data, inexperience regarding spin–orbit splitting, not considering the ratio of the peaks during deconvolution, and applying inappropriate background/baseline corrections may also lead to erroneous findings.

6 Conclusions and Future Outlook

XPS is a versatile technique and has been utilized in corrosion research for the past half-century. It provides both quantitative and qualitative analysis of surface chemistry, oxidation states of various elements, etc. In this chapter we covered how XPS has aided the development of novel smart coatings systems. The following main conclusions can be drawn that may be of benefit for the larger coatings systems and corrosion research community:

1. The unique findings of XPs analysis can be used to confirm the incorporation of various reinforcements present in the coatings developed for corrosion protection, especially when embedded in the matrix.
2. In the case of polymeric coatings, XPS can be deployed to confirm the presence, modification, and loading of nanocarriers. It is also efficiently used to verify the formation of passive filmed in the damaged part of smart self-healing polymeric coatings as a result of self-release of inhibitors and self-healing agents from the loaded nanocarriers.
3. XPS technique can be effectively used in the post-corrosion analysis of bare substrates and numerous families of coatings tested in various corrosive environments to properly understand the corrosion and protection mechanism.
4. Although XPS is a fascinating analytical technique, however, its utility can be further extended by considering its present limiting factors such as sample

size, plotting, and interpretation of obtained data. In addition, XPS can be integrated with other advanced techniques such as localized electrochemistry (SVET, LEIS, etc.) to enable important in situ analyses—that can further aid in the discovery and design of new corrosion protection systems. The size of the machinery has to be scaled up as most of the corrosion samples are more significant than the currently applicable specimen holder. Proper machining technology can be equipped with XPS so that the sample preparation can be wisely carried out by the operator keeping in mind the delicacy and requirement of the XPS machine. Argon bombardment may affect the corrosion product, which can be improved. As XPS is a compassionate technique, microscopic add-ins can be provided to focus on the area to be investigated.

5. In situ analysis are highly attractive to develop fundamental understanding of various corrosion processes. In this context, the utilization of XPS with localized electrochemistry techniques such as SVET or LEIS can advance the knowledge by providing breakthroughs that are otherwise difficult to achieve. Such an arrangement can provide opportunity for unique in-situ localized electrochemistry testing coupled with simultaneous compositional analysis, oxidation states of involved elements, etc. of the surface under investigation. The important findings thus obtained may be helpful in proper understanding of the respective corrosion mechanism and corrosion inhibition process.

Acknowledgements This book chapter was supported by NPRP11S-1226-170132 and NPRP13S-0120-200116 grants from the Qatar National Research Fund (QNRF) (a member of the Qatar Foundation). The authors are also thankful to Qatar University for their support through Grant-IRCC-2020-006.

References

1. C.S. Fadley, X-ray photoelectron spectroscopy: from origins to future directions. Nucl. Instruments Methods Phys. Res. Sect. A Accel. Spectrometers, Detect. Assoc. Equip. **601**, 8–31 (2009). https://doi.org/10.1016/j.nima.2008.12.189
2. B.V. Crist, XPS in industry—Problems with binding energies in journals and binding energy databases. J. Electron. Spectros. Relat. Phenomena. **231**, 75–87 (2019). https://doi.org/10.1016/j.elspec.2018.02.005
3. C.S. Fadley, X-ray photoelectron spectroscopy: progress and perspectives. J. Electron. Spectros. Relat. Phenomena. **178–179**, 2–32 (2010). https://doi.org/10.1016/j.elspec.2010.01.006
4. B. Díaz, J. Światowska, V. Maurice, M. Pisarek, A. Seyeux, S. Zanna, S. Tervakangas, J. Kolehmainen, P. Marcus, Chromium and tantalum oxide nanocoatings prepared by filtered cathodic arc deposition for corrosion protection of carbon steel. Surf. Coat. Technol. **206**, 3903–3910 (2012). https://doi.org/10.1016/j.surfcoat.2012.03.048
5. D.R. Baer, Y.-C. Wang, D.G. Castner, Use of XPS to quantify thickness of coatings on nanoparticles. Micros. Today. **24**, 40–45 (2016). https://doi.org/10.1017/S1551929516000109
6. G. Greczynski, L. Hultman, X-ray photoelectron spectroscopy: towards reliable binding energy referencing. Prog. Mater. Sci. **107**, 100591 (2020). https://doi.org/10.1016/j.pmatsci.2019.100591

7. D.R. Baer, G.E. McGuire, K. Artyushkova, C.D. Easton, M.H. Engelhard, A.G. Shard, Introduction to topical collection: Reproducibility challenges and solutions with a focus on guides to XPS analysis, J. Vac. Sci. Technol. A. **39**, 021601 (2021). https://doi.org/10.1116/6.0000873
8. S. Halkjær, J. Iversen, L. Kyhl, J. Chevallier, F. Andreatta, F. Yu, A. Stoot, L. Camilli, P. Bøggild, L. Hornekær, A.M. Cassidy, Low-temperature synthesis of a graphene-based, corrosion-inhibiting coating on an industrial grade alloy. Corros. Sci. **152**, 1–9 (2019). https://doi.org/10.1016/j.corsci.2019.01.029
9. Z.-Q. Zhang, L. Wang, M.-Q. Zeng, R.-C. Zeng, C.-G. Lin, Z.-L. Wang, D.-C. Chen, Q. Zhang, Corrosion resistance and superhydrophobicity of one-step polypropylene coating on anodized AZ31 Mg alloy. J. Magnes. Alloy. (2020). https://doi.org/10.1016/j.jma.2020.06.011
10. A. Stankiewicz, Z. Kefallinou, G. Mordarski, Z. Jagoda, B. Spencer, Surface functionalisation by the introduction of self-healing properties into electroless Ni-P coatings. Electrochim. Acta. **297**, 427–434 (2019). https://doi.org/10.1016/j.electacta.2018.12.026
11. L. Jiang, Z.Q. Chen, H.B. Lu, H.B. Ke, Y. Yuan, Y.M. Dong, X.K. Meng, Corrosion protection of NiNb metallic glass coatings for 316SS by magnetron sputtering. J. Mater. Sci. Technol. **79**, 88–98 (2021). https://doi.org/10.1016/j.jmst.2020.12.004
12. B. Li, D. Li, T. Mei, W. Xia, W. Zhang, Fabrication and characterization of boron nitride reinforced Ni–W nanocomposite coating by electrodeposition. J. Alloys Compd. **777**, 1234–1244 (2019). https://doi.org/10.1016/j.jallcom.2018.11.081
13. A. Bahgat Radwan, K. Ali, R.A. Shakoor, H. Mohammed, T. Alsalama, R. Kahraman, M.M. Yusuf, A.M. Abdullah, M. Fatima Montemor, M. Helal, Properties enhancement of Ni-P electrodeposited coatings by the incorporation of nanoscale Y_2O_3 particles. Appl. Surf. Sci. **457**, 956–967 (2018). https://doi.org/10.1016/j.apsusc.2018.06.241
14. M.H. Sliem, K. Shahzad, V.N. Sivaprasad, R.A. Shakoor, A.M. Abdullah, O. Fayyaz, R. Kahraman, M.A. Umer, Enhanced mechanical and corrosion protection properties of pulse electrodeposited NiP-ZrO_2 nanocomposite coatings, Surf. Coatings Technol. **403,** 126340 (2020). https://doi.org/10.1016/j.surfcoat.2020.126340.
15. J.K. Pancrecious, J.P. Deepa, V. Jayan, U.S. Bill, T.P.D. Rajan, B.C. Pai, Nanoceria induced grain refinement in electroless Ni-B-CeO_2 composite coating for enhanced wear and corrosion resistance of Aluminium alloy. Surf. Coatings Technol. **356**, 29–37 (2018). https://doi.org/10.1016/j.surfcoat.2018.09.046
16. B. Li, W. Zhang, Microstructural, surface and electrochemical properties of pulse electrodeposited Ni–W/Si3N4 nanocomposite coating. Ceram. Int. **44**, 19907–19918 (2018). https://doi.org/10.1016/j.ceramint.2018.07.254
17. O. Fayyaz, A. Khan, R.A. Shakoor, A. Hasan, M.M. Yusuf, M.F. Montemor, S. Rasul, K. Khan, M.R.I. Faruque, P.C. Okonkwo, Enhancement of mechanical and corrosion resistance properties of electrodeposited Ni–P–TiC composite coatings. Sci. Rep. **11**, 5327 (2021). https://doi.org/10.1038/s41598-021-84716-6
18. A.B. Radwan, R.A. Shakoor, Aluminum nitride (AlN) reinforced electrodeposited Ni–B nanocomposite coatings. Ceram. Int. **46**, 9863–9871 (2020). https://doi.org/10.1016/j.ceramint.2019.12.261
19. M. Jiang, L. Wu, J. Hu, J. Zhang, Silane-incorporated epoxy coatings on aluminum alloy (AA2024). Part 2 : Mechanistic investigations Visible Polymer coating, Corros. Sci. **92**, 127–135 (2015). https://doi.org/10.1016/j.corsci.2014.11.048
20. P. Wang, D.W. Schaefer, Why does silane enhance the protective properties of epoxy films? Langmuir **24**, 13496–13501 (2008). https://doi.org/10.1021/la8028066
21. K. Szczepanowicz, G. Mordarski, K. Podgórna, R.P. Socha, P. Nowak, T. Hack, Progress in organic coatings self-healing epoxy coatings loaded with inhibitor-containing polyelectrolyte nanocapsules, **84**, 97–106 (2015). https://doi.org/10.1016/j.porgcoat.2015.02.011
22. A. Khan, F. Ubaid, E. Fayyad, Z. Ahmad, P. Abdul Shakoor, F. Montemor, R. Kahraman, S. Mansour, M. Hassan, A. Hasan, A. Abdullah, Synthesis and properties of polyelectrolyte multilayered microcapsules reinforced smart coatings. J. Mater. Sci. **54** (2019). JMSC-D-19-01483. https://doi.org/10.1007/s10853-019-03761-9

23. A. Khan, M.H. Sliem, A. Arif, M.A. Salih, R.A. Shakoor, M.F. Montemor, R. Kahraman, S. Mansour, A.M. Abdullah, A. Hasan, Designing and performance evaluation of polyelectrolyte multilayered composite smart coatings. Prog. Org. Coatings. **137**, 105319 (2019). https://doi.org/10.1016/j.porgcoat.2019.105319
24. S. Habib, A. Khan, M. Nawaz, M.H. Sliem, R.A. Shakoor, R. Kahraman, A.M. Abdullah, A. Zekri, Self-healing performance of multifunctional polymeric smart coatings.3 Polymers (Basel). **11** (2019). https://doi.org/10.3390/polym11091519
25. B. Ramezanzadeh, Z. Haeri, M. Ramezanzadeh, A facile route of making silica nanoparticles-covered graphene oxide nanohybrids (SiO_2-GO); fabrication of SiO_2-GO/epoxy composite coating with superior barrier and corrosion protection performance. Chem. Eng. J. **303**, 511–528 (2016). https://doi.org/10.1016/j.cej.2016.06.028
26. M. Rostami, S. Rasouli, B. Ramezanzadeh, A. Askari, Electrochemical investigation of the properties of Co doped ZnO nanoparticle as a corrosion inhibitive pigment for modifying corrosion resistance of the epoxy coating. Corros. Sci. **88**, 387–399 (2014). https://doi.org/10.1016/j.corsci.2014.07.056
27. S. Deka, P.A. Joy, Electronic structure and ferromagnetism of polycrystalline Zn1−xCoxO ($0 \leq x \leq 0.15$), Solid State Commun. **134**, 665–669 (2005). https://doi.org/10.1016/j.ssc.2005.03.013
28. R. Mohammadkhani, M. Ramezanzadeh, S. Saadatmandi, Designing a dual-functional epoxy composite system with self-healing/barrier anti-corrosion performance using graphene oxide nano-scale platforms decorated with zinc doped-conductive polypyrrole nanoparticles with great environmental stability and non-t. Chem. Eng. J. **382**, 122819 (2020). https://doi.org/10.1016/j.cej.2019.122819
29. A. Khan, A. Hassanein, S. Habib, M. Nawaz, R.A. Shakoor, R. Kahraman, Hybrid halloysite nanotubes as smart carriers for corrosion protection (2020). https://doi.org/10.1021/acsami.0c08953
30. G. Xiao, S. Tan, Z. Yu, B. Dong, Y. Yi, G. Tian, H. Yu, S. Shi, CO_2 corrosion behaviors of 13Cr steel in the high-temperature steam environment. Petroleum **6**, 106–113 (2020). https://doi.org/10.1016/j.petlm.2019.12.001
31. D.M. Schlemper, H. Pezzin, Progress in organic coatings self-healing epoxy coatings containing microcapsules filled with different amine compounds—A comparison study **156** (2021)
32. E.B. Caldona, D.O. Wipf, D.W.S. Jr, Progress in organic coatings characterization of a tetrafunctional epoxy-amine coating for corrosion protection of mild steel **151** (2021)

Applications of Atomic Force Microscopy in Corrosion Research

Sultan Akhtar

Abstract The surface of the materials is often interacting with its surrounding. Corrosion is a chemical or/and electrochemical reaction of the material with its environment and degrades or damages it gradually. Corrosion research is the field to control and prevent corrosion by developing new tests and methods for observing corrosion damage under laboratory conditions and field situations. Microscopy approaches are used to determine corrosion reactions and rates by observing the surfaces of the corroded specimens. Atomic force microscopy (AFM) is a powerful tool to analyze the surface features and topography at a very high lateral resolution. The resolution of the AFM (fraction of nanometer) is better than the optical limit of diffraction. AFM machine is produced 2- and 3-dimensional (2D and 3D) images of the samples on the computer. These images are then used to study topographical structures, such as surface roughness, surface pits, cracks, and other features of the corroded specimen. There are several techniques of AFM that have been developed to understand the corrosion properties of the materials. For instance, scanning Kelvin probe force microscopy (SK-PFM) techniques are developed to produce Volta potential distribution maps, in-situ AFM is created to study the electrochemical processes, in-situ AFM scratching technique has been developed to stimulate the de-passivation of protective coatings.

Abbreviations

AFM	Atomic force microscopy
EC-AFM	Electrochemical atomic force microscopy
SPM	Scanning probe microscopy
SK-PFM	Scanning kelvin probe force microscopy
STM	Scanning tunneling microscopy
SEM	Scanning electron microscopy

S. Akhtar (✉)
Imam Abdulrahman Bin Faisal University, Dammam, Saudi Arabia
e-mail: suakhtar@iau.edu.sa

I. Toor (ed.), *Recent Developments in Analytical Techniques for Corrosion Research*,
https://doi.org/10.1007/978-3-030-89101-5_9

TEM	Transmission electron microscopy
OM	Optical microcopy
ECDPA	Ethyl-2-cyano-3-(4-(dimethylamino) phenyl) acrylate
ZnO	Zinc oxide
Si_3N_4 probe	Silicon nitride probe
SiO_2 probe	Silicon dioxide probe
$Fe_2O_3 \bullet xH_2O$	Rust
Fe^{+2} (eq)	Iron (aqueous)
Fe (s)	Iron (solid)
O_2	Oxygen
H_2O	Water
NaCl	Sodium chloride
IBM	International business machines corporation
2D	Two-dimensional
3D	Three-dimensional
μm	Micrometer (10^{-6} m)
nm	Nanometer (10^{-9} m)
N/m	Newton/meter

1 Introduction

The common name of corrosion is rust and is more specific for iron. However, corrosion is the degrading/weakening of a substance, especially a metal, and its properties due to a chemical reaction with the environment (rain, moisture air, sun, etc.) [1]. For example, silver tarnishes, copper, and brass turn to a bluish-green, and iron surface rusts due to corrosion reaction when exposed to air. Iron is the most important metal when subjected to corrosion. It should be noted that the surfaces of the materials especially the metals strongly interact with their environment and corrosion is the degradation of a material's surface due to interaction with its environment [2]. Corrosion occurs via an electrochemical process between metals which are good conductors of electricity and the environment. Most corrosion processes involve electrochemical reactions or a combination of physical/chemical processes. The rate of corrosion is greatly affected by the conditions of the environment. The study of the corrosion process is very important to design and engineer materials that can maintain high standards of safety in a harsh environment. Corrosion research is the area of research to control and prevent corrosion by developing new tests and methods for observing corrosion damage under laboratory conditions and field situations. Microscopy approaches are developed in order to determine corrosion reactions and rates by observing the surfaces of the corroded specimens. Atomic force microscopy (AFM) is a powerful tool to analyze the surface features of the specimens at a very high lateral resolution. AFM is a type of scanning probe microscopy (SPM), producing details in the form of electronic images with high-resolution, a fraction of nanometer which is far better than the optical limit of diffraction. AFM produces 2-dimensional (2D) and 3-dimensional (3D) images of the specimens on a computer [3, 4]. These images are then used to study the surface roughness, surface pits, cracks, and other features produced by corrosion species. Several techniques of AFM have been developed to obtain useful information to understand the corrosion behavior and corrosion properties of the materials. For instance, scanning kelvin-SPM (SK-PFM) techniques have been developed to generate the maps of the Volta potential distribution of the materials [5], In-situ AFM was created for characterization of electrochemical processes while the sample is immersed in a solution [4]. In-situ AFM scratching technique has been developed to stimulate the de-passivation of protective oxide or hydroxide film on metal in solution. This chapter focuses on a basic principle, imaging modes, advancement of AFM, and its applications in corrosion research.

2 Working Principle of AFM

The working principle of the AFM is given by G. K. Binning and his co-workers in 1986 and is based on the principle of STM (scanning tunneling microscope) [6, 7]. STM was presented by Binning et al. in 1981 where conductive samples are scanned and analyzed, while AFM allows the scanning of non-conductive samples as well [8, 9]. The basic principle of AFM is the interaction of the cantilever/tip assembly with the specimen. The cantilever/tip assembly is commonly known as a probe. A schematic representative of AFM working is shown in Fig. 1 [10–13]. The main parts of the AFM instruments include a cantilever with tip, sample holder, piezo scanner, photodiode laser, photodetector, and a computer. The more sensitive and important part of the AFM instrument is the cantilever/tip. The sharp tip of the cantilever is used to scan the sample in a line scan format. The samples on the sample holder are placed on a piezo controlled sample stage often known as piezo scanner. A piezo scanner is a mechanical device driven by piezoelectric actuator, which provides very precise x–y stage motion during sample scanning. A laser beam from the laser source is placed on the top-front of the cantilever to monitor the tip-deflection (up/down) due to surface height variation while scanning the sample. The sample usually has the higher and lower surface features. For example, the area 1 has the higher height of the sample than area 2 (Fig. 1a). A position sensitive photodetector is placed to register the tip deflections detected by a laser beam and consequently used to form an image.

In an AFM, a cantilever with a very sharp tip, only a few atoms thick is scanned the sample surface. The sample surface often consists of lower and higher surface features, see the arrangement of atoms in the lattices of areas 1 and 2 (Fig. 1b), where a small part of the sample is magnified to highlight the higher and lower parts of the sample. The construction of the cantilever tip is also shown (on top) by the blue colored atoms in the same scheme. It is shown that area 1 has two rows of atoms more than area 1 (total of four rows of atoms), hence considered as a higher surface part of

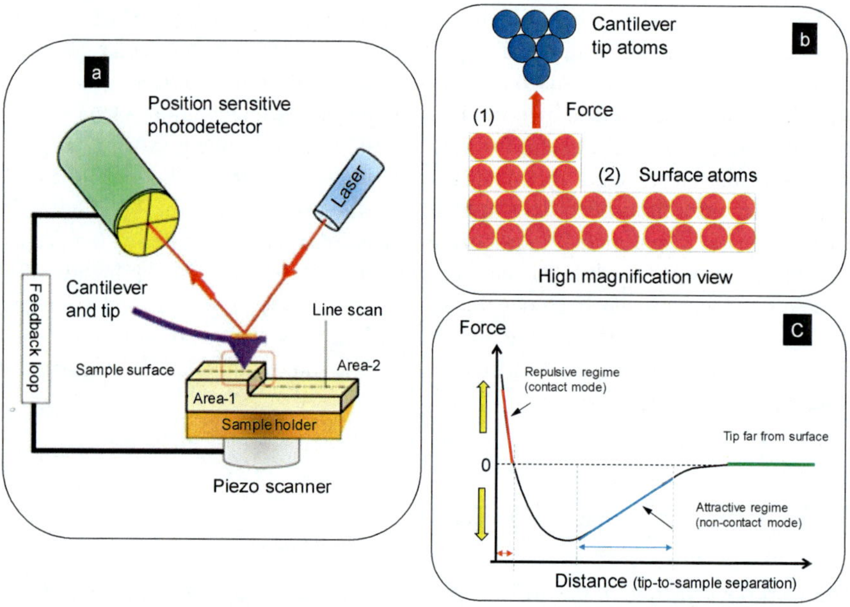

Fig. 1 **a** A schematic representation of the basic working principle of AFM and important parts of the AFM instrument; **b** a magnified view of the tip/sample setup taken from the red dotted rectangular area in "a", showing the tip atoms (blue) and the topographical features of the sample surface (purple). Area 1 shows the higher features while area 2 lower features of the sample surface. **c** A graphical relationship between the force and the sample/tip distance. The tip will deflect downward in the attractive regime (blue curve) due to attractive force and upwards in the repulsive regime (red curve) due to repulsive force as the tip scans the sample [14, 15]

the sample. As the tip approaches far from the surface, an attractive/repulsive force between the surface atoms and the tip resulted in the cantilever to deflect towards or away from the sample surface depending on the nature of the force, i.e., repulsive, or attractive. Depending on the tip-to-sample separation, two regimes are characterized, i.e. attractive regime, known as non-contact mode, and the repulsive regime is called contact mode of AFM. The types of two forces (repulsive and attractive) in AFM modes are demonstrated by a force-distance curve (see Fig. 1c) [14, 15]. In this curve, the attractive regime is highlighted by a blue line whereas the repulsive regime by a red line. In the non-contact mode, the forces are attracted, i.e. tip deflections are towards the surface while in the contact mode the forces are repulsive, and thus tip will be deflected away from the sample surface. In other words, if the attractive forces are downwards then the repulsive forces are in the upward direction, see the yellow arrows in the graph for both the forces (y-axis). Considering either mode, the higher parts of the sample will deflect the cantilever more than the lower parts due to attractive/repulsive forces.

The cantilever of the microscope can be considered as a spring. While scanning the sample, the force produced between the cantilever tip and the sample surface depends on the spring constant, i.e., cantilever constant or stiffness and the gap between the tip and the sample surface. The applicable force on the AFM probe can be expressed as follows using Hooke's law (Eq. 1):

$$F = k.x \tag{1}$$

In this equation, F is the force either repulsive or attractive, k is the cantilever constant and x is the cantilever deflection (upward or downward).

As the tip of the cantilever scans the sample, it deflects up and down according to the topographical properties of the sample. A laser beam adjusted or placed on the cantilever is reflected according to the cantilever deflection and collected by a position-sensitive photodetector. The displacement of the tip (x) for each deflection is measured and transformed in the form of a topographical 2D and 3D image. The images are then used to study the surface roughness, surface pits, cracks, and other features produced by the corrosion species. AFM is considered the high-resolution type of SPM, where its resolution is in the range of nanometer (10^{-9} m or billionth part of a meter), super passing the optical limit of the diffraction more than 1000 times. Using AFM, the surface characteristics of the sample can be scanned with a resolution range (scan area) from $100 \times 100\ \mu m^2$ to less than $1 \times 1\ \mu m^2$.

3 Imaging Modes of AFM

Though several approaches have been used to detect the cantilever bending the most used method was invented by Amer and Meyer. In this method, a laser beam is placed and adjusted onto the head of the cantilever (topside) for measuring the cantilever orientation. The cantilever fluctuations produced during the scanning are detected by a laser beam and measured by a position-sensitive photodetector. The output of the detector is transferred to a computer for producing the electronic 2D and 3D topographical image of the surface with atomic resolution and hence processing the data afterward for further analysis. It is noted that the cantilever is not only be deflected by contact forces employing tip/sample interaction while scanning the sample, but it could also be bend by far-range forces, i.e., van der Waal, magnetic, electric forces etc. In the process, the cantilever vibrates above the surface while scanning the sample as first proposed by Binning [9]. There are three modes of operation of the AFM depending on the tip-to-sample distance (gape) as demonstrated by Fig. 2 (plot and pictorial representation) [16]. They are given as follows:

(1) Contact mode: Direct contact of the tip with the surface
(2) Non-contact mode: Without touching the surface
(3) Tapping mode: Intermittent contact under vibration.

In the force-tip distance curve, two regimes of forces are characterized, attractive regime and repulsive regime (Fig. 2a). In the non-contact mode (orange area), the forces are attracted, i.e., the tip deflections are towards the surface while in the contact

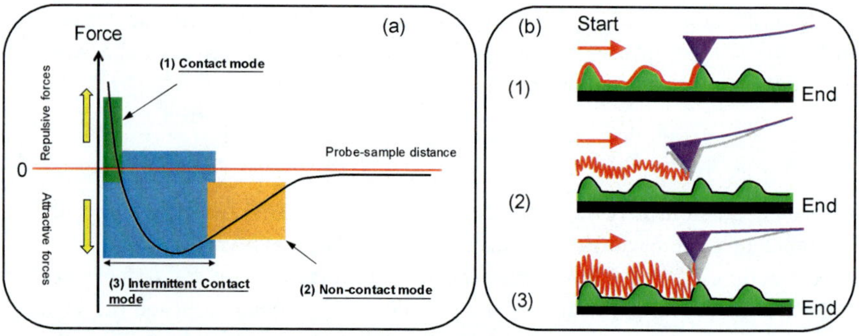

Fig. 2 **a** The force-tip distance curve describing the working modes of the AFM. The force between the tip and the sample surface is the function of the distance (defined by Hooke's law). Three kinds of modes are recognized: contact mode, non-contact mode, and intermittent contact mode. **b** A pictorial display of the three types of AFM modes showing the scanning of the sample with a possible position of the tip: (1) Contact mode, (2) Non-contact mode, and (3) Tapping mode [16]

mode (green area) the forces are repulsive, and thus tip will be deflected away from the sample surface.

In the contact mode (repulsive regime), we measured the repulsive forces between the cantilever tip and the sample surface. Feedback and feedback regulators are used to keep the force of the tip and cantilever constant during the scanning process. This mode of AFM is preferably used for hard samples and could produce images at high resolution, down to sub-Angstrom level. For non-contact mode, the attractive forces as known as van der Waals forces between the tip and the sample surface are measured. The process of scanning is non-destructive as the tip of the cantilever does not touch the surface while scanning the selected area of the sample. There is another technique for the formation of the images in the AFM, known as tapping mode or intermittent contact. In this mode of scanning, the tip of the cantilever oscillates with a very high frequency in the range of 5×10^4 to 5×10^5 cycles per second). Both the oscillation speed and amplitude are decreased due to loss of energy when the tip of the cantilever made any contact with the sample. The position of the tip and the possible deflections of the cantilever starts from the left-hand side of the sample are demonstrated in Fig. 2b. The deflections of the tip are highlighted by a red line for each mode of scanning. The summary of all the three modes of AFM operation, with description and advantages, is given in Table 1.

4 Development of AFM Technology

AFM instrument was invented by Gerd K. Binning in 1986 at IBM (International Business Machines Corporation) Zurich, Switzerland. The working principle of the AFM is based on STM (scanning tunneling microscope). By STM, only the conductive or conductive layers of samples are investigated. However, the AFM machine is

Table 1 Summary of the AFM modes [17]

(1) Contact mode	• Used silicon nitride (Si_3N_4) probes • Continuous physical contact between the tip and the sample (<0.5 nm) • Repulsive forces between the tip and the sample are produced (10^{-9} to 10^{-6} N) • With feedback force of the tip remains constant • Cantilever deflection keeps constant by Feedback regulation • High scan speed and high resolution • Samples are scanned in air and liquid
(2) Non-contact mode	• Used silicon (Si) probes with high spring constant (20–100 N/m) • Tip scans above the sample (1–10 nm), no physica contact appears • Attractive forces between the tip and the sample are produced (10^{-12} N) • Van der Waals forces are detected • Low resolution but no damage to the sample • Samples are scanned in a vacuum
(3) Tapping mode (Intermittent mode)	• Silicon (Si) probes with spring constant (2–50 N/m) • Intermittent and short contact between the tip and the sample (0.5–2 nm) • Both repulsive and attractive forces (tip-to-sample) are produced with force nearly (10^{-12} N) • The tip of the cantilever oscillates at frequency 50,000–500,000 cycles/sec • Tip oscillation amplitude reduces as the tip contacts the surface • Good resolution and minimal damage to the sample • Often used for biological specimens • Samples are scanned in air and liquid

equally applicable for non-conductive samples as well, e.g., biological and polymer samples could also be imaged with AFM. Binning was the first who reported that the cantilever with a sharp tip can be bent under interaction with sample surface by atomic forces [18]. Several methods were developed to measure the cantilever bending but the most useful and widely used method was invented by Amer and Meyer [19]. The schematic diagram along with the real AFM machine is shown in Fig. 3, showing the main parts, and demonstration of the AFM working method. The numbers in the parentheses are representing the main features of the AFM instrument. In AFM, a sharp tip is mounted onto the free end, the cantilevered head of the micromachined cantilever (1). The cantilever tip is utilized to scan the sample surface for 2D and 3D imaging. The cantilever is typically made of silicon (Si), borosilicate glass, or silicon nitride (Si_3N_4) with a tip radius of curvature from few nanometers to few tens of nanometers (~10^{-9} nm). Depending on the application, the surface of the AFM tip can be modified upon applying a coating of either gold (for the detection of covalent bonding), diamond coating to increase the wear resistance of the tip, or

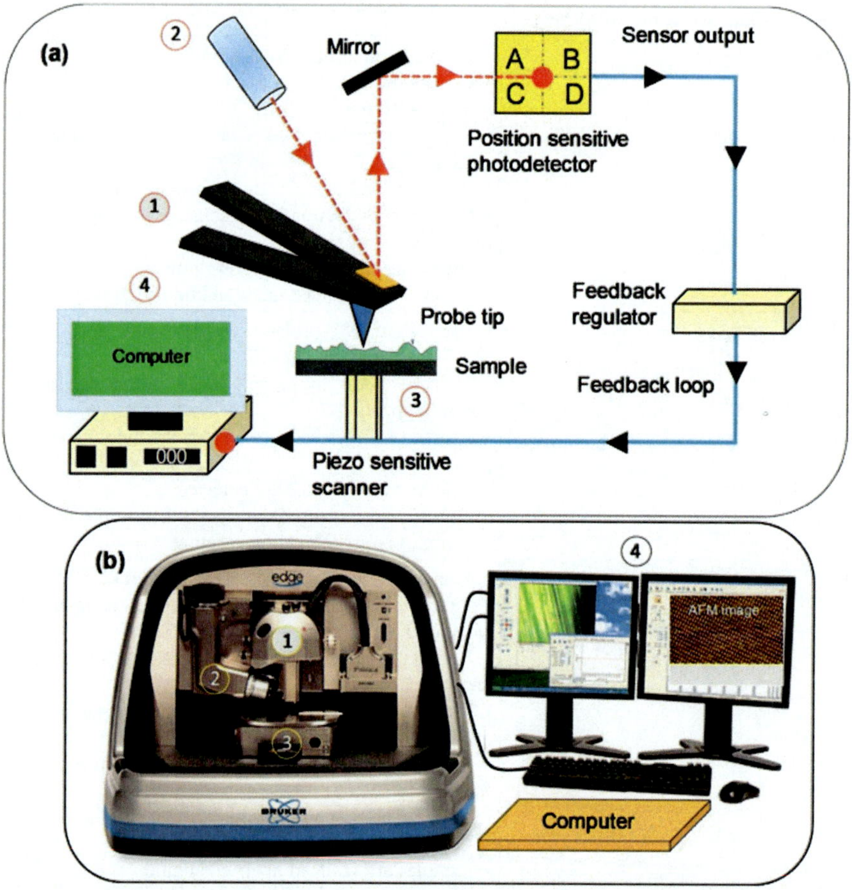

Fig. 3 Development of AFM technology. **a** Main parts of the AFM with their uses (only important parts are labeled). **b** Photo of the actual AFM machine (front view) and computer system where main components are labeled. Cantilever (1): spring which deflects as probe tip scans the sample surface; Probe tip: senses the sample properties and causes the cantilever to deflects; Laser diode (2): uses to detect the cantilever deflection; Position sensitive photodetector: measure the deflections of the cantilever by the laser beam; Feedback loop: controls z-directions; Piezo-scanner (3): position sample (x, y, z); Computer and displays (4): control system, perform data acquisition, display, and analysis. Credit: Image courtesy of Bruker Corporation (panel b) [24–26]

magnetic materials for the detection of magnetic properties) [20–23]. The coating of the cantilever surface utilizing reflected material (as shown by an orange rectangular disc on the cantilevered head) is further helped to increase the reflectance of the laser beam (2) from the cantilever surface and hence improved the overall deflection signal. The sample is mounted onto a sample stage equipped with a piezo-sensitive scanner having a piezo crystal (3) for control and precise movement of the sample while scanning. The piezo-sensitive scanner (3) controls (x, y, z) the movement of the sample and sample stage with micrometer precision. An XYZ dive permits the sample and sample stage to displace in x, y, and z-direction concerning tip (1).

When the sample is scanned via a piezo-sensitive scanner (3) of the AFM, the repulsion/attractive forces (interatomic forces) between the tip and the sample moves the tip that resulted in the bending of the cantilever upward or downward according to Hook's Law [15]. Due to the bending, the laser beam that was placed on the cantilevered head is reflected in variable orientation towards the position-sensitive photodetector for measuring the cantilever deflections at many points on a 2D surface. The most used photodetector consists of four sectioned electronic devices that can detect both longitudinal and torsion bending. The output signal of the detector is transferred to a computer (4) for making 2D and 3D images and for further data process and data analysis. The Cantilever can be bend both by direct contact forces and by far-ranging forces (van der Waals, chemical bonding, magnetic, electric, etc.). The gap between the tip and the sample surface defines the AFM operation, and there are three modes of operations are developed depending on the gap. One is known as static mode (contact mode) and the other two are dynamic modes of operation, also known as a non-contact and tapping mode. In the static mode, the tip is contacted to the sample surface while in the dynamic mode the cantilever just oscillates on the sample surface at a given frequency [9]. These modes of operations have shown some advantages and disadvantages, and their uses depend on the nature of the sample and the kind of information that we are looking for. For example, the contact mode is used when a high resolution is required. On the other hand, the tapping mode is beneficial for soft and biological samples, and the samples for this mode are scanned in air and liquid. Nearly, no sample damage is happed in this mode of operation. In addition, several other AFM related techniques have developed. For instance, Kelvin probe force microscopy (KPFM), also known as surface potential microscopy is developed to produce the maps of the Volta potential distribution across the surfaces. This technique is used to study the corrosion and pitting processes of the specimens [5, 27–29]. The working principle of the KPFM technique is based on the parallel plate capacitor. The first experiments using the KPFM technique were conducted by Lord Kelvin in 1898, whereas the technique was developed by William Zisman in the 1930s to measure the potential differences of dissimilar metals [30, 31].

In-situ AFM was developed for the study of electrochemical processes, and in-situ AFM scratching technique to stimulate the de-passivation of protective oxide or hydroxide film on metal in solution. In general, using AFM, 2D and 3D images can be produced on a computer which is then used to study the surface roughness, surface pits, cracks, and other features of the corroded specimens.

5 Uses of AFM in Corrosion Research

Corrosion is a chemical or/and electrochemical reaction of the material with its environment and degrades it. Iron is the most important metal when studied corrosion.

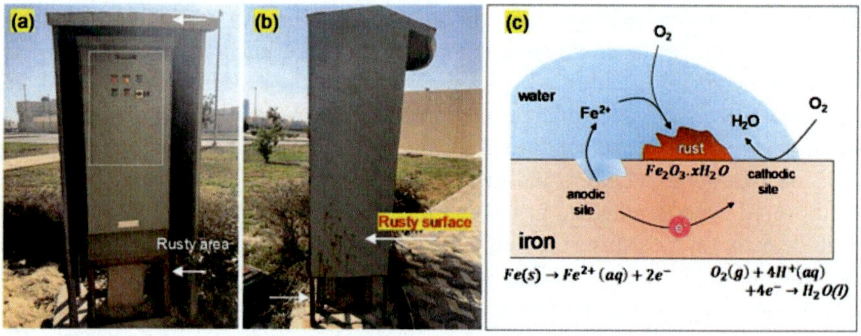

Fig. 4 Illustration of the corrosion process with a practical example. **a** Front-view and **b** side-view of the outdoor electric control panel with a protective cage. The areas of the electric control panel that are exposing directly to the environment showing a high tendency to rust (marked by white arrows) as compared to less exposed areas. **c** The development of rust over the surface of the iron is explained precisely. At the anodic site, iron is oxidized to Fe^{2+}(aq). O_2 is reduced to H_2O on the other side of the iron, known as the cathode. In the process, the electrons (e−) are moved from the anode to the cathode. H_2O is the solvent for the Fe^{2+}, behaves as a salt bridge. Subsequently, the Rust (Fe_2O_3•xH_2O) grows over iron by oxidation of $Fe^{2+}$by reacting with O_2 [2]

The surfaces of the materials especially the metals strongly interacted with the environment and degrade the materials due to this interaction. Corrosion occurs via an electrochemical process between metals which are good conductors of electricity and the environment. Most corrosion processes involve electrochemical reactions or a combination of physical/chemical processes. The rate of corrosion is greatly affected by the conditions of the environment. The development of corrosion (rust) over iron is illustrated by a practical example (outdoor electric control panel with an iron cage), exposing directly to the environment (water, rain, wind, sun, etc.) (Fig. 4). It was noticed that the portions of the electric control panel that are exposing directly to the environment (e.g., top, sides, and lower part especially the supporting legs) showing a high rate of corrosion as compared to those which are less exposed (i.e., area under the shed as marked with a white dotted rectangle) (see Fig. 4a, b). The highly rusted areas are marked by white arrows. The corrosion process develops the rust over the surface of the metallic objects and the process of corrosion over iron is summarized as follows (Fig. 3c) [16]: at the anodic site (on the left-hand side of the scheme), iron is oxidized to Fe^{2+}(aq), that is considered as a defect. Oxygen (O_2) is reduced to water (H_2O) on the other side of the iron bar (right-hand side), known as the cathode. In the process, the electrons (e^-) released at the anode are conducted by cathode courtesy of the electrically conductive nature of the iron. Water that is produced initially is the solvent for the Fe^{2+}, acts as a salt bridge. Subsequently, the Rust (Fe_2O_3•xH_2O) is formed over the surface of the iron by oxidation of $Fe^{2+}$by reacting with atmospheric O_2.

The study of the corrosion process is very important to design and engineer materials that can maintain high standards of safety in a harsh environment. Microscopy approaches are developed to determine corrosion reactions and rates by observing the surfaces of the corroded specimens. AFM is one of the techniques to study the surface chemistry of the materials at high lateral resolution. Using an AFM machine, 2D and 3D images of the specimens are produced on a computer for topographical studies of the materials.

AFM is considered as a standard microscopy technique for examining and evaluating the morphological features and measuring the surface structure in material science. AFM has shown the capability to acquire the images of the objects down to nanometer-scale resolution. The advantage of the AFM over the other microscopy techniques such as electron microscopy (transmission electron microscopy (TEM) and scanning electron microscopy (SEM)) is the relatively simple sample preparation: with little or no sample preparation is required for AFM visualization (see Table 2).

With the aid of the AFM, one can extract the information about the specimen, such as surface properties (topographical features, roughness, cracks, thickness, etc.), electrical properties (charge density), and magnetic properties (magnetic field), mechanical properties, etc. [32]. AFM can work under atmospheric pressure conditions (in air or liquids) without damaging the specimen as well as artificial conditions, such as, under drying conditions, coated specimens under vacuum or freezing. Almost all kinds of samples can be studied and investigated by AFM including, film coatings, composites, glass, polymers, plastics, steel, metals, graphene-based materials, synthetic and biological membranes, qualitative studies as well as quantitative studies including and real-time investigation of living cells. AFM is also used to visualize the single-stranded DNA/double-stranded DNA [33–35].

AFM is used in several studies to investigate the corrosion properties of the materials via the visualization of topographical features of the surface. Recently, a new

Table 2 Comparison of AFM with other microscopy techniques

Atomic force microscopy (AFM)	2D and 3D images are achieved at the same time Surface roughness profiles can be obtained For biological specimens: The ability to image the aqueous samples and no need for sample preparation (e.g. labeling, fixing, or coating, etc.) Large and thick specimens can be viewed
Optical microscopy (OM)	In general, 2D images are obtained using a special technique, known as the optical profilometry technique Large and thick specimens can be viewed For biological specimens: Dried samples can be viewed only
Electron microscopy (SEM and TEM)	In general, 2D images are obtained Surface roughness profiles are not available Specimens of certain dimensions can be viewed only especially thin specimens are needed for TEM to make them electronically transparent For biological specimens: Labeling, fixing, or coating is needed for imaging

technique is developedby coupling the AFM instrument with a electrochemical-cell, know as electrochemical atomic force microscopy (EC-AFM), where electrochemical cell consists of three electrodes (working, counter and reference electrodes) and a potentiostat [36]. By using this measurement setup, one can monitor the corrosion in an aqueous electrolyte and obtain an in-situ mapping of topographical features of the specimens. Furthermore, we can monitor the corrosion process by analyzing the AFM data acquired by higher-harmonic imaging mode. The method is helpful to extract some additional information about the specimen as it is sharpening the grain boundaries between the objects in heterogeneous samples. The method can be applied to identify the variations in steel structures after corrosion tests such as corrosion products, steel heterogeneity, chemical structures, especially iron oxides, physico-chemical properties as these features exhibited edges at which the higher harmonics are more intense [37]. The electrochemical atomic force microscopy (EC-AFM) is also applied to study the corrosion properties (e.g., corrosion at early-stage, surface dissolution, passive film and inhibitor properties, etc.) of the metallic materials by monitoring the surface and structural changes of the specimen surface in real-space conditions [36].

The following information could be obtained using EC-AFM techniques: surface passivation film and its structure when the specimen, carbon steel in the present example is submerged in a chemical solution (carbonate/bicarbonate) as studied by Li and his team fellows [38]. The passive film properties of 304L stainless steel in the acidic medium were also investigated. The ex-situ AFM results revealed that the surface topographical behavior of the steel substrate depends on the stability of the passive film at lower concentrations (0.1 M and 0.5 M) of nitric acid, while in-situ study revealed that the passive film could produce the particles of platelet-like shape and structure and these as-grown particles were aggregated at higher acid concentrations (0.6 M and 1 M) (*Padhy and co-workers*) [39].

Kreta group investigated the production of passive film of aluminum alloy (AA 2024-T3) during electrochemical experiments by exploring the surface changes (surface roughness, the formation of spiky surface, profile lines) using the EC-AFM imaging tools in NaCl environment. It was observed that the corrosion potential spikes were formed only for certain potential and vanished after the growth of a dense or thicker oxyhydroxide layer [40].

Nikhil and his group evaluated the corrosion inhibitor properties of zinc and ethyl-2-cyano-3-(4-(dimethylamino) phenyl) acrylate (ECDPA) composite for copper using EC-AFM. It was revealed by 2D and 3D EC-AFM imaging that the presence of ZnO nanoparticles increased the absorption of ECDPA molecules over copper (Fig. 5) [4]. SK-PMF technique is used to study the corrosion and pitting of the materials. P. Leblanc et al. applied to extract the information such as mapping of topography and Volta potential, volta potential maps after exposing the aluminum alloy (AA2024-T3) in sodium chloride (NaCl) solution [5].

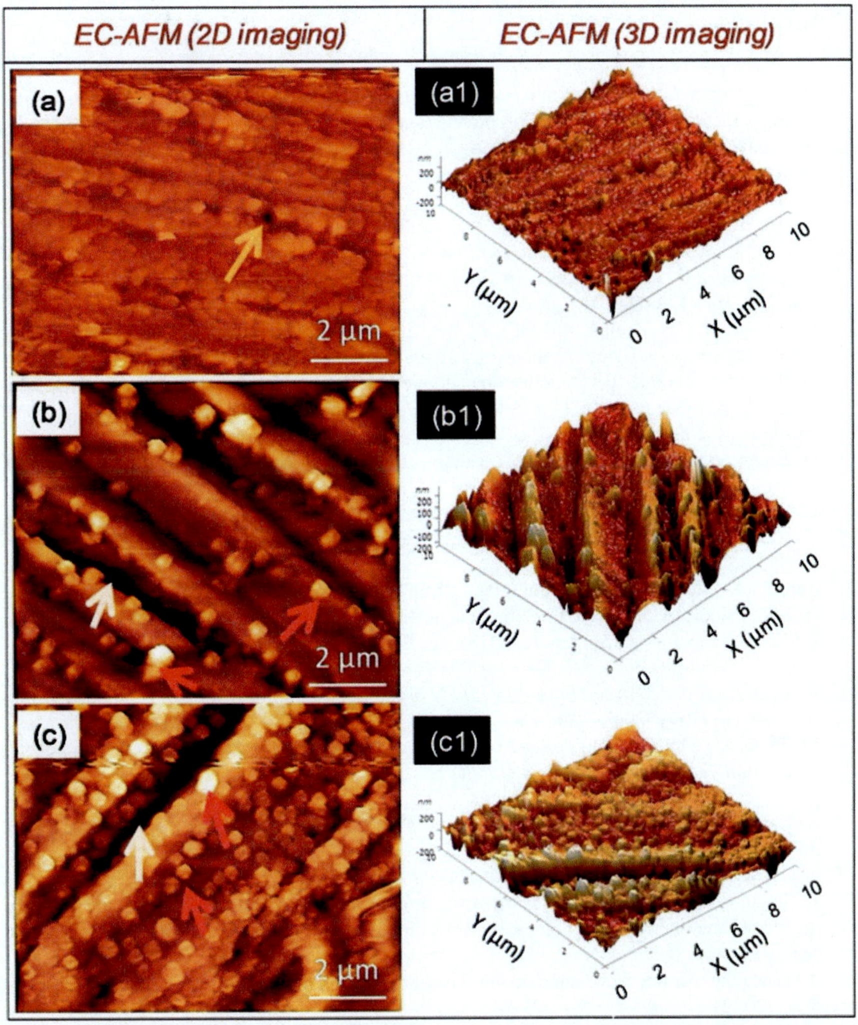

Fig. 5 Electrochemical- atomic force microscopy (EC-AFM) images (10 μm × 10 μm) of **a** pure copper, ECDPA inhibited, and **c** ECDPA/ZnO inhibited copper after 10 min immersion in HCl. **a–c** 2D and (a1-c1) corresponding 3D images. The no. of adsorbed particles was higher for composite (ECDPA/ZnO) than the individual ECDPA inhibitor. The arrows in the 2D images indicate a corrosion pit (a yellow arrow), adsorption/deposition of inhibitor molecules (red arrows), and scratches and deep pits (white arrows). Adapted with permission from [4]; copyright 2019 Elsevier Ltd.

References

1. P. Hammonds, Chapter 4 an introduction to corrosion and its prevention, in *Comprehensive Chemical Kinetics*, ed. by R.G. Compton (Elsevier, 1989), pp. 233–279
2. N.J. Tro, Electrochemistry. 19.9: Corrosion-Undesirable redox reactions, in *Chemistry: A Molecular Approach* (2019), pp. 19.9.1–19.9.5
3. A. Ibrahim et al., Effects of annealing on copper substrate surface morphology and graphene growth by chemical vapor deposition. Carbon **94**, 369–377 (2015)
4. G.J. Nikhil, R. Prakash, Composites of Donor-π-Acceptor type configured organic compound and porous ZnO nano sheets as corrosion inhibitors of copper in chloride environment. J. Mol. Liq. **280**, 160–172 (2019)
5. P. Leblanc, G.S. Frankel, A study of corrosion and pitting initiation of AA2024-T3 using atomic force microscopy. J. Electrochem. Soc. **149**(6), B239 (2002)
6. G.K. Bennig, Atomic force microscope and method for imaging surfaces with atomic resolution, in https://patents.google.com/patent/US4724318A/en, U.S. Patent, Editor. (Iinternational Business Machines Corporation, USA, 1988), p. 8
7. G. Binnig et al., Surface studies by scanning tunneling microscopy. Phys. Rev. Lett. **49**(1), 57–61 (1982)
8. G. Binnig, H. Rohrer, Scanning tunneling microscopy. Surf. Sci. **126**(1), 236–244 (1983)
9. G. Binnig, C.F. Quate, C. Gerber, Atomic force microscope. Phys. Rev. Lett. **56**(9), 930–933 (1986)
10. https://www.polifab.polimi.it/equipments/afm/
11. W. Sun, Principles of atomic force microscopy, in *Atomic Force Microscopy in Molecular and Cell Biology* (Springer, Singapore, 2018), pp. 1–28
12. https://www.azom.com/article.aspx?ArticleID=4437
13. K. Li et al., *Atomic Force Microscopy Study of the Interaction of DNA and Nanoparticles, in Advances in Experimental Medicine and Biology* (Springer, Dordrecht, 2014)
14. M. Ding, C. Shi, J. Zhong, 28—Atomic force microscopy for food quality evaluation, in *Evaluation Technologies for Food Quality* ed. by J. Zhong, X. Wang (Woodhead Publishing, 2019), pp. 715–741
15. B. Cappella, G. Dietler, Force-distance curves by atomic force microscopy. Surf. Sci. Rep. **34**(1), 1–104 (1999)
16. http://users.metu.edu.tr/chem355/assets/11-355-AFM.pdf
17. J. Zhong, From simple to complex: investigating the effects of lipid composition and phase on the membrane interactions of biomolecules using in situatomic force microscopy. Integr. Biol. **3**(6), 632–644 (2011)
18. Y.F. Dufrêne, Atomic force microscopy, a powerful tool in microbiology. J. Bacteriol. **184**(19), 5205 (2002)
19. G. Meyer, N.M. Amer, Novel optical approach to atomic force microscopy. Appl. Phys. Lett. **53**(12), 1045–1047 (1988)
20. O.H. Willemsen et al., Biomolecular interactions measured by atomic force microscopy. Biophys. J. **79**(6), 3267–3281 (2000)
21. K. Hasselbach et al., *High resolution magnetic imaging: MicroSQUID force microscopy.* J. Phys. Conf. Ser. **97**, 012330 (2008)
22. K.-H. Chung, D.-E. Kim, Wear characteristics of diamond-coated atomic force microscope probe. Ultramicroscopy **108**(1), 1–10 (2007)
23. X. Xu, A. Raman, *Comparative dynamics of magnetically, acoustically, and Brownian motion driven microcantilevers in liquids.* J. Appl. Phys. **102**(3), 034303 (2007)
24. M. Lin, Nanotechnology-based approaches for rapid detection of chemical and biological contaminants in foods (2012)
25. H. Yang et al., Application of atomic force microscopy as a nanotechnology tool in food science. J. Food Sci. **72**(4), R65–R75 (2007)
26. https://www.bruker.com/products/surface-and-dimensional-analysis/atomic-force-microscopes/dimension-edge/overview.html. (2020)

27. W. Melitz et al., Kelvin probe force microscopy and its application. Surf. Sci. Rep. **66**(1), 1–27 (2011)
28. M. Fujihira, Kelvin probe force microsopy of molecular surfaces. Annu. Rev. Mater. Sci. **29**(1), 353–380 (1999)
29. M. Nonnenmacher, M.P. O'Boyle, H.K. Wickramasinghe, Kelvin probe force microscopy. Appl. Phys. Lett. **58**(25), 2921–2923 (1991)
30. L. Kelvin, *V*. Contact electricity of metals. London, Edinb. Dublin Philos. Mag. J. Sci. **46**(278), 82–120 (1898)
31. W.A. Zisman, A new method of measuring contact potential differences in metals. Rev. Sci. Instrum. **3**(7), 367–370 (1932)
32. G. Meyer, N.M. Amer, Simultaneous measurement of lateral and normal forces with an optical-beam-deflection atomic force microscope. Appl. Phys. Lett. **57**(20), 2089–2091 (1990)
33. http://perso.univ-lemans.fr/~bardeau/IMMM-PEC/afm/afmintroduction.html
34. H. Yang, Y. Wang, Application of atomic force microscopy on rapid determination of microorganisms for food safety. J Food Sci **73**(8), 1750–3841 (2008)
35. A. Wu et al., Nanoscale structures of circle—MgCl2 constructed by plasmid DNA templates. Superlattices Microstruct. **37**(3), 151–161 (2005)
36. H. Chen et al., Application of electrochemical atomic force microscopy (EC-AFM) in the corrosion study of metallic materials. Materials **13**(3), 2020
37. S. Babicz et al., Corrosion process monitoring by AFM higher harmonic imaging. Measur. Sci. Technol. **28**(11), 114001 (2017)
38. Y. Li, Y.F. Cheng, Passive film growth on carbon steel and its nanoscale features at various passivating potentials. Appl. Surf. Sci. **396**, 144–153 (2017)
39. N. Padhy et al., Morphological and compositional analysis of passive film on austenitic stainless steel in nitric acid medium. Appl. Surf. Sci. **257**(11), 5088–5097 (2011)
40. A. Kreta et al., In situ electrochemical AFM, ex situ IR reflection–absorption and confocal Raman studies of corrosion processes of AA 2024–T3. Corros. Sci. **104**, 290–309 (2016)

Advancement in Corrosion Modeling and Prediction

A Review of Erosion-Corrosion Models for the Oil and Gas Industry Applications

Anass Nassef, Michael Keller, Shokrollah Hassani, Siamack Shirazi, and Kenneth Roberts

Abstract Carbon steel tubing and piping are susceptible to erosion-corrosion damage due to the erosive and corrosive nature of the flow in many situations in the oil and gas industry. The combined effect of solid particle erosion and corrosion can increase the metal degradation rate while simultaneously decreasing the efficiency of corrosion protection systems including iron-carbonate scale formation and chemical inhibition. These combined effects (synergies) can lead to higher corrosion rates, surface pitting, and material failure. This chapter discusses the development in the research of erosion-corrosion modeling for CO_2 environments under iron carbonate scale forming, non-scale forming and inhibited conditions. An overview is provided for erosion-corrosion and inhibited erosion-corrosion models. The erosion-corrosion model utilizes three steps approach based on existing mechanistic models for pure erosion and pure corrosion. The first step involves predicting erosion rate using the Sand Protection Pipe Saver (SPPS) erosion model. The second step involves predicting CO_2 corrosion using the mechanistic corrosion model called SPPS: CO_2. Finally, the combined erosion-corrosion rate is predicted for various environmental conditions based on presence and absence of iron carbonates scale. Frumkin adsorption isotherm is discussed as phenomenological model, which can be used to model the performance of imidazoline-based inhibitor and inhibited erosion-corrosion for different environmental conditions.

Keywords Carbon steel · CO_2 corrosion · Erosion · Iron carbonate scale ($FeCO_3$) · Erosion-corrosion · Inhibited erosion-corrosion · Sand · Calcium carbonate ($CaCO_3$) · SPPS

1 Introduction

In the oil and gas production industry, erosion-corrosion is a common problem in production and transportation facilities especially offshore pipelines downhole

A. Nassef (✉) · M. Keller · S. Hassani · S. Shirazi · K. Roberts
Department of Mechanical Engineering, College of Engineering and Natural Sciences, Erosion-Corrosion Research Center, The University of Tulsa, Tulsa, USA
e-mail: anass-nassef@utulsa.edu

I. Toor (ed.), *Recent Developments in Analytical Techniques for Corrosion Research*,
https://doi.org/10.1007/978-3-030-89101-5_10

components, fittings, valves, chokes, production manifolds, and process headers. Erosion-corrosion occurs due to erosive and corrosive environments. Erosion is the physical/mechanical removal of material by solid particles or liquid droplets, and corrosion is the electrochemical dissolution of metal surface through chemical and electrochemical reactions. The combined action between erosion and corrosion is known as erosion-corrosion.

In oil and gas production erosion-corrosion is caused predominately by the impact of entrained solid particles e.g. sand; $CaCO_3$ particles…etc., on carbon steel piping and equipment materials. CO_2 and H_2S are two common types of corrosion encountered in oil and gas production. However, CO_2 is the most commonly associated with erosion-corrosion. The combined effect of erosion and corrosion can increase metal degradation rate while negatively affecting corrosion mitigation films, such as removing a protective scales e.g. iron carbonate ($FeCO_3$); and iron sulfide (FeS), or stripping away an inhibitor film through solid particle erosion. These combined effects can lead to higher corrosion rates, pitting and material failure. The need to be able to predict erosion-corrosion and its effects on corrosion mitigation such scale iron carbonate scale and chemical inhibitors is particularly important when the wells are deep, off-shore or difficult to reach regions, because coupon testing is impractical and replacement costs are high.

In order to understand the nature and mechanisms of erosion-corrosion, it is necessary to have clear distinctions between erosion, corrosion, and erosion-corrosion. Erosion-corrosion results from the interaction between two components; the electrochemical (corrosion) and mechanical (erosion) processes. In erosion-corrosion, erosion and corrosion can either be independent in such case the metal loss rate due to erosion-corrosion is the sum of the two components, or synergistic in such case, erosion-corrosion results in significantly higher metal loss rates than the combined effects of erosion and corrosion acting separately. One common way of defining the contribution of erosion and corrosion and their interaction in the total erosion-corrosion mechanisms is given in Eq. (1): [1]

$$\begin{aligned}&\text{Erosion} - \text{Corrosion Rate}\\&= \text{Erosion Rate} + \text{Corrosion Rate} + \text{Effect of Erosion on Corrosion}\\&\quad + \text{Effect of Corrosion on Erosion Rate}\end{aligned} \tag{1}$$

In Eq. (1), "erosion rate" refers to the mechanical erosion rate of the meal that would occur under given conditions in absence of an electrochemical corrosion component, and corrosion rate refers to the corrosion rate of the metal that would occur under given conditions in the absence of mechanical component. The "effect of erosion on corrosion rate" refers to the increment, positive or negative, that the corrosion rate undergoes when the full effect of the mechanical component is considered, and the "effect of corrosion on erosion rate" refers to the increment, positive or negative, that the erosion rate undergoes when the full effect of corrosion is considered. The term "synergy" is often used to describe cases for which the sum of the interactions terms not equal to zero, that is, when erosion-corrosion in Eq. (1) is different from simply the sum of the erosion rate and corrosion rate [1].

In this chapter, the current understanding of the mechanisms and modeling that describe erosion, CO_2 corrosion, and erosion-corrosion are presented.

2 Solid Particles Erosion

The degradation of material that results from repeated impacts of solid particles, carried by a fluid, on the target material called solid particle erosion. When hard particles are entrained in a fluid flowing at any significant velocity, significant erosion rates may take place. This material loss phenomenon can be found in many industrial applications such as slurry transportation, aerospace industry, and oil and gas production. In oil and gas production, solid particles can be produced from direct release from reservoirs or from scale deposits on internal pipe walls. Such particles cause severe erosion damage to production tubing, pipelines, fittings and valves, mainly due to the presence of solid particles entrained in liquid and gas high velocity flow. Solid particle erosion is a serious concern to oil and gas companies not only because of metal loss it causes, but also because it has a major influence on corrosion and corrosion protection systems. Surface coverage by corrosion product films such as iron carbonate or magnetite; and by inhibitor films can be reduced through solid particle erosion leading to high corrosion rates. Thus, solid particle erosion has been a hot research topic for oil and gas producers.

The parameters that solid particle erosion depends on are commonly classified into three categories: particle impact information (impact speed and angle), properties of erodent particle and target material properties.

Particles moving at high velocities have more potential to degrade the surface than particles moving at lesser velocities, since high velocity particles have higher kinetic energies than low velocity particles. Finnie [2] showed that erosion rate (*ER*) is proportion to the square of the particle impact velocity (*Vp*). To date, many studies have been performed to find the velocity exponent (n), and most of the findings suggest a value between 2.0 and 3.0 [3, 4].

$$ER \propto Vp^n \tag{2}$$

Particle hardness, shape and size are important parameters in erosion. Mills et al. [5] showed that erosion in 90 degree tube bend is much higher for 230 μm particle size than for 70 μm particles.

Levy [6] found that erosion rates of 1020 steel is increased with increasing erodent particles hardness and also found that the erosion rate is very low for the soft and weak erodent materials such as calcite and apatite.

Levy also claimed that angular steel grit particles caused much sharper, deeper craters that increased the number of platelets (ductile extrusions); whereas, the spherical steel shot caused more shallow, rounded craters that did not produce platelets as effectively as the angular steel grit. Thus, the angular particles appear to be more effective in causing erosion of ductile materials than spherical particles [6]. Current experimental studies, however, revealed that there are some inter-relations between some of these parameters, and the angle function may depend on impact velocity and shape of the particle [2, 3].

Properties of target material including hardness, and ductility are also significant influential factors in erosion. Brittle materials such as iron carbonate exhibit different behavior than ductile materials such as carbon steel. Typical erosion behavior of ductile and brittle materials as function of particle impact angle is shown in Fig. 1 [4]. Erosion of ductile material increases with the impact angle up to a maximum point (approximately between 15 and 30 degrees) and then coasts down to a certain value at 90° impact angle. On the other hand, erosion of brittle materials increases with impact angle, and the maximum erosion is obtained at 90° impact angle.

Materials with different hardness were also found to have different impact velocity exponents, and thereby different erosion behaviors were obtained [7]. Thus, the erosion ratio (ER) which is the ratio of target mass loss to the mass of impinged particle is expressed as:

$$ER = \frac{Mass\ Loss\ of\ Target\ Material}{Mass\ of\ Erodent\ Particles} = f(Vp, \theta, Hv, \rho, D, Fs) \quad (3)$$

where: Vp and θ are particle impact speed and angle, Hv and ρ are target material hardness and density, D and Fs are particle size and sharpness factor, respectively.

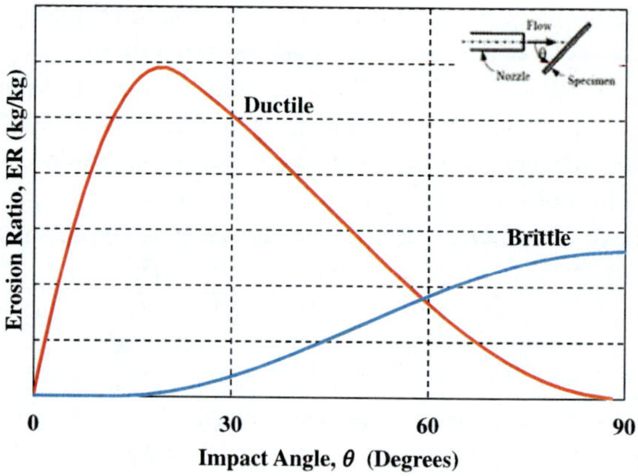

Fig. 1 Typical erosion behavior of ductile and brittle materials versus impact angle [4]

2.1 Erosion Modeling

In oil and gas industry, the American Petroleum Institute Recommended Practice 14E (API RP 14E) is used by some companies for guidance in the performance of many functions within the industry. The API RP 14E proposed a correlation for erosional velocity, V_e (in ft/s) for gas–liquid mixtures as follows,

$$V_e = \frac{c}{\sqrt{\rho_m}} \tag{4}$$

where c is an empirical constant and ρ_m is the gas/liquid mixture density in lb/ft^3. According to API RP 14E, for "solid-free fluids values of $c = 100$ for continuous service and $c = 125$ for intermittent service. In cases when corrosion is not anticipated, c-factor could raise up to $c = 250$. The origins for the development of API correlation is not clear, and many authors believe that the correlation was developed based on liquid impact erosion [8]. However, there is no experimental or theoretical evidence supporting this idea. The erosional velocity calculated from this correlation seems to be very conservative as compared to the experimental data from literature [9, 10].

Sand erosion has been studied extensively in the literature. Salama et al. [11] developed a correlation to predict the erosion rate in elbows as given in Eq. (5). Such correlation account for sand flow rate, flow velocity, pipe dimension and material properties in the calculation of erosion rate. Salama proposed a threshold velocity given by Eq. (6) for carbon steel material (P (Hardness) = 1.55×10^{05}) by solving this correlation for 10 mils per year erosion rate. Salama claimed that API RP 14E equation can only be used when sand flow rate (W) is approaching zero.

$$h = 93,000\frac{WV^2}{Pd^2} \tag{5}$$

$$V \cong \frac{4d}{\sqrt{W}} \tag{6}$$

where h is the erosion rate (mils per year, mpy), W is the sand flow rate (bbl/month), V is fluid flow velocity (ft/sec), P is the hardness (psi) and d is the pipe diameter (inches).

In addition to Salama's work, many researchers have attempted to develop erosion prediction equations and to improve the usefulness of the API 14E equation involving single-phase flow applications [12–18]. These models were developed based on limited experimental data and are only applicable for certain conditions. Salama [16] recently suggested a correlation, given in Eq. (7), for the API 14E erosional velocity to be used for sand-loaded fluid that could be applied to multiphase flow. A value of 400 for the C constant in Eq. (7) for solid-free noncorrosive fluids was suggested.

$$V_e = \frac{D\sqrt{\rho_m}}{20\sqrt{W}} \tag{7}$$

where: V_e is the erosional velocity limit in m/s. W is the sand flow rate in kg/day.

Shirazi and co-worker [19] at the Erosion/Corrosion Research Center (E/CRC) of the University of Tulsa developed more comprehensive and mechanistic model known as Sand Production Saver (SPPS). This model calculates the erosion rate for multiphase flows, for several flow geometries, and for several engineering materials. An equation was developed for calculating the erosion ratio as a function of particle shape, impingement velocity, impingement angle and the hardness of the target material. The E/CRC erosion ratio equation is given by Eq. (8).

$$ER = AF_s V_p^n F(\theta) \tag{8}$$

where *ER* is the erosion ratio defined as the ratio of the mass of target material removed to the mass of impacting particles; and A is an empirical constant depends on the properties of target material, which is calculated as a function of Brinell hardness (*BH*) of the material as in Eq. (9). Fs is a particle shape coefficient that equals to 1.0 for angular sharp particles, 0.53 for semi-rounded particles, or 0.2 for fully rounded sand particles; n is an empirical constant. E/CRC suggested that $n =$ 2.41 but others have suggested values between 1.5 and 3. V_p is the particle impact velocity. Computation Fluid Dynamics (CFD) and Particles Image Velocimetry (PIV) are two common methods currently used for obtaining particle impact velocity. $F(\theta)$ is the angle function that is given by Eq. (10). θ_0, *a, b, w, x, y, z* are target material dependent constant, which for carbon steel 1018 are equal to 15°, −38.4, 22.7, 1, 3.147, 0.3609, and 2.532 respectively.

$$A = 1559BH^{-0.59} \times 10^{-9} \tag{9}$$

$$F(\theta) = \begin{cases} a\theta^2 + b\theta & (for\theta \le \theta_0) \\ xcos^2\theta\sin(w\theta) + ysin^2\theta + z & (for\theta > \theta_0) \end{cases} \tag{10}$$

The E/CRC erosion ratio model undergoes a constant improvement and validation process through experimental data and CFD predations. Arabnejad and co-workers [20] at E/CRC improved the aforementioned model by adding semi-mechanistic erosion equation that utilizes empirical constants for different erodent particles. In this study, the angle function was modified, as it showed better agreement with broader range of experimental data. According to such model, the erosion ratio is calculated by Eqs. (11) and (12).

$$\mathrm{ER_C} = \begin{cases} \mathrm{C_1F_S}\frac{\mathrm{U^{2.41}\sin(\theta)[2K\cos(\theta)-\sin(\theta)]}}{\mathrm{2K^2}} & \theta < \tan^{-1}(\mathrm{K}) \\ \mathrm{C_1F_S}\frac{\mathrm{U^{2.41}COS^2(\theta)}}{2} & \theta > \tan^{-1}(\mathrm{K}) \end{cases} \tag{11}$$

$$ER_D = C_2 F_S \frac{(U \sin\theta - U_{tsh})^2}{2} \quad (12)$$

where C_1, C_2, K and U_{tsh} (m/s) are empirical constants for the erosion equation which for carbon steel 1018 are equal to 5.9e−8, 4.25e−8, 0.5 and 5.5, respectively. ER_C and ER_D are cutting and deformation erosions, respectively; while the total erosion is the summation of ER_C and ER_D [20].

Nassef and co-workers [21] at E/CRC studied the erosion behavior of iron carbonate ($FeCO_3$) scale formed on carbon steel in wake of CO_2 corrosion, and erosion model for predicting erosion magnitude of $FeCO_3$ scale was developed. Two kinds of erodent particles were investigated including sand and $CaCO_3$. A modified form of E/CRC model as given in Eq. (13) and (14) was fitted into dry erosion data as shown in Fig. 2.

$$ER(g/g) = K{*}F_s{*}V_p^{2.41}{*}F(\theta) \quad (13)$$

$$F(\theta) = (\sin\theta)^{n1} * \left(1 + Hv^{n3}(1 - \sin\theta)\right)^{n2} \quad (14)$$

where *ER* is erosion ration, and $F(\theta)$ is angle function and. Hv, K, F_s and V_p are material hardness, erosion factor, sharpness factor and particle impact velocity respectively which are equal to 3.089 GPa, 5.2e−7 for sand, 1.61e−7 for $CaCO_3$, 0.5 for

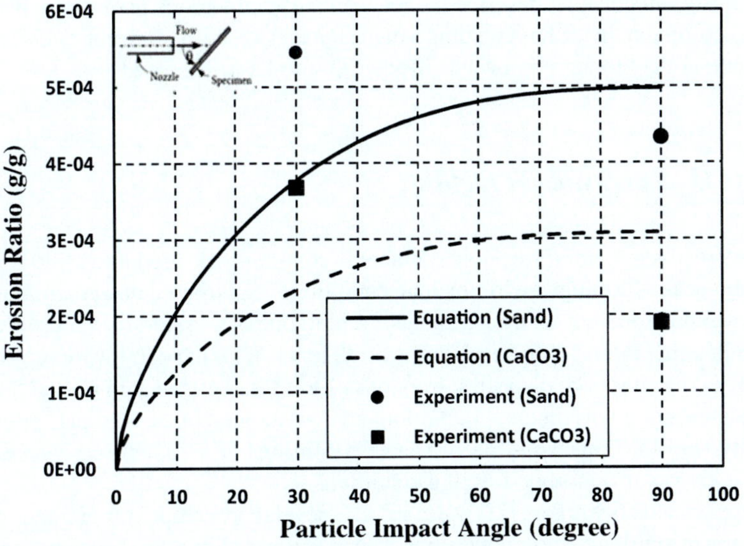

Fig. 2 Comparison of the erosion model and data for sand and $CaCO_3$ on iron carbonate scale under dry condition [21]

semi-rounded sand and 1 for sharp $CaCO_3$ and 23 m/s. The empirical constant n_1, n_2, n_3 used are equal to 0.65, 0.3, 0.4 respectively. The fitted erosion equation was then implemented into the CFD code to predict erosion for wet submerged condition. The predicted erosion magnitude was approximately one order-of-magnitude less than experimental data. Therefore, the K factor used in Eq. (13) had to be significantly modified (increased by a factor of 10) to obtain the final erosion model for iron carbonate scale in submerged liquid flow conditions.

3 CO_2 Corrosion

In oil and gas production, formation water containing carbon dioxide and other species such as hydrogen sulfides and other organic acids always accompany produced fluid. These species can negatively affect the integrity of carbon steel material through the corrosion process. Corrosion of carbon steel due to exposure to corrosive media containing CO_2 gas is known as CO_2 corrosion or sweet corrosion. Although this has been known for more than 100 years ago, internal corrosion of carbon steel pipelines and other facilities due to the exposure of CO_2 containing environments still represents a challenge to the oil and gas industry.

The importance and costly effects of CO_2 corrosion has led many investigators to address this type of corrosion and understand its mechanism. Many empirical, semi-empirical and mechanistic models were developed. A comprehensive review for these models is provided by Fosbøl et al. [22]. The review included more than 20 mechanistic models that describe the mechanisms and various parameters involved in CO_2 corrosion. In the next section, a mechanistic CO_2 corrosion model, commonly available in the literature known as SPPS-CO_2 model is discussed.

3.1 CO_2 Corrosion Modeling

CO_2 corrosion of carbon steel environment is commonly divided into scale-forming and non-scale forming environmental conditions. Environmental conditions that promotes precipitation of iron carbonate scale; known as scale-forming; is generally high temperature, 180°F to 200°F (82 °C to 93 °C), high pH, low flow velocity, and high concentration of species; iron ions (Fe^{+2}), and carbonate irons (CO_3). On the other hand, environmental conditions of low temperatures, low pH, high flow velocity, and low concentration of species; iron ions (Fe^{+2}), and carbonate irons (CO_3), known as non-scale forming conditions.

Dayalan and co-workers [23, 24] at E/CRC developed a model for predicting CO_2 corrosion in multiphase flow conditions. The model is steady-state model that predicts corrosion rate at equilibrium conditions for scale-forming, and non-scale forming conditions. For non-scale-forming conditions, the model describes CO_2 corrosion by four steps, listed in Table 1. The first step is the dissolution of CO_2 in the aqueous

Table 1 Schematic of steps in CO_2 corrosion process for non-scale-forming conditions [23]

Step I: Formation of reactive species in the bulk	*Step III: Electrochemical reactions at the surface*
$CO_2 + H_2O = H_2CO_3$ (15) $H_2CO_3 = H^+ + HCO_3^-$ (16) $HCO_3^- = H^+ + CO_3^{2-}$ (17)	Cathodic reactions $2H_2CO_3 + 2e = H_2 + 2HCO_3^-$ (21) $2HCO_3^- + 2e = H_2 + 2CO_3^{-2}$ (22) $2H^+ + 2e = H_2$ (23) Anodic iron dissolution $Fe = Fe^{+2} + 2e$ (24)
Step II: Transportation of reactants (bulk to surface)	*Step IV: Transportation of products (surface to bulk)*
$H_2CO_3(bulk) \rightarrow H_2CO_3(surface)$ (18) $HCO_3^-(bulk) \rightarrow HCO_3^-(surface)$ (19) $H^+(bulk) \rightarrow H^+(surface)$ (20)	$Fe^{+2}(Surface) \rightarrow Fe^{+2}(bulk)$ (25) $CO_3^{-2}(Surface) \rightarrow CO_3^{-2}(bulk)$ (26)

solution to form the various reactive species that plays role in the CO_2 corrosion reaction. The second step is the mass transfer of these reactant species from the bulk of the solution to the metal surface. The third step involves the electrochemical anodic dissolution of iron and cathodic reactions on steel surface. The fourth step involves mass transfer of the products of the corrosion reaction (Fe^{+2}, CO_3^{-2}) to the bulk of the solution. The overall rate of corrosion process is dependent on the rate at which each of the four steps take place for given environmental conditions.

Based on Table 1, three computational steps were created for CO_2 corrosion prediction program. The first step involves computation of the bulk concentrations of the various species which take part in the corrosion reaction. This step is conducted in the following manner.

The amount of CO_2 which dissolves in the aqueous phase at any given conditions of CO_2 partial pressure and temperature is calculated using Henry's Law.

$$CO_2(gas) = CO_2(liquid) \tag{27}$$

$$H = \frac{P_{CO2}}{[CO_2]} \tag{28}$$

where H is Henry's constant, P_{CO2} is the partial pressure of CO_2 in the gas phase, and $[CO_2]$ is the total concentration of CO_2 in the liquid phase.

The equilibrium constants for reactions (15) and (16), K_1 known as first dissociation constant, and K_2 known as the second dissociation constant for carbonic acid are given by:

$$K_1 = \frac{[H^+][HCO_3^-]}{[H_2CO_3]} \tag{29}$$

$$K_2 = \frac{[H^+][CO_3^{-2}]}{[HCO_3^-]} \tag{30}$$

where: $[H_2CO_3]$, $[HCO_3^-]$, $[H^+]$, and $[CO_3^{-2}]$ represent the concentration of each species in equilibrium with all other species. The proton (H^+) concentration is also controlled by water dissociation equilibrium,

$$H_2O = H^+ + OH^- \tag{31}$$

$$K_w = \frac{[H^+][OH^-]}{[H_2O]} \equiv [H^+][OH^-] \tag{32}$$

where the equilibrium constant, K_w is the ionic product of water. Also, the carbon mass balance is done as:

$$[CO_2] = [H_2CO_3] + [HCO_3^-] + [CO_3^{-2}] \tag{33}$$

and the charge balance (electroneutrality) is done as:

$$\sum_i z_i m_i = 0 \tag{34}$$

where z_i is the charge and m_i is the molarity of the ion i. Five mathematical equations are formulated and solved simultaneously from Eqs. (29)–(34) to determine the concentrations of $[H_2CO_3]$, $[HCO_3^-]$, $[H^+]$, $[CO_3^{-2}]$, $[H^+]$, $[OH^-]$. Newton–Raphson iterative method is used for solving non-linear equations.

The second step involves the computation of the mass transfer rates for transportation of reactants/products of corrosion reaction to/from the metal surface (computation of the mass transfer coefficients of the reactants and the products of the corrosion process). In this computation step, the mass transfer rate (V_{mt}) of reactant and product species are given by Eqs. (35) and (36) respectively:

$$V_{mt} = K_{mt}(C_b - C_s) \tag{35}$$

$$V_{mt} = K_{mt}(C_s - C_b) \tag{36}$$

where: K_{mt} is the mass transfer coefficient, and C_b and C_s are the bulk an surface concentrations of species respectively.

It is worth mentioning that during CO_2 corrosion process, the reactant species; $[H_2CO_3]$, $[HCO_3^-]$, and $[H^+]$ are transported from the bulk of the solution to the corroding steel surface. On the other hand, the product species; $[Fe^{+2}]$, and $[CO_3^{-2}]$

are transported in the opposite direction (from corroding steel surface to the bulk of solution). A hydrodynamic relation given by Chilton-Colburn [25], Eq. (37), was used to calculate the mass transfer coefficients (K_{mt}).

$$K_{mt} = \frac{fU(Sc)^{-2/3}}{8} \tag{37}$$

where f is the friction factor, U is the liquid velocity, and Sc is the Schmidt number.

The third and the last step involves the computation of the rates of anodic and cathodic electrochemical reactions at the steel surface. In this step, the model calculate the rate of an electrochemical reaction (V_{ec}) at the steel surface due to any species through Eq. (38).

$$V_{ec} = K_e C_s \tag{38}$$

where K_e is the electrochemical reaction rate constant and C_s is the surface concentration of the species.

The current due to cathodic reactions; Eqs. (21)–(23), and the anodic dissolution reaction of iron; Eq. (24) are calculated using Butler-Volmer equation:

$$i_c = nFK_{e,eq}C_s \exp \frac{-\propto nF(E_{corr} - E_{eq})}{RT} \tag{39}$$

for the cathodic reaction, and

$$i_a = nFK_{e,eq}C_s \exp \frac{(1-\propto)nF(E_{corr} - E_{eq})}{RT} \tag{40}$$

for the anodic reaction. Where i_c is the current density for the cathodic reaction, i_a is the current density of the anodic reaction, n is the number of electrons released or consumed during corrosion process, F is Faraday constant, $K_{e,eq}$ is the electrochemical reaction rate constant at temperature T and potential E_{eq},C_s is the surface concentration of the species undergoing electrochemical reaction, $\propto$ is the transfer coefficient ($\propto$= 0.5 for most reactions), E_{eq} is the equilibrium potential for the particular reaction, R is the gas constant, and T is the absolute temperature.

Three equations are produced from Eq. (39) for each cathodic reaction given by Eqs. (21)–(23), and one equation produced from Eq. (40) for the anodic reaction given by Eq. (24). Thus, the relations for currents i_{HCO_3}, $i_{HCO_3^-}$, i_{H^-} and $i_{Fe^{+2}}$ can be obtained.

Finally, the model calculate corrosion rate using Eq. (41). Such that, the corrosion current (i_{Corr}) is calculated by either adding up individual cathodic currents or anodic currents taking place at the steel surface. The i_{Corr} in amp/cm^2 and corrosion rate in mm/y.

$$Corrosion\ Rate = 1.16 \times 10^4 i_{Corr} \tag{41}$$

The concentrations of ferrous ions (Fe^{+2}) and carbonate ions (CO_3^{-2}) at the corroding carbon steel surface play a major role in CO_2 corrosion process. Iron carbonate ($FeCO_3$) corrosion product layer, also known as iron carbonate scale form on steel surface when the buildup in the concentrations of these two species exceeds the solubility limit of iron carbonate. It precipitates on steel surface according to reaction (42).

$$Fe^{2+} + CO_3^{-2} \rightleftharpoons FeCO_3(s) \tag{42}$$

Iron carbonate layer can reduce corrosion rate by providing a diffusion barrier for the species involved in the corrosion process and by covering a portion of steel surface and protect it against corrosion. Iron carbonate precipitation depending on precipitation rate can be protective or unprotective. When the precipitation rate is faster than the corrosion rate of underlying steel, a dense well attached and protective layer forms, sometimes very thin (~1–10 μm) but still protective. On the other, when the corrosion rate of underlying steel is faster than precipitation rate of iron carbonate, a porous and un-protective layer forms, which sometimes be very thick (10 m–100 μm) and still un-protective [26–28].

Many factors affect the formation of iron carbonate corrosion product layers. The most important one is water chemistry, and the concept of supersaturation, which is defined as:

$$F_{sat} = \frac{C_{Fe^{2+}} C_{CO_3^{-2}}}{K_{sp}} \tag{43}$$

where K_{sp} is the iron carbonate dissolution equilibrium constant [27, 29]. The scale will precipitate when F_{sat} value exceeds unity. The scale can be protective leading to low corrosion rate. This usually takes place at high pH (pH > 6), which is the key precondition for protective layer. It should be noted that this pH is refereeing to the local pH, at steel surface, which is always higher than pH in the bulk, often by one or two pH units. Temperature also has a major influence on iron carbonate scale formation. At high temperatures [T > 60 °C (140 °F)] precipitation proceeds at high rates yield dense, well-attached, and very protective surface layers, even at low supersaturation. On the other hand, low temperature [T < 60 °C (140 °F)] precipitation occurs at lower rates, which give rise to un-protective layers [26–28].

As mentioned earlier, the CO_2 corrosion model, SPPS-CO_2 developed by Daylan and co-workers [23] at E/CRC also accounts for the conditions where $FeCO_3$ scale is present on the surface. The presence of iron carbonate scale on steel works as a barrier for anodic and cathodic reactions leading lower rates of reactions on steel surface. Thus, additional steps were added to the model. The additional steps are cathodic electrochemical reactions on the scale surface, and the mass transfer process for reactants and products between the steel surface and the scale surface. The steps are

Table 2 Schematic of steps in CO_2 corrosion process for scale covered steel [23]

Step I: Formation of reactive species in the bulk	*Step V: Electrochemical reactions at the metal surface*
$CO_2 + H_2O = H_2CO_3$ (44) $H_2CO_3 = H^+ + HCO_3^-$ (45) $HCO_3^- = H^+ + CO_3^{2-}$ (46)	Cathodic reactions $2H_2CO_3 + 2e = H_2 + 2HCO_3^-$ (56) $2HCO_3^- + 2e = H_2 + 2CO_3^{-2}$ (57) $2H^+ + 2e = H_2$ (58) Anodic iron dissolution $Fe = Fe^{+2} + 2e$ (59)
Step II: Transportation of reactants (bulk to scale surface (sc))	*Step VI: Transportation of products (metal surface (me) to scale surface (sc))*
$H_2CO_3(bulk) \rightarrow H_2CO_3(sc)$ (47) $HCO_3^-(bulk) \rightarrow HCO_3^-(sc)$ (48) $H^+(bulk) \rightarrow H^+(sc)$ (49)	$Fe^{+2}(me) \rightarrow Fe^{+2}(sc)$ (60) $CO_3^{-2}(me) \rightarrow CO_3^{-2}(sc)$ (61)
Step III: Transportation of reactants (scale surface (sc) to metal surface (me))	*Step VII: Transportation of products (scale surface (sc) to bulk)*
$H_2CO_3(sc) \rightarrow H_2CO_3(me)$ (50) $HCO_3^-(sc) \rightarrow HCO_3^-(me)$ (51) $H^+(sc) \rightarrow H^+(me)$ (52)	$Fe^{+2}(sc) \rightarrow Fe^{+2}(bulk)$ (62) $CO_3^{-2}(sc) \rightarrow CO_3^{-2}(bulk)$ (63)
Step IV: Electrochemical reactions at the scale surface (sc)	
$2H_2CO_3 + 2e = H_2 + 2HCO_3^-$ (53) $2HCO_3^- + 2e = H_2 + 2CO_3^{-2}$ (54) $2H^+ + 2e = H_2$ (55)	

listed in Table 2 [23]. Additional parameters are considered in the model including: area of steel surface available for anodic and cathodic reactions; the kinetics of cathodic reactions on the surface of scale; additional mass transfer process through scale; and concentrations of various species at the scale surface in addition to their concentrations at the steel surface.

Under any given environmental conditions, when the scale formation has reached a state of equilibrium (steady state), the following conditions can be satisfied.

$$\begin{aligned} &Rate\,of\,mass\,transfer\,of\,reactants(from\,bulk\,to\,scale\,surface)\\ &= Sum\,of\,the\,rates\,of\,cathodic\,reactions\\ &\quad(on\,steel\,surface\,and\,scale\,surface)\\ &= Rate\,of\,anodic\,reaction\\ &= Rate\,of\,mass\,transfer\,of\,products\\ &\quad(from\,steel\,surface\,to\,bulk) \end{aligned} \tag{64}$$

As in the previous part of the model (non-scale forming conditions), the computations for CO_2 corrosion prediction according to this model are conducted in three steps. The first two steps are identically the same at described above. However, many additional electrochemical reactions and mass transfer processes were included in the third step. The mathematical equations formulated in the third step as the following 20 equations:

$$I_{H_2CO_3,me} = A_{me}2FK_{H_2CO_3,me}[H_2CO_3]_{me}10^{-3}EXP\left[\frac{-\beta F}{RT}(E_{corr} - E^{\circ}_{H_2CO_3})\right] \tag{65}$$

$$I_{HCO_3^-,me} = A_{me}2FK_{HCO_3^-,me}\left[HCO_3^-\right]_{me}10^{-3}EXP\left[\frac{-\beta F}{RT}\left(E_{corr} - E^{\circ}_{HCO_3^-}\right)\right] \tag{66}$$

$$I_{H^+,me} = A_{me}2FK_{H^+,me}\left[H^+\right]_{me}10^{-3}EXP\left[\frac{-\beta F}{RT}(E_{corr} - E^{\circ}_{H^+})\right] \tag{67}$$

$$I_{Fe} = A_{me}2FK_{Fe}\left[HO^-\right]_{me}10^{-3}EXP\left[\frac{(2-\beta)F}{RT}(E_{corr} - E^{\circ}_{Fe})\right] \tag{68}$$

$$I_{H_2CO_3,sc} = A_{sc}2FK_{H_2CO_3,sc}[H_2CO_3]_{sc}10^{-3}EXP\left[\frac{-\beta F}{RT}(E_{corr} - E^{\circ}_{H_2CO_3})\right] \tag{69}$$

$$I_{HCO_3^-,sc} = A_{sc}2FK_{HCO_3^-,sc}\left[HCO_3^-\right]_{sc}10^{-3}EXP\left[\frac{-\beta F}{RT}\left(E_{corr} - E^{\circ}_{HCO_3^-}\right)\right] \tag{70}$$

$$I_{H^+,sc} = A_{sc}2FK_{H^+,sc}\left[H^+\right]_{sc}10^{-3}EXP\left[\frac{-\beta F}{RT}(E_{corr} - E^{\circ}_{H^+})\right] \tag{71}$$

$$I_{corr} = I_{Fe} = \left(I_{H_2CO_3,me} + I_{HCO_3^-,me} + I_{H^+,me}\right) + (I_{H_2CO_3,sc} + I_{HCO_3^-,sc} + I_{H^+,sc}) \tag{72}$$

$$\begin{aligned}
&\left(I_{H_2CO_3,me} + I_{HCO_3^-,me} + I_{H^+,me}\right) + (I_{H_2CO_3,sc} + I_{HCO_3^-,sc} + I_{H^+,sc}) \\
&= 2FM_{H_2CO_3,l}10^{-3}\{[H_2CO_3]_B - [H_2CO_3]_{sc}\} \\
&\quad + FM_{HCO_3^-,l}10^{-3}\{[H_2CO_3]_B - [H_2CO_3]_{sc}\} \\
&\quad + FM_{H^+,l}10^{-3}\{\left[H^+\right]_B - \left[H^+\right]_{sc}\}
\end{aligned} \tag{73}$$

$$\begin{aligned}
&I_{H_2CO_3,me} + I_{HCO_3^-,me} + I_{H^+,me} \\
&= 2FM_{H_2CO_3,sc}10^{-3}\{[H_2CO_3]_{sc} - [H_2CO_3]_{me}\} \\
&\quad + FM_{HCO_3^-,sc}10^{-3}\{[H_2CO_3]_{sc} - [H_2CO_3]_{me}\}
\end{aligned}$$

$$+ FM_{H^+,sc}10^{-3}\{[H^+]_{sc} - [H^+]_{me}\} \quad (74)$$

$$I_{Fe} = 2FM_{Fe^{+2},l}10^{-3}\{[Fe^{+2}]_{sc} - [Fe^{+2}]_B\} \quad (75)$$

$$I_{Fe} = 2FM_{Fe^{+2},sc}10^{-3}\{[Fe^{+2}]_{me} - [Fe^{+2}]_{sc}\} \quad (76)$$

$$\begin{aligned} &M_{H_2CO_3,l}10^{-3}\{[H_2CO_3]_B - [H_2CO_3]_{sc}\} \\ &\quad + M_{HCO_3^-,l}10^{-3}\{[H_2CO_3]_B - [H_2CO_3]_{sc}\} \\ &\quad + M_{CO_3^-,l}10^{-3}\{[CO_3^{-2}]_B - [CO_3^{-2}]_{sc}\} = 0 \end{aligned} \quad (77)$$

$$\begin{aligned} &M_{H_2CO_3,sc}10^{-3}\{[H_2CO_3]_{sc} - [H_2CO_3]_{me}\} \\ &\quad + M_{HCO_3^-,sc}10^{-3}\{[H_2CO_3]_{sc} - [H_2CO_3]_{me}\} \\ &\quad + M_{CO_3^-,sc}10^{-3}\{[CO_3^{-2}]_{sc} - [CO_3^{-2}]_{me}\} = 0 \end{aligned} \quad (78)$$

$$K_1 = \frac{[H^+]_{me}[HCO_3^-]_{me}}{[H_2CO_3]_{me}} \quad (79)$$

$$K_2 = \frac{[H^+]_{me}[CO_3^{-2}]_{me}}{[HCO_3^-]_{me}} \quad (80)$$

$$K_w = [H^+]_{me}[OH^-]_{me} \quad (81)$$

$$K_1 = \frac{[H^+]_{sc}[HCO_3^-]_{sc}}{[H_2CO_3]_{me}} \quad (82)$$

$$K_2 = \frac{[H^+]_{sc}[CO_3^{-2}]_{sc}}{[HCO_3^-]_{sc}} \quad (83)$$

$$K_w = [H^+]_{sc}[OH^-]_{sc} \quad (84)$$

where: I refers to the current due to the cathodic reduction of H_2CO_3, HCO_3^-, H^+ species and current due to anodic dissolution of iron. E° is the standard potentials, and E_{corr} is the corrosion potential. The "K" with chemical species as subscript refers to the standard electrochemical reaction rate constants. Subscripts me, and sc refer to at metal (carbon steel) and at the scale. Subscripts B refers to bulk of the solution and l refers to liquid. A is the fraction of the surface available for the electrochemical reactions and has value 0–1. The terms M refer to the mass transfer coefficients of the respective species. F is the Faraday constant, β is the transfer coefficient, R is the universal gas constant, and T is the temperature in Kelvin. K_1 and K_2 are the equilibrium constants for the first and second dissociation of carbonic acid. K_w is the ionic product of water and is the equilibrium constant for the dissociation of water.

Sun et al. [29] developed a correlation for calculating K_{sp} of iron carbonate scale as a function of temperature and ionic strength, given by Eq. (85). This correlation was incorporated into SPPS: CO_2 model in a later study [30], so it accounts for changing temperature and ionic strength on the solubility limit for precipitation and growth of iron carbonate ($FeCO_3$) scale.

$$\log K_{sp} = -59.350 - 0.041T_k - \frac{2.196}{T_k} + 24.572 \times \log T_k + 2.51I^{0.5} - 0.657 \quad (85)$$

where: T_k is temperature in Kelvin and I is the ionic strength in molarity, which can be calculated from Eq. (86).

$$I = \frac{1}{2}\sum_i C_i Z_i^2 \quad (86)$$

where C_i is the concentration of different species in the aqueous solution, all of which are in mol/L, and Z_i are the species charges.

In a recent work by Al-Mutahar and co-workers at E/CRC [28], SPPS: CO_2 model was modified by incorporating equations that account for scale kinetics such as scale precipitation rate, scale growth rate, and diffusion rate of reactants and products through the scale.

The scale precipitation rate was account for using Yean's model [31]. Yean's model was recast to include the solubility product of K_{sp}, CO_3^{-2} concentration, and diffusion boundary layer thickness, h_{bl} [28]. The value of h_{bl} is approximated by pipe roughness. The final form of the precipitation model used in SPPS: CO_2 is given by Eq. (87).

$$\dot{m}\left(\frac{moleFe^{+2}}{m^2 s}\right) = \rho_{liq} h_{bl} \frac{k}{[CO_3^{-2}]}\left([Fe^{+2}][CO_3^{-2}] - K_{sp}\right) \quad (87)$$

where $\dot{m}$ is the precipitation rate of iron. The solubility product, K_{sp}, was given by Eq. (85).

The change in scale thickness was considered in the model using Eq. (88) [28]. This equation calculates the change in scale thickness (Δh) at different time steps as function of the scale precipitation rate, the scale density, and porosity.

$$\Delta h = \frac{\dot{m}\Delta t}{\rho_{FeCO_3}(1 - \varepsilon)} \quad (88)$$

where $\dot{m}$ is the precipitation rate in (kg/m^2.s), ρ_{FeCO_3} is the density of the solid phase of the scale in (kg/m^3), Δt is the time step in seconds and ε is the scale porosity.

The diffusion through iron carbonate scale was account for using Eq. (89). This equation calculates the effective diffusion coefficient (D_{eff}) as function of the diffusion coefficient in the bulk of the solution (D_{bulk}), porosity (ε), tortuosity (τ), and an empirical constant (m) as given by Eq. (89).

$$D_{eff} = D_{bulk}\frac{\varepsilon^m}{\tau} \tag{89}$$

4 Erosion-Corrosion

Erosion-corrosion in oil and gas production systems is caused predominately by solid particles impacting interior of the pipe in a corrosive fluid. Solid particles are usually silica sand, but they could be other solid particles such as calcium carbonate ($CaCO_3$). Erosion-corrosion deterioration of carbon steel in CO_2 environments in presence of solid particles is a common threat to oil and gas production and transportation facilities. The combined effect of erosion and corrosion can decrease the efficiency of corrosion protection methods such as iron-carbonate scale formation or chemical inhibition and cause pitting and severe corrosion. Oil and gas will increasingly be produced from deeper wells in more difficult to reach regions, and higher corrosive and erosive situations will be expected because of higher temperatures, pressures and also higher solid particles production rates [32]. The importance of erosion-corrosion in oil and gas industry has encouraged the E/CRC at the University of Tulsa to conduct research in this area [32–45]. The next section discusses the erosion-corrosion mitigation and modeling.

4.1 Erosion-Corrosion Modeling

As described by Eq. (1), the mechanisms of erosion-corrosion can be explained by understanding the interaction between erosion and corrosion, known as synergism. This results in higher metal loss rates than the sum of the independent erosion and corrosion mechanisms. For carbon steel material, in CO_2 environments, erosion-corrosion mechanisms are described for two conditions; scale-forming and non-scale forming, which have different degrees of synergism. For bare carbon steel erosion and corrosion under non-scale-forming conditions, the interaction between erosion and corrosion mechanisms, if any, is negligible [43–45]. In such cases, erosion-corrosion can be modeled as the summation of bare-metal corrosion and bare metal erosion. In the absence of any synergistic effect between erosion and corrosion, the total penetration rate due to erosion-corrosion can be calculated simply as the summation of individual components. On the other hand, under scale-forming conditions, the interaction between erosion and corrosion processes is very significant and cannot be neglected. Formation of iron carbonate scale provides protection, and reduces the corrosion rate to very small values. However, the presence of solid particles such as sand or $CaCO_3$ in the flow can deteriorate the scale, and increase corrosion to unacceptable levels. Under such conditions the erosion-corrosion is a function of three components: scale formation, under-scale corrosion, and scale erosion [28].

Shadley and co-workers [36–38] at the E/CRC studied erosion-corrosion of iron carbonate scale covered carbon steel in a CO_2 saturated flow with sand. Thirteen sets of environmental conditions were tested using a flow loop with a direct impingement condition. The highest erosivity that the iron carbonate scale could withstand before being partial or totally removed was identified for each set of the environmental conditions. The highest erosivity was defined as the Erosion-Corrosion Resistance (ECR) of the scale. The erosivity can be defined as the erosion rate of the exposed material under the given flow, particle, and geometry conditions, which was calculated for low carbon steel material using previously mentioned SPPS erosion model [19]. The ECR data collected was used to develop a correlation for predicting ECR as a function of temperature, pH and F_{sat}, given by Eq. (43). The ECR correlation is given by Eq. (90).

$$ECR\left(\frac{mm}{y}\right) = C_0 F_{sat}^{C1} f_T f_{pH} \tag{90}$$

where f_T; the temperature function and f_{pH}; the pH function is given by Eqs. (91) and (92) respectively. C_0 through C_7 are empirical constants given in Table 3.

$$\log(f_T) = C_2 e^{C_5(T - C_4)} \tag{91}$$

$$\log\left(f_{pH}\right) = C_5 \tan[C_6(pH - C_7) \tag{92}$$

Using ECR correlation along with E/CRC models mentioned previously for erosion [19] and CO_2 corrosion [23, 24], Shadley et al. [36, 37] developed a model for prediction of threshold velocity (V_{th}) and program was used to predict the threshold velocity for erosion- corrosion penetration rate for carbon steel and protective iron carbonate scale. The threshold velocity (V_{th}) is defined as the maximum flow velocity that the scale can tolerate at a given environmental condition before being partially or totally removed. Thus, for any set of environmental and erosion conditions, threshold velocities can be found in two steps. First, for a given set of test conditions, results

Table 3 Empirical constants in Eqs. (90–92)

C_0(mm/y)	C_1	C_2	C_3(°C^{-1})	C_4	C_5	C_6	C_7
1.038×10^{-3}	1.594	0.7748	0.0504	65.56	3.629	1.560	5.650

from the E/CRC mechanistic model [23, 24] for CO_2 corrosion can be used to plot ECR as a function of flow velocity. Second, erosion parameters in the E/CRC mechanistic erosion model [19] can be used to calculate and plot erosivity as a function of flow velocity. The intersection of the two curves defines the threshold velocity [36, 37]. This model also calculates corrosion rate which can be high or low depending on whether the intended flow velocity is above or below the threshold velocity. Figure 3 shows prediction of threshold velocity for 3.5-in elbow for multiphase flow at a temperature of 57°C (135°F); 64.7$psia$ CO_2 pressure, 5.5 pH, 10 ppm Fe^{+2}; 30 Ibs/day particles rate; 150 μm, sand particle size; 1.2 m/s liquid superficial velocity; and 5 m/s gas superficial velocity. For this case, the predicted threshold velocity is greater than the superficial gas velocity. This indicates that the erosivity is smaller than the ECR. In such case, the model assumes that iron carbonate is able to form throughout the piping interior. On the other hand, when the operating velocity is higher than the threshold velocity, the erosivity is higher than ECR. In these cases, the model assumes that the scale is not able to adhere to piping surfaces, and pitting or uniform corrosion occurs.

Al-Mutahar and co-workers at E/CRC [28] developed a procedure for predicting erosion-corrosion penetration rate known as SPPS-EC by integrating the three E/CRC models; SPPS: CO_2, SPPS and Threshold Velocity model. The first step of SPPS-EC is to check for iron carbonate scale formation. For non-scale forming conditions, the erosion-corrosion penetration rates are assumed to be equal to the summation of erosion and bare metal corrosion rates. For the conditions where scale forms, erosion rate of scale, erosion rate of bare steel, and the rate of scale formation have to be found. If the formation rate of the scale is smaller than the erosion rate, then it is assumed that the scale cannot maintain itself and erosion-corrosion rate is found by the summation of the corrosion rate and erosion rate of bare steel. In such case the flow velocity is greater than the threshold velocity. However, if the formation rate is

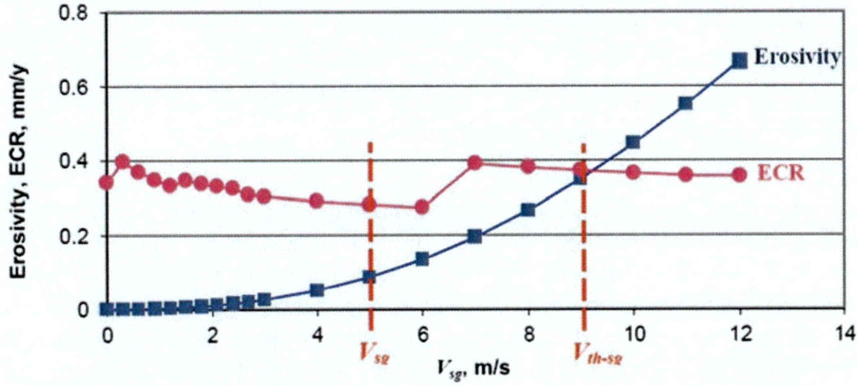

Fig. 3 Prediction of threshold velocity for 3.5-in elbow for multiphase flow at a temperature of 57°C (135°F); 64.7$psia$ CO_2 pressure, 5.5 pH, 10 ppm Fe^{+2}; 30 Ibs/day particles rate; 150 μm, sand particle size; 1.2 m/s liquid superficial velocity; and 5 m/s gas superficial velocity [28]

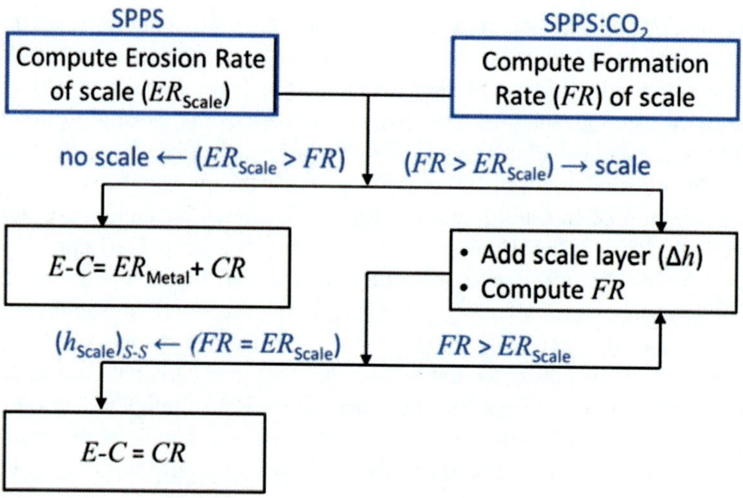

Fig. 4 Computation steps of erosion-corrosion model (SPPS-EC)

greater than the erosion rate of rate of scale, the scale thickness, h, is incremented until the rate of scale formation is equal to the erosion rate of the scale. At this point, the scale thickness is maintained and it is equal to the steady-state scale thickness. Once the steady-state scale thickness is reached, the corrosion rate calculated for an iron carbonate scale of thickness, h, and the erosion rate of the underlying steel is set equal to zero [28]. In this case the flow velocity is below the threshold velocity, and erosion-corrosion penetration rate is assumed to be equal to corrosion rate at the steady state thickness, h. A flow chart illustrating the computation procedure of the erosion-corrosion model is shown in Fig. 4. Where: ER_{Scale} is the erosion rate of scale, ER_{Metal} is the erosion rate of bare steel, h_{Scale} is the scale thickness, and $(h_{Scale})_{s-s}$ is the steady-state scale thickness.

4.2 *Erosion-Corrosion Inhibition*

Chemical inhibitors are commonly utilized to mitigate erosion-corrosion. The performance of corrosion inhibitors, however, are also influenced by flows containing particulates [46–52]. Therefore, prediction of chemical inhibitor effectiveness when solid particles are being produced particularly important. Modeling this behavior is critical when the wells are deep, or off-shore, because coupon testing is impractical and replacement costs are high. McMahon and co-workers [49] claimed that large amounts of corrosion inhibitor can be lost from the bulk solution by adsorption of inhibitor onto the surfaces of produced sand particles. It was also found that this effect only becomes significant for high concentrations of sand and small particle sizes, especially for oil-wetted particles [49]. However, Ramachandran and co-workers

[50] claimed that the effect of inhibitor loss due to sand adsorption is small for industrial corrosion inhibitors used under field conditions. Tandon and co-workers [51] alleged that for CO_2 corrosion under non-scale forming conditions with sand, inhibitor concentrations of 100 ppm and above appeared not to be affected much by the presence of small amounts of sand (<1% wt). Neville and co-workers [52] found that inhibitor that are effective in a CO_2 saturated static brine are not necessarily effective under erosion-corrosion conditions.

Hassani and co-workers at E/CRC [47] have investigated the performance of corrosion inhibitors under sand production for total brine solution, in which it was reported that sand erosion decreases inhibitor efficiency. This decrease in inhibitor performance was attributed to the reduction in surface coverage of the physically adsorbed inhibitor layer on steel surfaces. On the other hand, in a CO_2 saturated solution with 60% oil, Nassef and co-investigators at E/CRC [48] showed that the inhibitor is more effective in minimizing the corrosion part of erosion-corrosion than for the same system without sand. LPR date for this observation is shown in Fig. 5. Such improvement in inhibitor efficiency was likely the result of immediate removal of iron carbide by action of the sand. A layer of iron carbide has been reported to be covering steel surface for the inhibited corrosion system with no sand. They found that this layer of iron carbide (Fe_3C) impairs inhibitor performance by not allowing enough inhibitor to reach the steel surface below the iron carbide [48]. It is worth mentioning that the complete mechanism of this process is not fully understood. More details about such observations as well as inhibited erosion-corrosion modeling are covered in the next section.

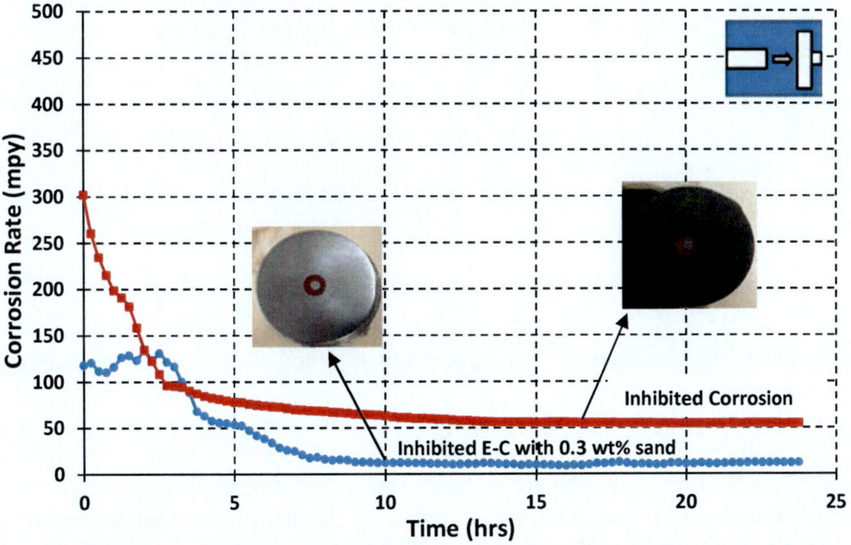

Fig. 5 LPR data comparing inhibited corrosion with no sand and corrosion component of inhibited erosion-corrosion with 0.3% wt, sand concentration for 40% water cut and 200 ppm inhibitor concentration (2 gallons of brine solution) [48, 53]

4.3 Inhibited Erosion-Corrosion Modeling

The interaction of the inhibitor with the metal surface is generally explained by chemical-physical adsorption phenomena of the inhibitor molecule to a metal surface. The adsorbed inhibitor layer retards the electrochemical reaction at the metal surface.

Different inhibitor adsorption phenomenological models have been developed for characterizing the mechanism of corrosion inhibition such as Flory–Huggins [54] Frumkin, Temkin, Langmuir, Freundlich, Bockris-Swinkels, Hill-de Boer, Parsons, Damaskin-Parsons, and Kastening-Holleck models [55]. Such models have the same general form given by Eq. (93).

$$K_{\frac{a}{d}} C_{inh} = G(\theta, n) e^{-f\theta} \tag{93}$$

where $G(\theta, n)$ is a configurational factor, C_{inh} is the inhibitor concentration, θ is the fraction of surface coverage by inhibitor, f represents the interaction between adsorbed inhibitor molecules, and n is the ratio of the inhibitor surface coverage to that of water.

Inhibitor adsorption isotherms can be developed based on corrosion rate measured at different inhibitor concentrations. The adsorption isotherm for a given inhibitor depends on the interaction of an inhibitor molecule with the metal surface, and on the mutual interaction between adsorbed inhibitor molecules. In the Langmuir isotherm, the adsorption/desorption of inhibitor molecules is considered while the mutual interaction between inhibitor molecules is negligible [55]. Based on Langmuir isotherm, there is a linear relationship between inhibitor concentration and theta/(1-theta), as per Eq. (86). The proportionality constant $K_{\frac{a}{d}}$ is known as the adsorption/desorption constant.

$$\frac{\theta}{(1-\theta)} - K_{\frac{a}{d}} C_{inh} \tag{94}$$

The surface coverage by inhibitor is calculated by Eq. (95).

$$\theta = 1 - \frac{CR_{\theta}}{CR_{\theta=0}} \tag{95}$$

CR_{θ} is the measured inhibited corrosion rate and $CR_{\theta=0}$ is the uninhibited corrosion rate.

In the Frumkin isotherm, however, the mutual interaction between adsorbed inhibitor molecules is important and both mutual interaction as well as the effect of adsorption/desorption of inhibitor molecules are taken into consideration [55]. Based on the Frumkin isotherm, there is an exponential relationship between the inhibitor concentration and the surface coverage by inhibitor, and the isotherm equation is given by Eq. (96).

$$K_{\frac{a}{d}} C_{inh} = \left(\frac{\theta}{1-\theta}\right) e^{-f\theta} \quad (96)$$

where $K_{\frac{a}{d}}$ is the adsorption/desorption constant, f is an interaction parameter, θ is the surface coverage of the inhibitor, and C_{inh} is the concentration of the inhibitor. In a mechanistic interpretation, f can either be a negative or positive value depending on the degree of intermolecular interaction between inhibitor molecules that are absorbed on a metal surface. A negative value for f is used when the adsorbed molecules attract each other, and a positive value is used when the molecules repel each other. The surface coverage,θ is defined as in Eq. (95).

Based on studies at E/CRC [47, 53], both Langmuir and Frumkin isotherms were found to be good models for developing a correlation for prediction of corrosion rate as a function of inhibitor concentration. Frumkin isotherm, however, was found to be more representative model than Langmuir isotherm, because it considers the interaction between adsorbed inhibitor molecules as well as the adsorption/desorption of inhibitor molecules. The Frumkin isotherm also resulted in a better fit with experimental data [47, 53].

Figure 6 shows Frumkin isotherm fit to inhibited corrosion versus inhibited erosion-corrosion data collected by Hassani [47]. These experimental data was collected in a CO_2 saturated flow loop with direct impingement geometry under

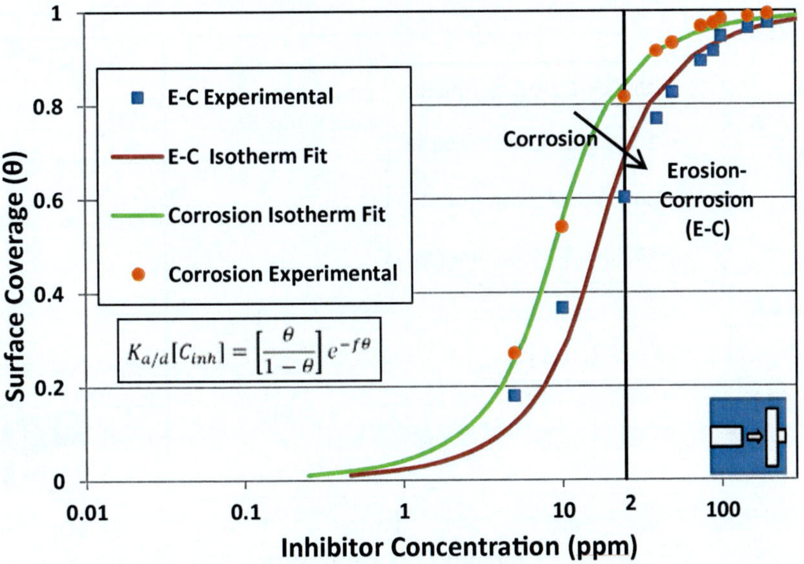

Fig. 6 Inhibitor performance for corrosion versus erosion-corrosion under non-scale forming conditions [Temperature of 57 °C (135 °F), 20 psia CO_2 gas, 32.6 ft/s (9.9 m/s) flow velocity, 0.3% wt sand concentration, 150 μm particle size, and varies concentrations of imidazoline-based inhibitor] [47]

non-scale forming conditions [Temperature of 57 °C (135 °F), 20 psia CO_2 gas, 32.6 ft/s (9.9 m/s) flow velocity, 0.3% wt sand concentration, 150 μm particle size, and varies concentrations of imidazoline-based inhibitor]. As shown in Fig. 6, addition of sand reduced inhibitor efficiency. This can obviously be demonstrated by the black vertical line drawn at 25 ppm inhibitor concentration. The surface coverage with no sand is higher than coverage with sand.

Under same environmental conditions discussed in Fig. 6, Nassef [48, 53] found that addition of oil into inhibited erosion-corrosion system, resulted in improvement of inhibitor efficiency. Figure 7 shows Frumkin isotherm fit to inhibited erosion-corrosion for 100% water cut [47] versus 40% water cut. The mechanism of this added protection due to presence of oil phase was explained by the changes in the wettability of the carbon steel surface [56]. Adsorption of inhibitor on steel surface, makes steel surface more hydrophobic, thereby increasing oil wettability. Increasing oil wettability of the surface, makes it harder for water to come into direct contact with the steel. Decreasing the likelihood of corrosion.

Nassef and co-workers at E/CRC [53, 57–59] also studied the performance of imidazoline-based inhibitor under iron carbonate scale forming conditions [Temperature of 80 °C (175 °F), 20 psia CO_2 gas, 14.0 ft/s (4 m/s) flow velocity, and 0.4% wt particles concentration]. Two kinds for particles were investigated including 150 μm sand and 250 μm calcium carbonate ($CaCO_3$) particles using CO_2 saturated flow loop under direct impingement configuration. Figure 8 shows Frumkin isotherm fit

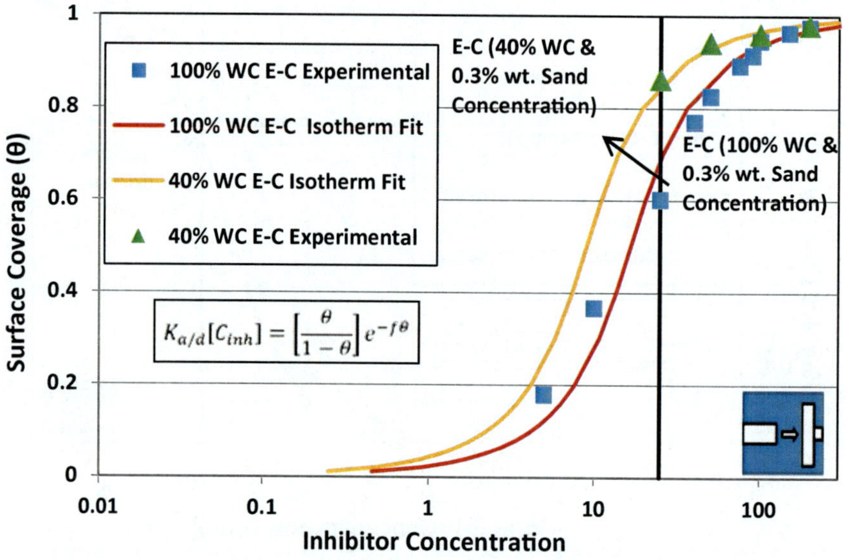

Fig. 7 Inhibitor performance for 40% versus 100% water cut with 0.3% wt sand concentration. Data for 100% water cut is from Hassani and co-worker [47] and data for 40% water cut is from Nassef and co-workers [53]. Solid lines represent fits for the Frumkin isotherm model at each water cut. Symbols represent experimentally measured surface coverage

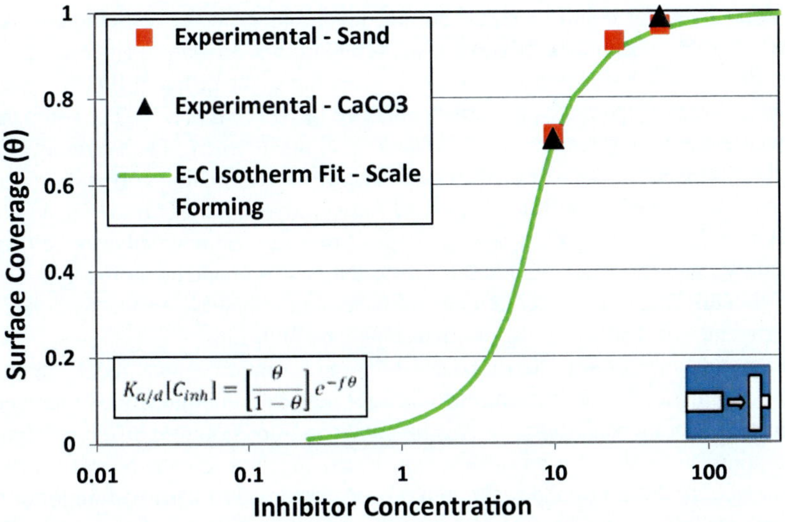

Fig. 8 Inhibitor performance for sand versus $CaCO_3$ particles. Solid line represents fit for the Frumkin isotherm model with both kinds of particles. Symbols represent experimentally measured surface coverage [53, 57–59]

for sand and $CaCO_3$ particles. In Fig. 8 , it can be pointed out that surface coverage by inhibitor of $CaCO_3$ particles system is almost matching with the coverage measured for sand. Thus, both kinds of particles, which represent different errosivities resulted in comparable inhibitor effectiveness.

5 Conclusion

Erosion-corrosion is a complex phenomenon to model. It becomes more complicated when corrosion process resulted in formation of corrosion products such as iron carbonate scale. This chapter discusses a procedure for predicting erosion-corrosion penetration rate known as SPPS: E-C by integrating the three E/CRC models including SPPS: CO_2, SPPS and Threshold velocity model. This procedure predicts erosion-corrosion based on the environment tendency to form iron carbonate scale. For environmental conditions that are favorable for scale formation, erosion-corrosion is modeled for two scenarios based on the rate of scale formation, and scale erosion rate. In case when scale formation is smaller than scale erosion rate, it is assumed that the scale cannot adhere to steel surface, and erosion-corrosion rate equals to the summation of corrosion rate and erosion rate of base steel. In case when scale formation is greater than scale erosion rate, however, erosion-corrosion is approximated by the corrosion rate when scale is protecting steel surface. On the other hand for conditions that are not favorable for scale formation, erosion-corrosion rate is assumed to be equal to the summation of erosion and bare steel corrosion. Erosion-corrosion measurements obtained through a different E/CRC research projects, were compared to predictions obtained by SPPS: E-C [28]. The comparison showed encouraging results. The surface condition in terms of scale removal was predicted correctly for most of the cases. The predicted erosion-corrosion rates for those cases were within a factor of three of measured data [28]. This model, however, can be modified by adding new scale deposition rate, and new values of diffusion boundary layer thickness obtained for different environmental conditions to SPPS-CO_2 model. Thus, the model can be validated, and enhanced further by collecting experimental data covering varies environmental conditions.

Frumkin adsorption isotherm is discussed as phenomenological model, which can be used to model the performance of imidazoline-based inhibitor for different environmental conditions. Different environmental conditions showed different inhibitor adsorption properties. Based on Frumkin isotherm, the presence of sand particles can reduce inhibitor efficiency through physical removal of adsorbed inhibitor film of steel surface as well as through partial adsorption of inhibitor on solid particles. Frumkin isotherm also indicted that addition of oil into non-oil system at same erosivity and environmental conditions improved inhibitor efficiency. This added protection in the presence of an oil phase took place as a result of oil wetting enhanced by the presence of inhibitor, i.e. inhibitor makes the steel surface hydrophobic, thereby favoring an oil layer near the surface rather than a water layer. Research discussed in this chapter, also show that Frumkin isotherm can be modified based on erosivity caused by sand and $CaCO_3$ particles. The modified Frumkin isotherm showed that both kinds of particles resulted in comparable surface coverage by inhibitor, thus comparable inhibitor performance [53]. Varies modified Frumkin adsorption isotherms can be obtained to cover different environmental conditions, erosivities, and different kinds of inhibitors.

The mechanistic models developed by E/CRC including SPPS, SPPS: CO_2, SPPS: E-C, and the modified Frumkin isotherms can be used together to predict total erosion-corrosion rate occurs in oil and gas production industry at varies operating conditions. Erosion-corrosion and inhibitor performance modeling is an important part of solid production management during operation in the oil and gas industry. It can be used to predict erosion-corrosion rates, inhibitor performance and required dosage of inhibitor in difficult to reach regions such as offshore or deep wells when coupon testing is impractical. Erosion-corrosion and Inhibitor performance modelling is also used to guide and improve inspection and maintains activities, as a result the cost of inspection, maintenance, and material failure can be reduced.

References

1. S. Shirazi, B. McLaury, J. Shadley, K.P. Roberts, E.F. Rybicki, H.E. Rincon, S. Hassani, F.M. Al-Mutahar, G.H. Al-Aithan, Erosion-corrosion in oil and gas pipelines, in *Oil and Gas Pipelines:*

Integrity and Safety Handbook. First edition (John Wiley & Sons Inc., 2015), p. 399
2. I. Finnie, Erosion of surfaces by solid particles. Wear **3**(2), 87–103 (1960)
3. A. Mansouri, H. Arabnejad, S. Shirazi, B. McLaury, A combined CFD/experimental methodology for erosion prediction. Wear **332, 333**, 1090–1097 (2015)
4. H. Arabnejad, Development of erosion equations for solid particle and liquid droplet impact, in *Mechanical Engineering* (The University of Tulsa, Tulsa, OK, 2015)
5. D. Mills, J. Mason, Particle size effects in bend erosion. Wear **44**(2), 311328 (1977)
6. A. Levy, Solid particle erosion and erosion-corrosion of materials. ASM Int. (1995)
7. Y.I. Oka, M. Matsumura, T. Kawabata, Relationship between surface hardness and erosion damage caused by solid particle impact. Wear **162164**, 688–695 (1993)
8. M. Salama, E. Venkatesh, Evaluation of API RP 14E erosional velocity limitations for offshore gas wells, in *Offshore Technology Conference* (1983)
9. A. Thiruvengadam, S.L. Rudy, Experimental and analytical investigations on multiple liquid impact erosion. Hydronaut. INC laurel MD (1969)
10. D.W.C. Baker, K.H. Jolliffe, D. Pearson, The resistance of materials to impact erosion damage. Philos. Trans. Royal Soc. London. Ser. A. Math. Phys. Sci. **260**(1110), 193–203
11. M. Salama, E. Venkatesh, *Evaluation of Erosional Velocity Limitations in Offshore Gas Wells, 15th Annual OTC*. (Houston, Texas, 1983), 2–5 May, OTC Number 4485
12. M. Salama, E. Venkatesh, *Evaluation of Erosional Velocity Limitations in Offshore Gas Wells, 15th Annual OTC* (Houston, Texas, 1983), May 2–5, OTC Number 4485
13. S. Shirazi, B. McLaur, J. Shadley, E. Rybicki, Generalization of the API RP 14E guideline for erosive services. J. Pet. Technol. (Distinguished Author Series), 693–698 (1995)
14. B. McLaury, S. Shirazi, Generalization of API RP 14E for erosive service in multiphase production, in *SPE Annual Technical Conference and Exhibition* (Society of Petroleum Engineers, Houston, TX, 1999), 2, p. 10
15. K. Jordan, Erosion in multiphase production of oil & gas, in *Corrosion 98, Paper No. 58, NACE International Annual Conference* (San Antonio, 1998)
16. M. Salama, Erosional velocity limits for water injection systems, in *Corrosion 93, Paper No. 62, NACE International Annual Conference* (1993)
17. A. Bourgoyne, Experimental Study of erosion in diverter systems due to sand production. Paper presented at the 1989 SPE/IADC Drilling Conference, New Orleans, SPE/IADC 18716, 1989
18. S. Svedeman, K. Arnold, Criteria for sizing multiphase flow lines for erosive/corrosive service. Paper presented at the 1993 SPE Conference, Houston, SPE 26569
19. S. Shirazi, B. McLaury, J. Shadley, E. Rybicki, Generation of the API RP 14E guidelines for erosive services. J. Petrol. Technol. **47**, 693–698 (1995)
20. H.K. Arabnejad, Development of erosion equations for solid particle and liquid droplet impact. Ph.D. Dissertation, Department of Mechanical Engineering, The University of Tulsa, OK, 2015
21. A. Nassef, N. Banazadeh-Neishabouri, M. Keller, K. Roberts, E. Iski, E. Rybicki, S. Shirazi, Comparison of erosion resistance of iron carbonate protective layer with calcium carbonate particles versus sand. ADIPEC 2017, Paper No. 188531-MS SPE International, Abu Dhabi, UAE, 2017)
22. P. Fosbøl, K. Thomsen, E. Stenby, Improving mechanistic CO_2 corrosion models, in *Corrosion 2009* (NACE International, Atlanta, GA, 2009)
23. E. Dayalan, J. Shadley, E. Rybicki, S. Shirazi, F. De Moraes, CO_2 corrosion prediction in pipe flow under $FeCO_3$ scale-forming conditions, in *Corrosion/98* (NACE International, San Diego, CA, 1998), p. 137
24. E. Dayalan, G. Vani, J. Shadley, S. Shirazi, E. Rybicki, Modeling CO_2 corrosion of carbon steels in pipe flow, in *Corrosion/95* (NACE International, Houston, TX, 1995)
25. T. Chilton, A. Colbum, Mass transfer (absorption) coefficients, prediction from data on heat transfer and fluid friction. Ind. Eng. Chem. **26**(11), 1183–1187 (1934)
26. A. Kahyarian, M. Achour, S. Nesic, CO_2 corrosion of mild steel, in *Trends in Oil and Gas Research and Technologies*, ed. by A. El-Sherik (Woodhead Publishing, Cambridge, USA, 2017), pp. 149–190

27. F. Farelas, B. Brown, S. Nesic, Iron carbide and its influence on the formation of protective iron carbonate in CO_2 corrosion of mild steel, in *CORROSION 2013*, Paper No. 2291 (NACE International, Houston TX, 2013)
28. F. Al-Mutahar, Erosion-corrosion for carbon steel in sweet production with sand: modeling and experiments, in *Mechanical Engineering, 2012* (University of Tulsa, Tulsa, OK, 2012)
29. W. Sun, S. Nesic, R.C. Woollam, The effect of temperature and ionic strength on iron carbonate (FeCO3) solubility limit. Corros. Sci. **51**(6), 1273–1276 (2009)
30. S. Hassani, K. Roberts, S. Shirazi, J. Shadley, E. Rybicki, C. Joia, Flow Loop study of chloride concentration effect on erosion, corrosion and erosion-corrosion of carbon steel in CO_2 saturated systems, in *Corrosion 2011* (Houston, TX, 2011)
31. S. Yean, H. Saiari, A. Kan, M. Tomson, *Ferrous Carbonate Nucleation and Inhibition, SPE International Oilfield Scale Conference, Aberdeen* (Society of Petroleum Engineers, UK, 2008)
32. K. Tummala, *Effect of Sand Production on Chemical Inhibition of CO2 Corrosion of Carbon Steel for Iron Carbonate Scale Forming Conditions at High Flow Rates, in Mechanical Engineering* (University of Tulsa, Tulsa, 2009)
33. M. Tandon, K.P. Roberts, J.R. Shadley, S. Ramachandran, E.F. Rybicki, V. Jovancicevic, Flow loop studies of inhibition of erosion-corrosion in CO_2 environments with sand, in *Corrosion. 2006* (San DIego, CA, 2006)
34. K. Dave, *Effect of Sand Production on Corrosion Inhibition in the Oil and Gas Industry, in Mechanical Engineering* (Univesity of Tulsa, Tulsa, 2007)
35. K. Dave et al., Effect of a corrosion inhibitor for oil and gas wells when sand is produced. (2008), p. 085701–0857014
36. J.R. Shadley, S.A. Shirazi, E. Dayalan, E.F. Rybicki, Prediction of erosion-corrosion penetration rate in a carbon dioxide environment with sand. Corrosion **54** (1998)
37. J.R. Shadley, S.A. Shirazi, E. Dayalan, E.F. Rybicki, Prediction of erosion-corrosion penetration rate in a CO_2 Environment with Sand, in*Corrosion. 1998* (NACE International, 1998)
38. J.R. Shadley, S.A. Shirazi, E. Dayalan, M. Ismail, E.F. Rybicki, Erosion-corrosion of a carbon steel elbow in a carbon dioxide environment, in *CORROSION. 1995* (NACE International, 1995)
39. B.S. McLaury, S.A. Shirazi, J.R. Shadley, E.F. Rybicki, How erosion-corrosion patterns in a choke change as material losses in the choke progress, in *CORROSION. 2006* (NACE International, 2006)
40. J. Chen, J.R. Shadley, E.F. Rybicki, Activation/re-passivation behavior of 13Cr in CO_2 and sand environments using a modified electrochemical noise technique, in *CORROSION 2002* (NACE International, 2002)
41. J. Chen, J.R. Shadley, H. Rincon, E.F. Rybicki, Effects of temperature on erosion-corrosion of 13Cr, in *CORROSION. 2003* (NACE International, 2003)
42. H.E. Rincon, J.R. Shadley, K.P. Roberts, E.F. Rybicki, Erosion-corrosion of corrosion resistant alloys used in the oil and gas industry, in *CORROSION. 2008* (NACE International, 2008)
43. H.E. Rincon, Testing and prediction of erosion-corrosion for corrosion resistant alloys used in the oil and gas production industry, in *Mechanical Engineering* (University of Tulsa, Tulsa)
44. S. Hassani, K. Roberts, S. Shirazi, J. Shadley, E. Rybicki, C. Joia, *Flow Loop Study of Chloride Concentration Effect on Erosion, Corrosion and Erosion-Corrosion of Carbon Steel in CO2 Saturated Systems, Corrosion 2012* (NACE International, Salt Lake City, UT, 2012a)
45. S. Hassani, K. Roberts, S. Shirazi, J. Shadley, E. Rybicki, C. Joia, *Characterization and Prediction of Chemical Inhibition Performance for Erosion-Corrosion Conditions in Sweet Oil and Gas Production, Corrosion 2012* (NACE International, Salt Lake City, UT, 2012b)
46. K.C. Tummala, K. Roberts, J. Shadly, E. Rybicki, S. Shirazi, Effect of sand production and flow velocity on corrosion inhibition under scale forming conditions, in *CORROSION 2009*, Paper No. 09474 (NACE International, Houston, TX, 2009)
47. S. Hassani, K. Roberts, S. Shirazi, J. Shadley, E. Rybicki, C. Joia, Characterization and prediction of chemical inhibition performance for erosion-corrosion conditions in sweet oil and gas production. Corrosion **68**(10), 885–896 (2012)

48. A.S. Nassef, M. Keller, K. Roberts, E. Iski, E. Rybicki, S. Shirazi, Erosion-corrosion of low carbon steel inhibition in oil-brine-sand flow, in *CORROSION. 2016*, Paper No. 7426 (NACE International, Houston, TX, 2016)
49. J. McMahon, J. Martin, L. Harris, Effect of sand and interfacial adsorption loss on corrosion inhibitor efficiency, in *CORROSION 2005*, Paper No. 05274 (NACE International, Houston, TX, 2005)
50. S. Ramachandran, Y. Ahn, K. Bartrip, Further advances in the development of erosion corrosion inhibitors, in *CORROSION 2003*, Paper No. 05292 (NACE International, Houston, TX, 2003)
51. M. Tandon, J. Shadley, E. Rybicki, K. Roberts, S. Ramachandran, V. Jovancicevic, Flow loop studies of inhibition of erosion-corrosion in CO_2 environment with sand, in *CORROSION 2006*, Paper No. 06597 (NACE International, Houston, TX, 2006)
52. A. Neville, H.C. Wang, Erosion-corrosion mitigation by corrosion inhibitors—an assessment of mechanisms. Wear **267**(1–4), 195–203 (2009)
53. A.S. Nassef, Erosion-corrosion and inhibited erosion-corrosion of carbon steel in sweet production with $CaCO_3$ versus sand particles, 2019, in *Mechanical Engineering* (The University of Tulsa, OK, 2019)
54. B. Ateya, B. El-Anadouli, F. El-Nizamy, The adsorption of thiourea on mild steel. Corros. Sci. **24**(6), 509–515 (1984)
55. H. Kaesche, *Corrosion of Metals: Physicochemical Principles and Current Problems*, 1st ed. (Springer, Berlin, Germany, 2003)
56. L. Chong, S. Richter, S. Nešić, How do inhibitors mitigate corrosion in oil-water two-phase flow beyond lowering the corrosion rate. Corrosion **70**(09), 958966 (2014)
57. A. Nassef, M. Keller, K. Roberts, E. Iski, E. Rybicki, S. Shirazi, Comparison of inhibited erosion-corrosion with $CaCO_3$ particles versus sand. in *CORROSION 2018*, Paper No. 11093 (NACE International, Houston, TX, 2018)
58. A. Nassef, M. Keller, K. Roberts, E. Iski, E. Rybicki, S. Shirazi, Inhibited CO_2 corrosion with $CaCO_3$ particles versus sand, in *MECC 2018*, Paper No. 12527 (NACE International, Houston, TX, 2018)
59. A. Nassef, M. Keller, K. Roberts, E. Iski, E. Rybicki, S. Shirazi, Inhibited erosion-corrosion of carbon steel in sweet production with $CaCO_3$ particles versus sand, CORROSION 2019, Paper No. 13433 (NACE International, Houston, TX, 2019)

Effect of Hydrogen and Defects on Deformation and Failure of Austenitic Stainless Steel

Eugene Ogosi, Amir Siddiq, Umair Bin Asim, and Mehmet E. Kartal

Abstract Austenitic stainless steels are used for a variety of applications and could suffer degradation of properties when exposed to hydrogen. The performance of these steels are also dependent on crystallographic texture which in practice is a factor influenced by manufacturing processes. A study has been performed using a crystal plasticity based finite element model to understand the effect of crystal orientation with respect to loading direction for FCC single crystals in both hydrogenated and non-hydrogenated environment. The purpose of the study is to understand the effect of crystal orientation on how hydrogen influences plastic deformation and void growth. Simulations have been performed for a variety of stress triaxilaities, Lode parmeters and hydrogen concentrations. It is observed that initial crystal orientation has a varied effect on the influence hydrogen has on plastic deformation and void growth. Hydrogen in trap distribution at various stages of the deformation process was also found to be influenced by intial crystal orientation. Hydrogen affects the evolution of crystal rotation during deformation but was not found to significantly affect the general pattern of crystal orientation evolution.

Keywords Crystal plasticity · Crystal orientation · Hydrogen embrittlement · Void growth · Plastic deformation

1 Introduction

Manufacturing processes are used to form steel into useful products. The mechanical and corrosion resistant properties are known to be strongly dependent on the

E. Ogosi
Apache North Sea, Prime Four Business Park Kingswells, Aberdeen AB15 8PU, UK

E. Ogosi · A. Siddiq (✉) · M. E. Kartal
School of Engineering, University of Aberdeen, Aberdeen AB24 3FX, UK
e-mail: amir.siddiq@abdn.ac.uk

U. B. Asim
Department of Materials Science and Engineering, Texas A&M University, College Station, TX 77843, USA

I. Toor (ed.), *Recent Developments in Analytical Techniques for Corrosion Research*,
https://doi.org/10.1007/978-3-030-89101-5_11

texture of the material inherited during manufacture [1]. The crystallographic texture of steel has been found to affect hydrogen resistant properties [2, 3]. Hydrogen has a deleterious effect on steel and since the nineteenth century, its effect have been subject to many studies and reviews [4–7]. Specifically, experimental work has been performed on hydrogen embrittlement of austenitic stainless steel [8, 9]. The hydrogen enhanced localized plasticity (HELP) theory is used to explain how hydrogen, in steel undergoing deformation, enhances dislocation motion at a microscopic scale by elastic shielding of dislocation interactions with other dislocations and microstructural structures [7, 10]. In face centred cubic (FCC) metals, with relatively low hydrogen difussivity, this has been observed to lead to local dislocation pile ups [11, 12] and increased plastic flow properties [13]. Crystal orientation in relation to the rolling, normal and transverse directions have been found to influence materials resistance to hydrogen induced cracking for both face centred cubic (FCC) and body centred cubic (BCC) crystal materials [14, 15]. Crystal orientation parallel to the normal direction was found to improve resistance to hydrogen induced cracking [14, 15]. Potirniche et al. investigated the effect of crystal orientation on void growth of an FCC crystals without hydrogen influence [16]. They performed 2D finite element simulations using a crystal plasticity based model to understand the effects of crystal orientation and loading biaxiality on void characteristics. Results from these studies showed that crystal orientation had a significant effect on void growth. Asim et al. [17] previously observed that crystal orientation has significant impact on the void growth and that this effect reduces with increasing triaxiality. Activation of slip systems plays an important part in the dependence of plastic deformation on loading direction and crystal orientation. It was reported by Michler and colleagues [18] that austenite phase stability is not a guarantee that stainless steel is resistance to hydrogen embrittlement. They performed experiments on stable austenitic stable steels and observed hydrogen embrittlement even when there was no evidence of phase transformation to martensite. Crystal orientation was found to affect plastic slip localization, void growth and its shape evolution. Hua et al. performed experimental studies to understand the effect of crystal orientation on hydrogen diffusion using specialized microscopy and electron back scatter diffraction (EBSD) [19]. They found that that hydrogen diffusion had a strong dependence on crystal orientation. Hydrogen mobility in 304 stainless steel crystal grains were found to be higher in (111) oriented grains when compared with (001) and (101) oriented grains. Depending of crystal orientation, it was observed that hydrogen affected the plastic deformation differently [20]. This was explained by the difference in deformation mechanism viz a viz slip and twinning experienced and the effect of applied stress on splitting of dislocations. In this work, we investigate the effects of initial crystal orientation on how hydrogen affects the properties of austentic stainless steels and study the effects of hydrogen on crystallographic rotation evolution. Simulations have been performed for a variety of stress triaxialities (0.8, 1, 1.5, 2 & 3) and Lode parameters ($-1, 0$ & 1). The model formulation is presented in Sect. 2. Methodology is explained in Sect. 3. Results are presented in Sect. 4. Conclusions are presented in Sect. 5.

2 Crystal Plasticity Model with Hydrogen Effects

2.1 Crystal Plasticity Theory

A crystal plasticity model [21, 22] has been extended to include hydrogen influence [23]. A brief summary of the model formulation is presented. Deformation gradient is expressed as a product of its elastic F^e, hydrogen $\boldsymbol{F}^h$ and elastic $\boldsymbol{F}^e$ components [24];

$$\mathbf{F} = \mathbf{F}^e\mathbf{F}^h\mathbf{F}^{\mathbf{p}} \tag{1}$$

Marin's formulation [21] with a hydrogen component is expressed as:

$$\mathbf{F} = \mathbf{V}_e\mathbf{F}^*, \quad \mathbf{F}^* = \mathbf{R}^e\mathbf{F}^h\mathbf{F}^p \tag{2}$$

$\boldsymbol{V}_e$ is elastic stretch and $\boldsymbol{R}^e$ is rotation. The velocity gradient $\boldsymbol{l}$ in the final deformed state is given by;

$$\boldsymbol{l} = \dot{\boldsymbol{F}}\boldsymbol{F}^{-1} \tag{3}$$

$\dot{\boldsymbol{F}}$ and $\boldsymbol{F}^{-1}$ are the rate of change deformation gradient and its inverse. Velocity gradient $(\tilde{\boldsymbol{L}})$ in an intermediate state with elastic stretch unloaded is given by:

$$\tilde{\boldsymbol{L}} = \boldsymbol{V}^{e^{-1}}\boldsymbol{l}\boldsymbol{V}^e = \boldsymbol{V}^{e^{-1}}\boldsymbol{V}^e + \tilde{\boldsymbol{L}}^* \tag{4}$$

$\boldsymbol{V}^{e-1}$ is the inverse of stretch.

$$\widetilde{\mathbf{L}}^* = \mathbf{R}^e\mathbf{R}^{eT} + \mathbf{R}^e\widehat{\mathbf{L}}^h\mathbf{R}^{eT} + \mathbf{R}^e\mathbf{F}^h\overline{\mathbf{L}}^p\mathbf{F}^{h-1}\mathbf{R}^{eT} \tag{5}$$

$\dot{\boldsymbol{R}}^e$ is the rotation change rate and $\boldsymbol{R}^{eT}$ is the transpose of $\boldsymbol{R}^e$. The plastic part of the velocity gradient $\overline{\boldsymbol{L}}^p$, is given by:

$$\overline{\mathbf{L}}^p = \sum \dot{\gamma}^\alpha \overline{s}^\alpha \otimes \overline{\mathbf{m}}^\alpha \tag{6}$$

$\overline{\boldsymbol{s}}^\alpha$ is the direction and $\overline{\boldsymbol{m}}^\alpha$ is the normal component. Combing (5) and (6):

$$\widetilde{\mathbf{L}}^* = \tilde{\Omega}^e + \widehat{\mathbf{L}}^h + \sum_{\alpha=1}^{N} \dot{\gamma}^\alpha \widetilde{\mathbf{s}}^\alpha \otimes \mathbf{m}^\alpha \tag{7}$$

$\tilde{\boldsymbol{\Omega}}^e = \dot{\boldsymbol{R}}^e\boldsymbol{R}^{eT}$ is elastic spin. $\dot{\gamma}^\alpha$ is the rate shear strain.
Sofronis provides an expression for the hydrogen deformation gradient [25]:

$$F^h = \left(1 + \frac{(c - c_o)\lambda}{3}\right) I \tag{8}$$

c_o and c are the initial and final hydrogen concentrations. λ is $\frac{\Delta V}{V_m}$, ΔV is volume change and V_m is atomic volume.

The hydrogen part of the velocity gradient $\hat{L}^h$ is:

$$\hat{L}^h = \dot{F}^h \cdot F^{h-1} = \frac{1}{3}\left[\frac{3\lambda}{3 + (c - c_0)}\right](c)\dot{c}I \tag{9}$$

The Second Piola–Kirchhoff stress tensor, $\tilde{S}$ is:

$$\tilde{S} = \widetilde{\mathbb{C}^e} : \widetilde{E^e} \tag{10}$$

$\tilde{\mathbb{C}}^e$ and $\tilde{E}^e$ are the elasticity and Green–Lagrange strain tensors. Deformation tensor splits into a symmetric, $\tilde{D}$, and skew, $\tilde{W}$ part:

$$\tilde{D} = V^{eT} d V^e = \tilde{E}^e + \left[\mathrm{sym}\left(\tilde{C}^e \tilde{\Omega}^e\right) + \sum_{\alpha=1}^{N} \dot{\gamma}^\alpha \mathrm{sym}\left(\tilde{C}^e \tilde{Z}^\alpha\right)\right] \tag{11}$$

$$\tilde{W} = V^{eT} w V^e = \mathrm{skew}\left(V^{eT} \dot{V}^e\right) + \left[\mathrm{skew}\left(\tilde{C}^e \tilde{\Omega}^e\right) + \sum_{\alpha=1}^{N} \dot{\gamma}^\alpha \mathrm{skew}\left(\tilde{C}^e \tilde{Z}^\alpha\right)\right] \tag{12}$$

where $\tilde{C}^e = R^e \tilde{C}^e R^{eT}$ and $\tilde{Z}^\alpha = \tilde{s}^\alpha \otimes \tilde{m}^\alpha$.

Plastic slip evolution is given as:

$$\dot{\gamma}^\alpha = \dot{\gamma}_0^\alpha \left[\frac{|\tau^\alpha|}{\kappa_s^\alpha}\right]^{\frac{1}{m}} \mathrm{sign}(\tau^\alpha) \tag{13}$$

$\dot{\gamma}^\alpha$ is strain rate inα, $\dot{\gamma}_0^\alpha$ is the reference strain rate, κ_s^α is the current crystal strength of α, τ^α is resolved stress and m is rate sensitivity. The slip system hardens as follows;.

$$\dot{\kappa}_s^\alpha = h_0 \left(\frac{\kappa_{s,S}^\alpha - \kappa_s^\alpha}{\kappa_{s,S}^\alpha - \kappa_{s,0}^\alpha}\right) \sum_{\alpha=1}^{N} |\dot{\gamma}^\alpha|, \kappa_{s,S}^\alpha = \kappa_{s,S0}^\alpha \left[\frac{\sum_\alpha |\dot{\gamma}^\alpha|}{\dot{\gamma}_{s0}}\right]^{1\backslash m'} \tag{14}$$

$\dot{\kappa}_s^\alpha$ is the current hardening rate, h_0 is a reference coefficient for hardening, $\kappa_{s,S}^\alpha$ is the saturation strength value and $\sum_\alpha |\dot{\gamma}^\alpha|$ is accumulated slip.

$\kappa_s^\alpha(t = 0)$ is the critical resolved shear stress (CRSS) of each slip system. $\kappa_{s,0}^\alpha$, $\kappa_{s,s0}^\alpha$, $\dot{\gamma}_{s,0}^\alpha$ and m' are other plastic property defining material parameters.

2.2 Incorporation of Hydrogen Effects

It has been proposed by Oriani that hydrogen in steel will either reside in normal interstitial lattice sites (NILS) or in traps and remain in equilibrium [26]. It has been discussed previously that diffusion of hydrogen in austenitic stainless steel (FCC) is slow and as such the concentration of hydrogen at material points can be assumed to remain the same during deformation, although there is a transfer of hydrogen atoms from NILS to trap sites [23]. Evidence supporting the constant hydrogen theory has been presented by Schebler [27]. Discussion on the incorporation of hydrogen influence into the model has been discussed previously [23, 28] so only a summary is given here. We express the concentration of hydrogen $C_{i,bulk}$ at a given material point as

$$C_{i,bulk} = C_L + C_{i,traps} \tag{15}$$

C_L is the concentration in NILS and $C_{i,traps}$ is the hydrogen in traps before deformation.

Hydrogen in traps C_T during deformation increases as there is a transfer from C_L to traps.

$$C_T = \theta_T \psi N_T \tag{16}$$

θ_T is hydrogen occupancy in traps, ψ is the number of sites per trap and N_T represent the number of traps per lattice site. N_T is:

$$N_T = \frac{\sqrt{3}}{a_{fcc}} \rho \tag{17}$$

a_{fcc} is lattice parameter. Evolution of dislocation density,$\dot{\rho}$ is:

$$\int_0^t \dot{\rho} dt = \left(k_1 \sqrt{y}\right) \int_0^t /\dot{\gamma}/ dt \tag{18}$$

$\dot{\gamma}$ is rate of change of strain. k_1 is a measure of immobile dislocations and $\sqrt{y}$ is the average dislocation separation length as proposed by Estrin et al. [29]. Krom's formulation [30] is used to determine C_T:

$$C_T = \frac{1}{2}\left[\frac{N_L}{K_T} + C_{Total} + N_T - \sqrt{\left(\frac{N_L}{K_T} + C_{Total} + N_T\right)^2 - 4N_T C_{Total}}\right] \tag{19}$$

The terms H_i and H_f are used to capture the hydrogen effect as previously discussed [23].

$$\kappa_{h,0}^{\alpha} = \kappa_{s,0}^{\alpha} * (1 + \mathrm{H_i} C_{initial}) \tag{20}$$

$\kappa_{s,0}^{\alpha}$ is crystal strength with no hydrogen. $C_{initial}$ is hydrogen traps before plastic deformation and defined as per Caskey's relationship [31];

$$C_{initial} = \mathrm{f}\, \mathrm{C_L} e^{18400} / (\mathrm{RT}) \tag{21}$$

f is atoms per unit length of dislocation. 18,400 J/mol is bonding energy.

Crystal strength evolution in (14) is altered to include hydrogen effect as follows:

$$\dot{\kappa}_{s}^{\alpha} = h_0 \left(\frac{\kappa_{s,S}^{\alpha} - \kappa_{s}^{\alpha}}{\kappa_{s,S}^{\alpha} - \kappa_{s,0}^{\alpha}} \right) \sum_{\alpha=1}^{N} |\dot{\gamma}^{\alpha}| (1 + \mathrm{H_f C_T}) \tag{22}$$

Stress triaxiality and Lode parameter are used to quantify stress states [32, 33] as previously discussed by the authors [17]. Lateral displacements are controlled using a multipoint constraint (MPC) subroutine in ABAQUS finite analysis software [34] to keep stress triaxialities and Lode parameter constant at each iteration of the simulations. The relationships in Sects. 2.1 and 2.2 are implemented in a user material subroutine (UMAT) and ABAQUS as discussed in the next chapter.

3 Methodology

A representative volume element (RVE) is constructed using reduced integrated elements in ABAQUS finite element software. Sample element with an embedded void is shown in Fig. 1 along with the relevant boundary conditions. The embedded void represents a void which has already nucleated as part of the fracture process. The analysis is done using ABAQUS finite element analysis software [34].

The relationship between strain, stress, void growth, void size, stress triaxiality and initial crystal orientation have been covered previously [17, 35]. Only a summary is given here. Void growth is tracked by the term;

$$Normalised\ void\ fraction\ (NVF) = \frac{f}{f_0},\ f = \frac{V_{void}}{V_{RVE}} \tag{23}$$

Initial void volume fraction, f_0 is given by $(4/3)\pi r^3/s^3$ where r is sphere radius, f is void volume fraction, and s is side length. V_{void} is the void volume and $\mathrm{V_{RVE}}$ is the sum of solid material and void volume. Loading is done by applying a positive displacement in the x direction and the sides of the RVE are constraint to remain straight. In Fig. 1, $\mathrm{F_{Top}}$, $\mathrm{F_{Front}}$, and $\mathrm{F_{Left}}$ are mobile while $\mathrm{F_{bottom}}$, $\mathrm{F_{back}}$ and $\mathrm{F_{right}}$ are constraint to be fixed during displacement. Lateral displacement is adjusted to keep stress triaxilities constant by applying a multipoint constraint (MPC) subroutine. Using the technique discussed by Tekoglu [36], the stress triaxiality (X) and Lode

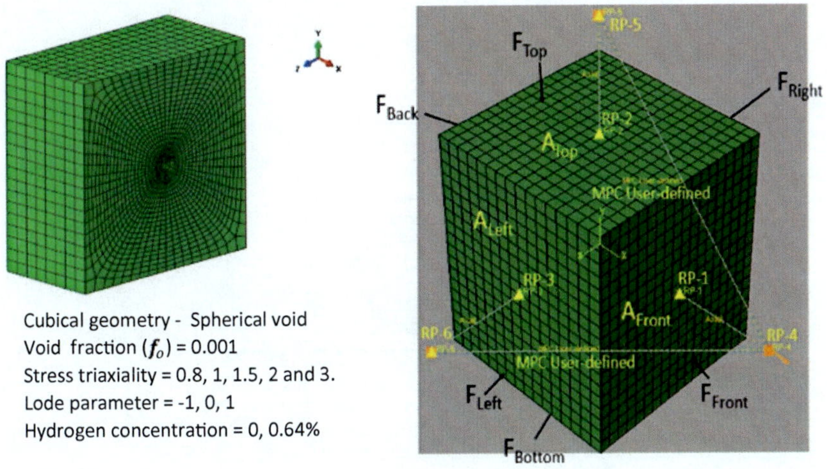

Fig. 1 RVE sample showing boundary conditions

parameter (L) is held constant by satisfying;

$$\rho_{11} = \frac{\sum_{11}}{\sum_{22}}; u_x = \rho_{11}\frac{A_{Front}}{A_{Top}}u_y \tag{24}$$

$$\rho_{33} = \frac{\sum_{33}}{\sum_{22}}; u_z = \rho_{33}\frac{A_{Left}}{A_{Top}}u_y \tag{25}$$

Current RVE volume is

$$V_{RVE} = (s + u_x) * \left(s + u_y\right) * (s + u_z) \tag{26}$$

The value of each element ($V_{e,i}$) is stored and summed to obtain

$$V_{solid} = \sum_{i=1}^{N} V_{e,i} \tag{27}$$

Current value of void volume fraction (f) is then obtained by

$$f = \frac{V_{RVE} - V_{Solid}}{V_{RVE}} \tag{28}$$

Crystal orientation is represented using Euler angles. Euler angles are three angles used to represent the rotation of an object in space. Here, we have applied the Kock's convention in depicting these angles. Consider a crystal (small cube) shown in relation to a fixed global coordinate axis (big cube) in Fig. 2. We apply a displacement in

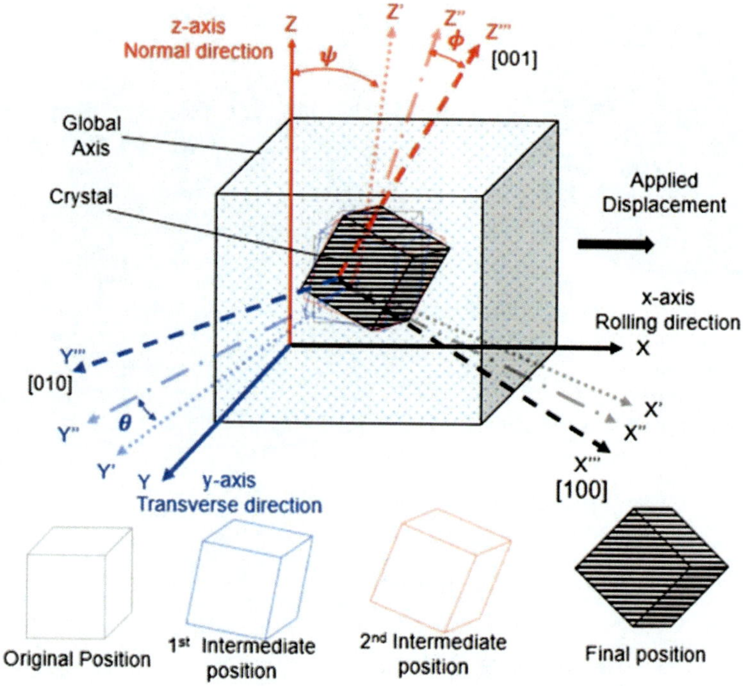

Fig. 2 Representation of crystal orientation in relation to a fixed global axis

the x direction or rolling direction and this impacts the position of the crystal in relation to a fixed global coordinate. The first angle ψ represents a rotational angle in the z-axis from the original position (Z) to a first intermediate position (Z′). The second angle θ represents rotation and displacement in the y-axis from the first intermediate position (Y′) to a second intermediate position (Y″). The third angle ϕ represents rotation and displacement in the z-axis from the second intermediate position (Z″) to the final position (Z‴).

Table 1 summarises the various cases which have been considered. A variety of initial crystal orientations, Lode parameters and stress triaxialitied have been considered. However, for the sake of brevit, only selected results (highlighted in light grey) are presented to discuss pertinent findings and observation.

Table 2 shows the various initial crystal orientations and their corresponding stress states. Euler angles in relation to applied displacement considered are illustrated in Fig. 3.

Using the model described, simulations are performed for a range parameters. Results obtained are discussed in Sect. 4.

Table 1 Study cases

Initial crystal orientation		Ori 0			Ori A			Ori B			Ori C			Ori D			Ori E			Ori F		
Lode parameter		−1	0	1	−1	0	1	−1	0	1	−1	0	1	−1	0	1	−1	0	1	−1	0	1
Stress triaxiality	0.8	x	x	x	x	x	x	x	x	x	x	x	x	x	x	x	x	x	x	x	x	x
	1	x	x	x	x	x	x	x	x	x	x	x	x	x	x	x	x	x	x	x	x	x
	1.5	x	x	x	x	x	x	x	x	x	x	x	x	x	x	x	x	x	x	x	x	x
	2	x	x	x	x	x	x	x	x	x	x	x	x	x	x	x	x	x	x	x	x	x
	3	x	x	x	x	x	x	x	x	x	x	x	x	x	x	x	x	x	x	x	x	x

Table 2 Crystal orientations and Euler angles

Euler angles	ψ	θ	ϕ
Orientation 0 (Ori 0)	0°	0°	0°
Orientation A (Ori A)	5°	5°	5°
Orientation B (Ori B)	30°	30°	30°
Orientation C (Ori C)	45°	45°	30°
Orientation D (Ori D)	60°	60°	90°
Orientation E (Ori E)	75°	90°	120°
Orientation F (Ori F)	145°	120°	180°

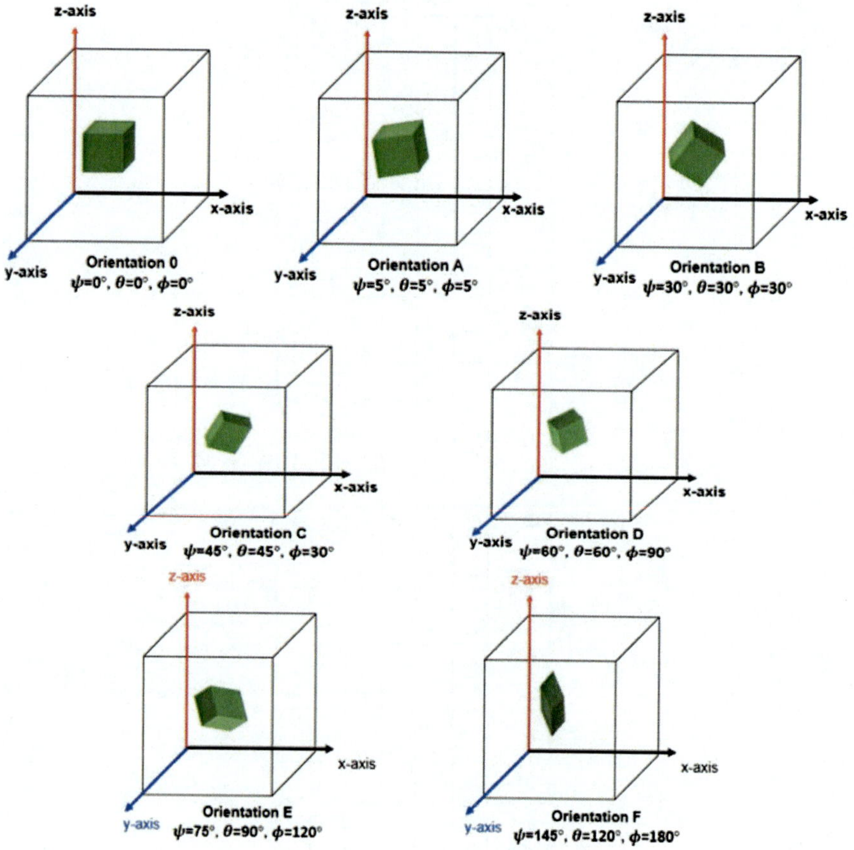

Fig. 3 Illustration of study cases with intial crystal orientation shown in relation to applied displacement

4 Results and Discussion

4.1 *Effect of Intial Crystal Orientation and Hydrogen on Equivalent Stress*

Figure 4 shows the equivalent stress vs equivalent strain curves for different initial crystal orientations. It can be seen that the presence of hydrogen shifts the graph upwards so that the stress value required to transist from elastic to plastic deformation increases. This has been observed previously and is discussed in other works [23, 27]. Table 3 presents an average percentage change equivalent stress due to the presence of hydrogen for the various initial crystal orientations considered. The change in equivalent stress $\Delta(\sigma)$(expressed as a percentage) due to hydrogen is computed using the relationship:

$$\Delta(\sigma) = \frac{\sigma_{hydrogen} - \sigma_{Nohydrogen}}{\sigma_{Nohydrogen}} \times 100\% \tag{29}$$

$\sigma_{hydrogen}$ is the average equivalent stress for hydrogenated RVE samples and $\sigma_{Nohydrogen}$ is the average equivalent stress for non hydrogenated RVE samples. $\sigma_{hydrogen}$ and $\sigma_{Nohydrogen}$ are computed up to a equivalent strain value of 0.24 for initial crystal orientations 0, A, B, C, D and E. The maximum equivalent strain obtained for initial crystal orientation F during simulation was 0.155 and values were computed up to this strain.

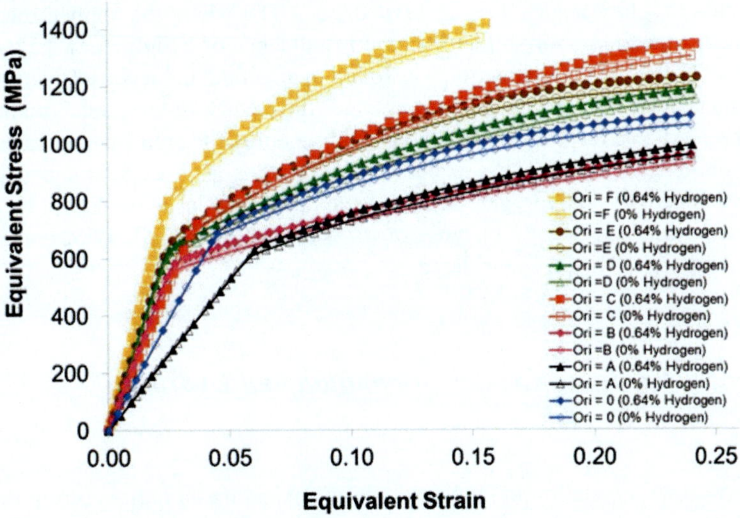

Fig. 4 Stress strain curves for different initial crystal orientations, viz. stress triaxiality = 1, lode parameter = −1 $f_0 = 0.001$ and hydrogen content = 0.64%

Table 3 Influence of hydrogen on equivalent stress vs strain relationship at different initial crystal orientations. Stress triaxiality = 1, Lode Parameter = −1 $f_0 = 0.001$ and hydrogen content is 0.64%

	Equivalent stress in (MPa) $\sigma_{hydrogen}$	Equivalent stress (MPa) $\sigma_{No\,hydrogen}$	Change in equivalent stress $\Delta(\sigma)$ (%)	Equivalent strain
Orientation 0 (Ori 0)	987	954	3.52	0.24
Orientation A (Ori A)	856	824	3.94	0.24
Orientation B (Ori B)	665	643	3.38	0.24
Orientation C (Ori C)	1231	1194	3.10	0.24
Orientation D (Ori D)	1055	996	5.84	0.24
Orientation E (Ori E)	1101	1068	3.15	0.24
Orientation F (Ori F)	1221	1181	3.44	0.15

The results show a variation in percentage increase in equivalent stress due to hydrogen of between 3.1% to 5.84% for different intitial crystal orientation. The highest change 5.84% was found for initial crystal orientation D (Euler angles $\psi = 60°$, $\theta = 60°$, $\phi = 90°$) and the least change in equivalent was 3.1% for initial crystal orientation C (Euler angles $\psi = 45°$, $\theta = 45°$, $\phi = 30°$). For an initial crystal orientation 0 (Euler angles $\psi = 0°$, $\theta = 0°$, $\phi = 0°$) where the applied direction is penpedicular to the normal, an increase in influence of hydrogen of 3.52% was observed. Initial crystal orientation was found to affect the influence of hydrogen in increasing equivalent stresses and its magnitude differently. Initial crystal orientation was observed to alter the slope of the elastic modulus for both hydrogenated and non hydrogenated samples. As explained elsewhere, plastic work is expended by dislocation activity so hydrogen is not expected to affect elastic modulus [23]. This observation was consistent for a range of stress triaxialities (0.8, 1, 1.5, 2 and 3) and Lode parameters (-1, 0, 1) considered.

4.2 *Effect of Intial Crystal Orientation and Hydrogen on Void Growth*

Figure 5 shows a variation in the void fraction depending on initial crystal orientations. The results are consistent with the observations reported by Asim et al. [17]. They found that void growth had a strong dependence on initial crystal orientation. Although not clearly observable from the graphs there is a shift in the graph

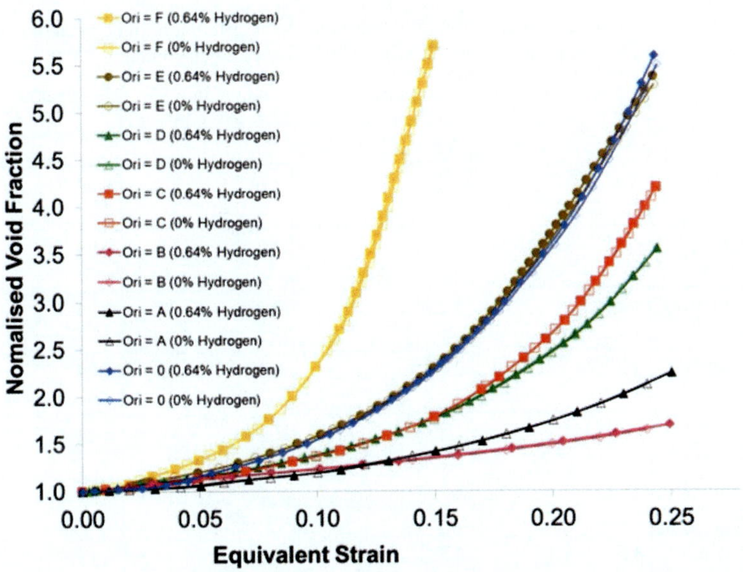

Fig. 5 Normalised void fraction for different initial crystal orientations, viz. stress triaxiality = 1, lode parameter = −1 $f_0 = 0.001$ and hydrogen content = 0.64%

left for hydrogenated samples which indicates that hydrogen promotes void growth. This has been discussed by the authors elsewhere [28]. Table 4 presents an average percentage change in normalised void fraction due to the presence of hydrogen for the various initial crystal orientations considered. The average change in normalised void fraction ΔNVF (expressed as a percentage) due to hydrogen is computed using the relationship:

$$\Delta NVF = \frac{NVF_{hydrogen} - NVF_{\text{No hydrogen}}}{NVF_{\text{No hydrogen}}} \times 100\% \tag{30}$$

$NVF_{hydrogen}$ is the average normalised volume fraction for hydrogenated RVE samples and $NVF_{Nohydrogen}$ is the average normalised volume fraction for non hydrogenated RVE samples. $NVF_{hydrogen}$ and $NVF_{Nohydrogen}$ are computed up to a equivalent strain value of 0.24 for initial crystal orientations 0, A, B, C, D and E. The maximum equivalent strain obtained for initial crystal orientation F was 0.155.

The results show a percentage increase in nominal void fraction due to the presence of hydrogen up to 2.7% for different initial crystal orientation. The strongest influence due to hydrogen was 2.7% and this was observed for an initial crystal orientation D (Euler angles $\psi = 60°$, $\theta = 60°$, $\phi = 90°$). For an initial crystal orientation 0 (Euler angles $\psi = 0°$, $\theta = 0°$, $\phi = 0°$) where the applied displacement direction is penpedicular to the normal, a percentage increase in influence of hydrogen in relation to nominal void fraction was 2.0%. It is noted that the effect of hydrogen

Table 4 Influence of hydrogen on normalised void fraction at different initial crystal orientations. Stress triaxiality = 1, lode parameter = −1 $f_0 = 0.001$ and hydrogen content is 0.64%

	Normalised void fraction $NVF_{hydrogen}$	Normalised void fraction $NVF_{No\,hydrogen}$	Change in NVF ΔNVF (%)	Equivalent strain
Orientation 0 (Ori 0)	3.45	3.38	2.0	0.24
Orientation A (Ori A)	1.61	1.60	0.6	0.24
Orientation B (Ori B)	1.25	1.24	0.4	0.24
Orientation C (Ori C)	2.79	2.77	0.7	0.24
Orientation D (Ori D)	2.31	2.24	2.7	0.24
Orientation E (Ori E)	2.61	2.55	2.4	0.24
Orientation F (Ori F)	3.6	3.52	2.3	0.15

varies depending on the initial crystal orientation. This observation was consistent for a range of stress triaxialities (0.8, 1, 1.5, 2 and 3) and lode parameters (−1, 0, 1) considered.

4.3 *Effect of Initial Crystal Orientation on Hydrogen in Traps Distribution During Deformation*

To understand the effect of initial crystal orientation on the distribution of hydrogen in traps, contour plots for representative equivalent strains have been derived. Plots of hydrogen in trap distribution for selected equivalent strain values for a variety of initial crystal orientations are shown in Fig. 6. It is observed that the distribution and magnitude of hydrogen in traps had a strong dependence on initial crystal orientation. In all cases considered hydrogen in traps was found to be highest in the immediate vicinity of the void as previously reported [28]. Initial crystal orientations 0, A and B showed a lower maximum value of hydrogen concentrated around the void relative to initial crystal orientations C, D, E and F. A close scrutiny of the Euler angles of these initial crystal orientations, show that initial crystal orientations A and B have a lesser deviation (in terms of Euler angles ψ, θ and ϕ) from initial crystal orientation 0 and initial crystal orientations C to F having a higher deviation. The difference in concentration of hydrogen in trap around the void may offer an explanation on the effect intial crystal orientation has on hydrogen influence on local plastic deformation in the area around the void and overall void growth. However, the relationship between hydrogen in trap concentration around the void and its influence on void growth due to the effect of initial crystal orientation appears to be more complex. As presented in Table 4, initial crystal orientations B reported the lowest change in normalised void fraction compared to other initial crystal orientations (i.e. 0.4%) and also had the least concentration of hydrogen in traps in the vicinity of the void. However, initial crystal orientation C reported a relatively low change in normalised void fraction (i.e. 0.7%), but had a high concentration of hydrogen in trap distribution around the void compared with other cases considered. It is worthwile

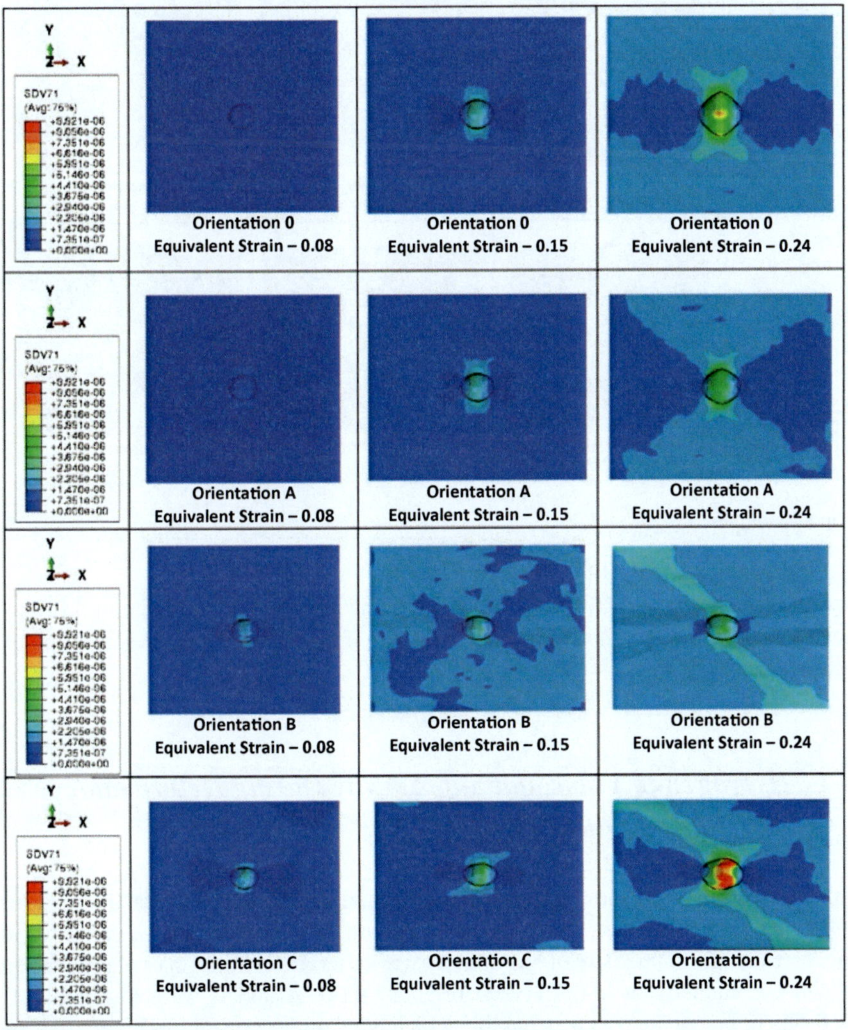

Fig. 6 Contour plots showing hydrogen in traps distribution for different initial crystal orientations viz. stress triaxiality = 1, lode parameter = -1, $f_0 = 0.001$ and hydrogen content is 0.64%

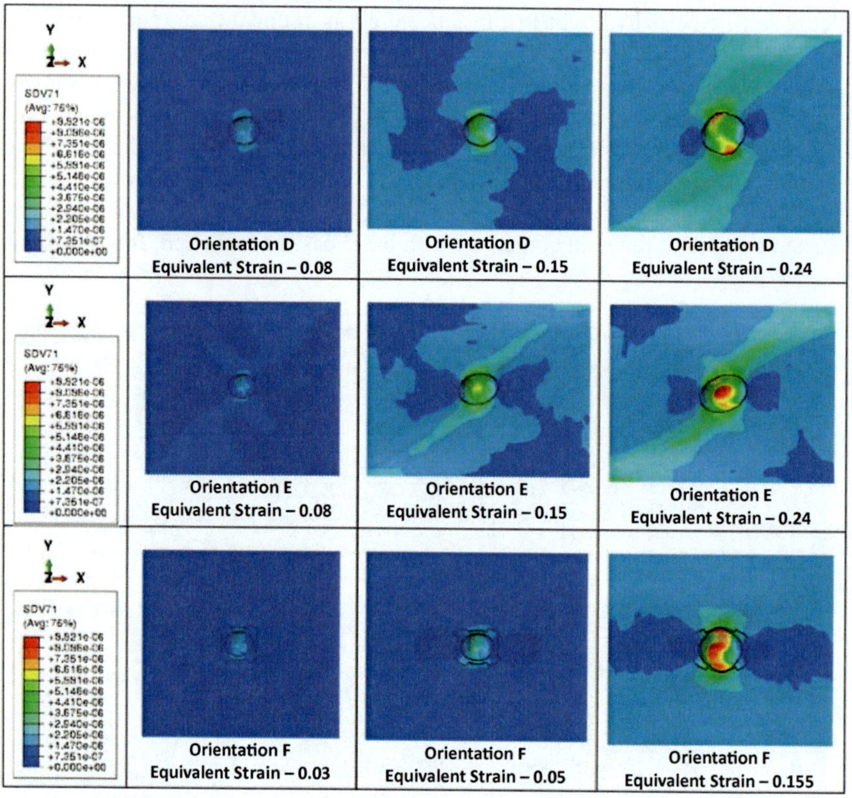

Fig. 6 (continued)

noting that the shape of the voids for the cases considered are varied and this may have an addition effect in void growth evolution if hydrogen influence is affected by void shape.

4.4 *Hydrogen in Traps and Slip Activity During Deformation for Different Initial Orientations*

Contour plots in Fig. 7 show that slip activity for the various initial crystal orientations follow the same pattern and intensity as the amount of hydrogen in traps presented in Fig. 6. The highest intensity of slip activity was observed in initial crystal orientation C and the intensity in initial crystal orientations 0, A and B were lower than C to F. There is evidence of a strong dependency of initial crystal orientation on the distribution of slip acivity which concentrates around the void.

4.5 Crystal Orientation Evolution in the Presence of Hydrogen

The evolution of crystal orientation for different stress triaxialities and initial crystal orientations have been studied previously. It was found that the magnitude and direction of crystal rotation was dependent on stress triaxialities and initial crystal orientations [17]. The crystal orientation evolution for hydrogenated and non hydrogenated

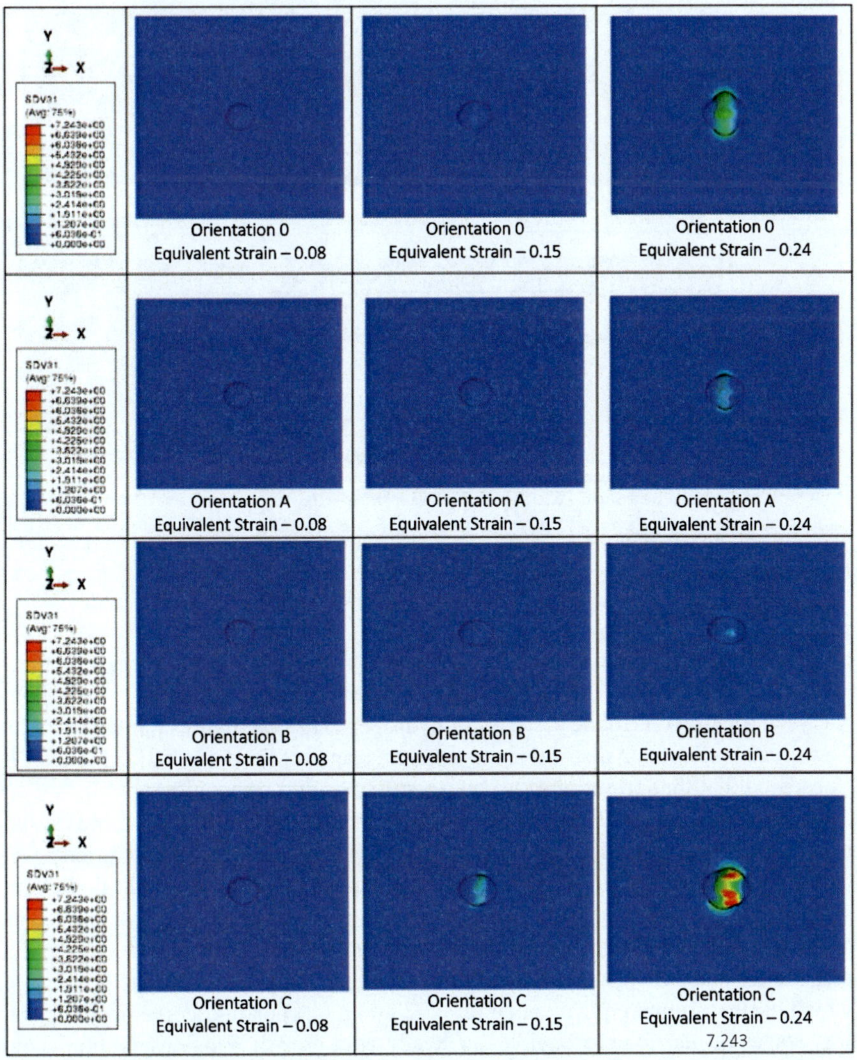

Fig. 7 Contour plots showing slip activity (SDV31) for different initial crystal orientations viz. stress triaxiality = 1, lode parameter = −1, $f_0 = 0.001$ and hydrogen content = 0.64%

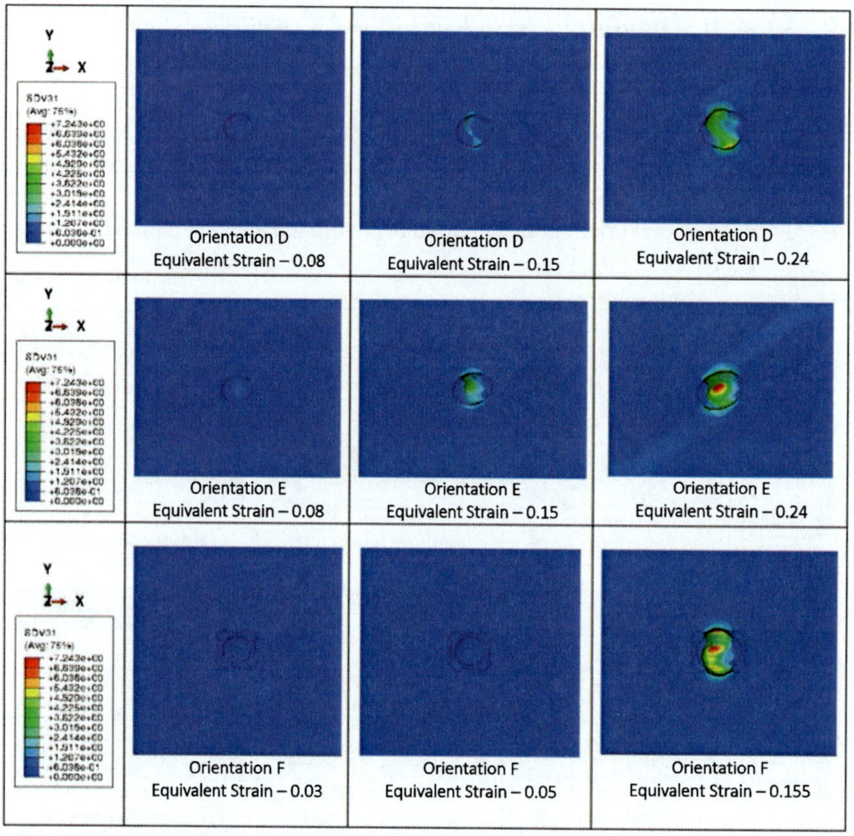

Fig. 7 (continued)

samples have been investigated here. The stress triaxiality and initial crystal orientation are kept the same with RVE samples studied to observe if there are any effect on crystal rotation due to the presence of hydrogen. The pole figures plotted in Fig. 8 show crystal orientation distribution at different stages of the deformation process. It is observed that that presence of hydrogen stifled crystal orientation initially. At an equivalent strain value of 0.07, the RVE sample with no hydrogen show more crystal rotation with respect to the sample exposed to hydrogen indicating this orientation inhibiting effect of hydrogen. However, at an equivalent strain of 0.08, this effect changes and hydrogen appears to encourage crystal orientation. This can be seen as the RVE sample exposed to hydrogen shows more crystal rotation with respect to the sample not exposed to hydrogen. A reference to Figs. 4 and 5 indicate that this change (which occurs around equivalent strain 0.07 and 0.08) position corresponds to the early stages of plastic deformation and steady void growth. This observation show a similar characteristic to previously discussed phenomenon where hydrogen was found to initially impede void growth and beyond a certain equivalent strain value promote void growth [28]. For equivalent strains 0.09, 0.15 and 0.24, the pole figures show that crystal orientation is largely similar with subtle changes in magnitude. The pattern of crystal rotation remains broadly similar with no significant effect due to hydrogen. It can be inferred that hydrogen has an effect on the evolution of crystal

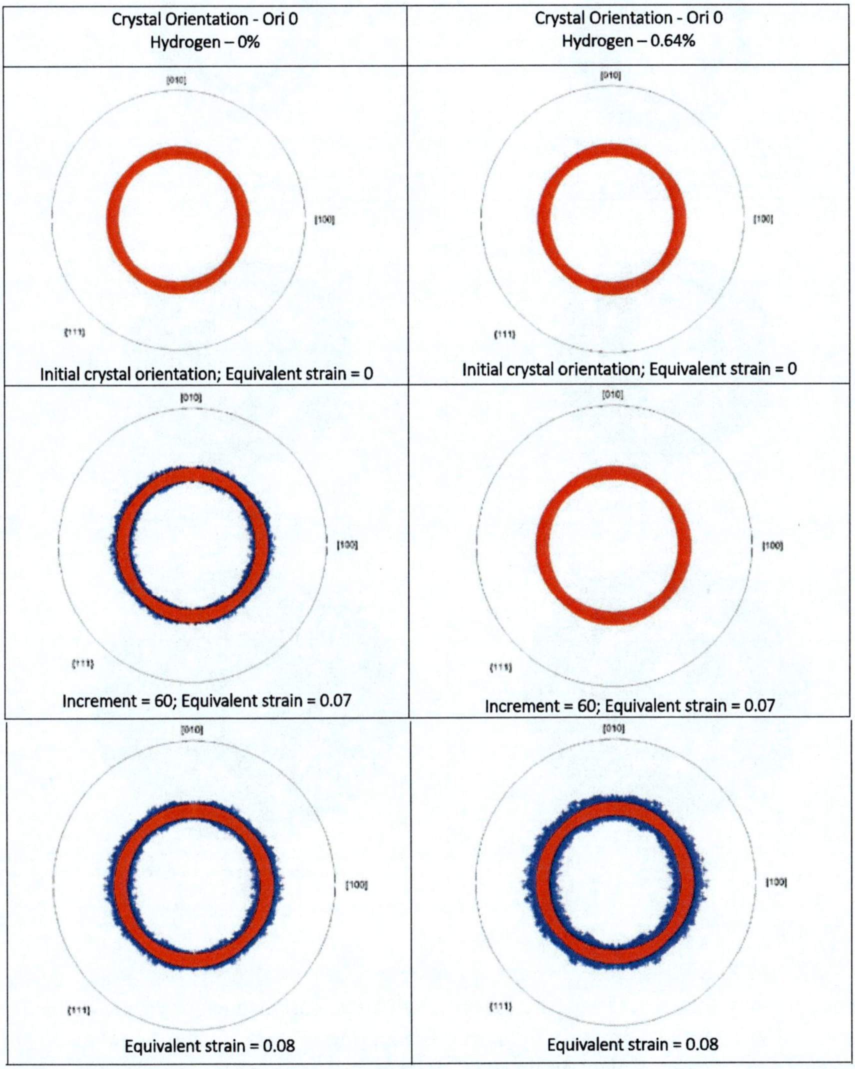

Fig. 8 Pole figures showing progression of crystal orientation for hydrogenated and non hydrogenated FCC single crystal viz. stress triaxiality = 1, lode parameter = $-1, f_0 = 0.001$ and hydrogen content = 0.64%

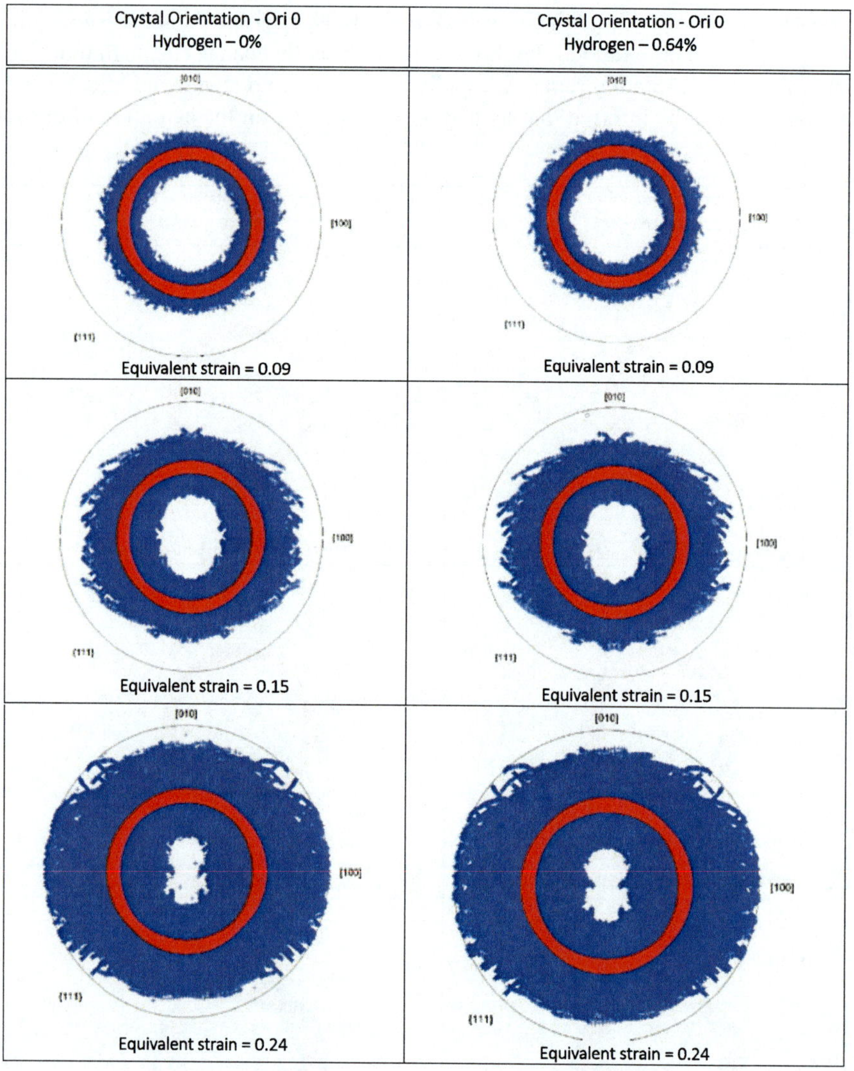

Fig. 8 (continued)

orientation with respect to plastic deformation at the early stages of plastic deformation and void growth. Initially hydrogen is found to impede crystal orientation, but beyond a certain equivalent strain (0.07–0.08 in this study case), this is reversed.

5 Conclusions

The effect of crystal orientation on hydrogen influence on void growth in a single crystal of austenitic stainless steel has been studied using a crystal plasticity formulation. There have been finite element based studies into the effect of lattice orientation on void growth without hydrogen influence previously [16, 17] and experimental evidence that hydrogen distribution and mechanisms are affected by crystal orientation [19, 20]. The following conclusions have been drawn from this study:

1. Initial crystal orientation relative to applied displacement is found to affect the influence of hydrogen on plastic deformation and void growth. This effect was found for a variety of stress triaxialities and Lode parameters considered.
2. The distribution and magnitude of hydrogen in traps, especially in the vicinity of the void, was also found to have a strong dependence on initial crystal orientation. The variation in hydrogen in trap distribution around the void provides evidence that there is an effect of intial crystal orientation on hydrogen influence on fracture processes.
3. The evolution of crystal orientation during plastic deformation is affected by the presence of hydrogen although there was no evidence that the general pattern of rotation is significantly affected.

6 Future Challenges and Opportunities

The use of models to predict material performance provide the advantage of replicating experimental and real life conditions. Specifically for predicting hydrogen damage, the knowledge of hydrogen transport to the metal surface, entry into the metal, transport within its microstructure, its interaction with crystal defects and microstructure and modification of material properties due to hydrogen are complex phenomena [37] and a good model should account for these. For metals with a FCC crystal structure, it has already been explained that bulk hydrogen transport is limited due to the slow diffusive properties of hydrogen in austenite. For these materials, the interaction of hydrogen with microstructural features and the effects on the material properties is more relevant. A predictive framework based on FEM which incorporates the effects of stress states and crystal orientations for metals with FCC crystal structure will be of practical use to engineers. Suggestion for future work and areas of improvements are as follows:

1. The development of an analytical relationship that captures the effect hydrogen has on fracture processes influenced by strain, stress triaxialities, Lode parameter and crystal orientation will be beneficial.
2. Hydrogen influence is believed to be linked to activation of individual slip systems which occur differently depending on initial crystal orientation. This will have relevance in understanding why certain orientations favour a higher

hydrogen influence on mechanical properties compared with others and should be further investigated.

References

1. S.I. Wright, D.P. Field, Recent studies of local texture and its influence on failure. Mater. Sci. Eng. A **257**(1), 165–170 (1998)
2. V. Venegas, F. Caleyo, T. Baudin, J.H. Espina-Hernández, J.M. Hallen, On the role of crystallographic texture in mitigating hydrogen-induced cracking in pipeline steels. Corros. Sci. **53**(12), 4204–4212 (2011)
3. M. Masoumi, C.C. Silva, H.F.G. de Abreu, Effect of crystallographic orientations on the hydrogen-induced cracking resistance improvement of API 5L X70 pipeline steel under various thermomechanical processing. Corros. Sci. **111**, 121–131 (2016)
4. T. Graham, On the occlusion of hydrogen gas by metals, Proc. R. Soc. London, 422–427 (1868)
5. W. Johnson, On some remarkable changes produced in iron and steel by the action of hydrogen and acids. R. Soc. London **14**(2), 168–179 (1874)
6. S.P. Lynch, Progress towards the understanding of mechanisms of hydrogen embrittlement and stress corrosion cracking. NACE Corros. 2007 Conf. Expo (07493), 1–55 (2007)
7. I.M. Robertson, H.K. Birnbaum, P. Sofronis, Hydrogen effects on plasticity **15**(09) (2009)
8. M. Hatano, M. Fujinami, K. Arai, H. Fujii, M. Nagumo, Hydrogen embrittlement of austenitic stainless steels revealed by deformation microstructures and strain-induced creation of vacancies. Acta Mater. **67**, 342–353 (2014)
9. Y. Mine, T. Kimoto, Hydrogen uptake in austenitic stainless steels by exposure to gaseous hydrogen and its effect on tensile deformation. Corros. Sci. **53**(8), 2619–2629 (2011)
10. P. Birnbaum, H. K., Sofronis, "Hydrogen-enhanced localized plasticity—a mechanism for hydrogen-related fracture," *Mater. Sci. Eng. A*, vol. 176, no. 1–2, pp. 191–202, 1994.
11. Y. Yagodzinskyy, T. Saukkonen, S. Kilpeläinen, F. Tuomisto, H. Hänninen, Effect of hydrogen on plastic strain localization in single crystals of austenitic stainless steel. Scr. Mater. **62**(3), 155–158 (2010)
12. Y. Yagodzinskyy, E. Malitckii, T. Saukkonen, and H. Hanninen, Hydrogen-induced strain localization in austenitic stainless steels and possible origins of their hydrogen embrittlement, in *2nd International Conference on Metals and Hydrogen*, 2014, May, pp. 203–213 (2014)
13. D.P. Abraham, C.J. Altstetter, Hydrogen-enhanced localization of plasticity in an austenitic stainless steel. Metall. Mater. Trans. A **26**(11), 2859–2871 (1995)
14. G.P. Potirniche et al., Role of crystallographic texture on the improvement of hydrogen-induced crack resistance in API 5L X70 pipeline steel. Int. J. Hydrogen Energy **42**(3), 14786–14793 (2017)
15. M. Béreš et al., Role of lattice strain and texture in hydrogen embrittlement of 18Ni (300) maraging steel. Int. J. Hydrogen Energy **42**(21), 14786–14793 (2017)
16. G.P. Potirniche, J.L. Hearndon, M.F. Horstemeyer, X.W. Ling, Lattice orientation effects on void growth and coalescence in fcc single crystals. Int. J. Plast **22**(5), 921–942 (2006)
17. U. Asim, M.A. Siddiq, M. Demiral, Void growth in high strength aluminium alloy single crystals: A CPFEM based study, Model. Simul. Mater. Sci. Eng. **25**(3), 035010 (2017)
18. T. Michler, C. San Marchi, J. Naumann, S. Weber, M. Martin, Hydrogen environment embrittlement of stable austenitic steels. Int. J. Hydrogen Energy **37**(21), 16231–16246 (2012)
19. Z. Hua, B. An, T. Iijima, C. Gu, J. Zheng, The finding of crystallographic orientation dependence of hydrogen diffusion in austenitic stainless steel by scanning Kelvin probe force microscopy. Scr. Mater. **131**, 47–50 (2017)
20. E.G. Astafurova et al., Hydrogen-enhanced orientation dependence of stress relaxation and strain-aging in Hadfield steel single crystals. Scr. Mater. **136**, 101–105 (2017)

21. E.B. Marin, On the formulation of a crystal plasticity model. Sandia National Laboratories (2006)
22. A. Siddiq, S. Schmauder, Simulation of hardening in high purity niobium single crystals during deformation. Steel Grips J. Steel Relat. Mater. **3**(4), 281–286 (2005)
23. E.I. Ogosi, U.B. Asim, M.A. Siddiq, M.E. Kartal, Modelling hydrogen induced stress corrosion cracking in austenitic stainless steel. J. Mech. **36**(2), 213–222 (2020)
24. R. Hill, J.R. Rice, Constitutive analysis of elastic-plastic crystals at arbitrary strain. J. Mech. Phys. Solids **20**(6), 401–413 (1972)
25. H.K. Birnbaumt, P. Sofronis, Mechanics of the hydrogen-dislocation-impurity interactions-I. Increasing shear modulus. J. Mech. Phys. Solids **43**(1), 49–90 (1995)
26. R.A. Oriani, Hydrogen embrittlement of steels. Annu. Rev. Mater. Sci. **8**(1), 327–357 (1978)
27. G. Schebler, On the mechanics of the hydrogen interaction with single crystal plasticity. University of Illinois (2011)
28. E. Ogosi, A. Siddiq, U.B. Asim, M.E. Kartal, Crystal plasticity based study to understand the interaction of hydrogen, defects and loading in austenitic stainless steel single crystals. Int. J. Hydrogen Energy
29. Y. Estrin, H. Mecking, A unified phenomenological description of work hardening and creep based on one-parameter models. Acta Metall. **32**(1), 57–70 (1984)
30. A. Krom, Numerical modelling of hydrogen transport of steel (1998)
31. G.R.J. Caskey, Hydrogen solubility in austenitic stainless steels. Scr. Metall. **34**(2), 1187–1190 (1981)
32. T. Luo, X. Gao, On the prediction of ductile fracture by void coalescence and strain localization. J. Mech. Phys. Solids **113**, 82–104 (2018)
33. C. Tekoglu, J.W. Hutchinson, T. Pardoen, On localization and void coalescence as a precursor to ductile fracture. Philos. Trans. R. Soc. A Math. Phys. Eng. Sci. **373**(2038) (2015)
34. Dassault Systèmes Simulia Corp, "ABAQUS 6.18." Providence, p. 2018 (2018)
35. U.B. Asim, M.A. Siddiq, M.E. Kartal, Representative volume element (RVE) based crystal plasticity study of void growth on phase boundary in titanium alloys. Comput. Mater. Sci. **161**, 346–350 (2019)
36. C. Tekoglu, Representative volume element calculations under constant stress triaxiality, lode parameter, and shear ratio. Int. J. Solids Struct. **51**(25–26), 4544–4553 (2014)
37. O. Barrera, D. Bombac, Y. Chen, T.D. Daff, P. Gong, D. Haley, Understanding and mitigating hydrogen embrittlement of steels : a review of experimental, modelling and design progress from atomistic to continuum. J. Mater. Sci. **53**(9), 6251–6290 (2018)

Molecular Modeling for Corrosion Inhibitor Design

Ime Bassey Obot and Abduljabar Q. Alsayoud

Abstract Different corrosion measurement techniques have been used in determining the efficiency of corrosion inhibitors. These techniques such as weight loss are tedious and time consuming and do not provide any information on the mechanism of interaction of the corrosion inhibitor with the metal surface at the molecular level. With the advancement in computer hardware and software development, interactions between a metal surface and corrosion inhibitor can be computed and quantify. Such information is important in the design and development of new and improved corrosion inhibitors. Molecular modeling technique such as molecular dynamics (MD) simulation has become a powerful tool for the design and mechanistic study of corrosion inhibitors in recent times. In this chapter, a brief introduction and theory of MD simulation are presented. Some applications of MD simulation in the design of corrosion inhibitor for the most important industrial metals such as steel, copper and aluminum are highlighted. Strengths and limitations of MD simulation to corrosion inhibitor design are discussed.

Keywords Molecular modeling · Corrosion inhibitors · Molecular dynamics simulations · Forcefields · Corrosion · Metals

1 Introduction

Over the years, the global cost of corrosion has significantly increased. In 2016, the global corrosion cost was estimated at US $2.5 trillion amounting to 3.4% of the global GDP [1]. In 2002, The U.S. Federal Highway Administration (FHWA) released a detailed study on the direct costs linked with metallic corrosion in nearly

I. B. Obot (✉)
Centre of Research Excellence in Corrosion, King Fahd University of Petroleum and Minerals, Dhahran 31261, Saudi Arabia
e-mail: obot@kfupm.edu.sa

A. Q. Alsayoud
Department of Mechanical Engineering, King Fahd University of Petroleum and Minerals, Dhahran 31261, Saudi Arabia

I. Toor (ed.), *Recent Developments in Analytical Techniques for Corrosion Research*,
https://doi.org/10.1007/978-3-030-89101-5_12

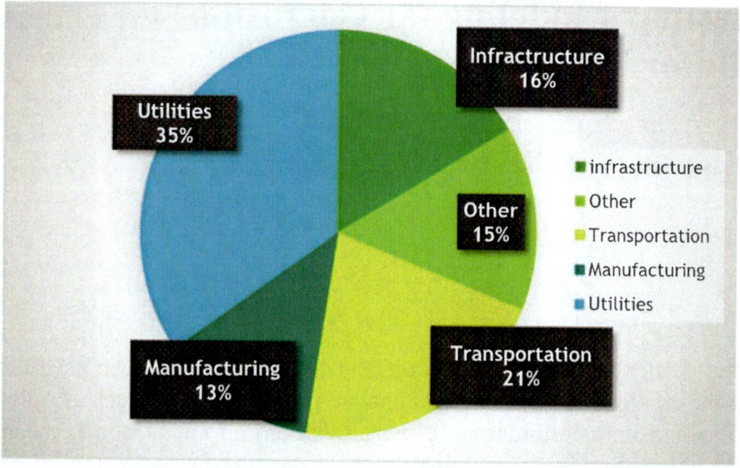

Fig. 1 Shows the cost of corrosion (in percentages) for the different industrial sector of the US economy in 2002 Adapted and redrawn from Ref. [2]

every U.S. industry, from infrastructure and transportation to production and manufacturing. The outcome of the study showed that the total direct cost of corrosion in the US was $276 billion which correspond to 3.1% of the US GDP [2]. This study illustrates without a doubt that mitigating corrosion is of extreme importance. The study also showed that the utilities sector which includes gas distribution pipelines, electrical transmission lines and potable water lines has the biggest share among other sectors such as infrastructure, transportation, manufacturing, and others. Figure 1 shows the cost of corrosion (in percentages) for the different industrial sector of the US economy in 2002.

Corrosion protection is an important aspect in the field of engineering. Corrosion protection methods include the selection of corrosion resistant materials, protective coatings, cathodic protection, and the use of corrosion inhibitors. Corrosion inhibitors are cost effective and are the most practical approach used in corrosion mitigation most especially in aqueous environments [3–5]. Corrosion inhibitors are chemicals that can decrease the rate of metals dissolution when present in a minimal concentration in aggressive environments [6]. Corrosion inhibitors can be classified into inorganic or organic chemical compounds. The organic corrosion inhibitors contain heteroatoms like sulphur, oxygen, nitrogen, and phosphorus in their molecular structure which serves as binding or adsorption sites as they interact with the metal surface. Figure 2 shows a typical classification of corrosion inhibitors and their electrochemical mechanism of actions.

Traditionally, evaluation of corrosion inhibition performance is primarily conducted experimentally using standard laboratory technique such as weight loss assessment, linear polarization, potentiodynamic polarization, and electrochemical impedance spectroscopy etc. [7–10]. However, these experimental methods are costly and time-consuming, and are often deficient in elucidating inhibition mechanisms.

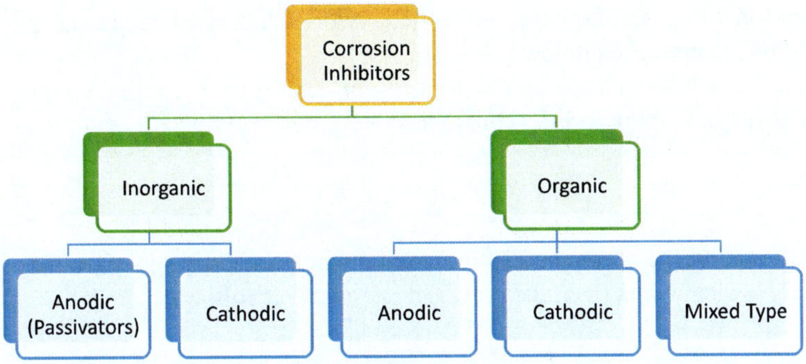

Fig. 2 Typical classification of corrosion inhibitors

Given the tremendous advancements in the area of computer programming and software development, researchers are beginning to embrace computer simulation methods as an essential tool for understanding the mechanism of how corrosion inhibitors adsorbed on metal surfaces with the overall aim of designing better corrosion inhibitors [11, 12]. These computer simulation methods operate based on mathematical and theoretical principles to provide deeper insights (on an atomic and molecular scale) into the complex interfacial processes controlling the metal corrosion and corrosion protection phenomena [13].

Molecular modeling methods can be classified into three major categories, namely: ab initio electronic structure calculations, semi-empirical methods, and molecular mechanics (or atomistic simulations). Atomistic simulations also called force field methods or molecular mechanics are purely empirical methods and are based upon the principles of classical physics, they rely on force fields with embedded empirical parameters to predict properties of molecules. The two main categories of the classical atomistic simulations are the Molecular Dynamics (MD) and Monte Carlo (MC) simulations techniques [13, 14]. Molecule dynamics simulations are usually performed under some well-defined (fixed) thermodynamic states, such as a constant number of particles (N), volume (V) and energy (E). These thermodynamic states defined the ensemble of the system under investigation.

In this chapter, a brief introduction and theory of molecular dynamics simulation (MDS) are presented. Some applications of MDS in the design of corrosion inhibitor for the most important industrial metals such as steel, copper and aluminum are highlighted. Finally, the strengths and limitations of MDS to corrosion inhibitor design are discussed.

2 Theory of Molecular Dynamics Simulation

The MD simulation is also called the deterministic simulation. It provides the actual trajectory of a system by simulating the time evolution of the system through solving the Newton's classical equation of motion. A brief theory of molecular dynamics simulations is presented below [13, 15].

For a particle (say an atom), ***i***, with mass, $\boldsymbol{m_i}$, the force, $\boldsymbol{F_i}$, acting on the particle can be given according to Eq. (1a) [15]:

$$F_i = m_i a_i(t) \tag{1a}$$

The force on the atom can be computed at each time step based on the relative potion of all the atoms from the potential energy, ***V*** as follows:

$$F_i(t) = -\Delta V(\mathrm{r}(t)) \tag{1b}$$

Then by integrating the Newton's law of motion Eq. (2)., successive configurations of the system can be produced which yield the trajectory; specifying how the coordinates and velocities of the atoms in the system vary as a function of time.

$$\frac{dV}{dr_i} = m_i \frac{dr_i}{dt^2} \tag{2}$$

The integration of the Newton's law of motion involves the Taylor series expansion of the positions (***r***), velocities (first derivative with respect to time, ***v***) and accelerations (second derivative with respect to time, ***a***) of the particle.

$$r(t + \delta t) = r(t) + v(t)\delta t + \frac{1}{2}a(t)\delta t^2 + \cdots \tag{3}$$

$$v(t + \delta t) = v(t) + a(t)\delta t + \frac{1}{2}b(t)\delta t^2 + \cdots \tag{4}$$

$$a(t + \delta t) = a(t) + b(t)\delta t + \cdots \tag{5}$$

If the acceleration of an atom is known, then its velocity in the next time step can be predicted and calculated. Positions and velocities of the atom in the next step can be calculated from the atom positions, velocities, and accelerations at any previous instant in time. This can be achieved using any of the simulation algorithms like the Verlet algorithm or the Leapfrog algorithm.

The Verlet algorithm uses positions and accelerations at a time ***t*** and the positions from time ***t* – *δt*** to calculate new positions at time ***t* + *δt***.

$$r(t + \delta t) = r(t) + v(t)\delta t + \frac{1}{2}a(t)\delta t^2 \tag{6}$$

$$r(t - \delta t) = r(t) + v(t)\delta t + \frac{1}{2}a(t)\delta t^2 \tag{7}$$

Adding both equations yields Eq. (8).

$$r(t+\delta t) = 2r(t) - r(t-\delta t) + a(t)\delta t^2 + \ldots \tag{8}$$

In the Leapfrog algorithm, the velocities of the particle are calculated at half time step intervals; first, at time $\boldsymbol{t + \frac{1}{2}\delta t}$, and then used to calculate the particle position at another time $\boldsymbol{t + \delta t}$. It has an advantage that the velocities can be explicitly calculated. However, a major disadvantage is that the velocities cannot be calculated at the same time as the positions; making it difficult to evaluate the total energy (which is the sum of the potential and kinetic energies) at any point in time. The velocities at time, $\boldsymbol{t}$, can be estimated by the relationship in Eq. (9).

$$v(t) = \frac{1}{2}\left[v\left(t - \frac{1}{2}\delta t\right) + v\left(t + \frac{1}{2}\delta t\right)\right] \tag{9}$$

In MD simulations, three key steps are employed: Energy minimization, dynamic simulation and prediction of desired properties. Figure 3 schematically illustrates the key steps in MD cycle. Energy minimization serves to find the lowest energy of a system as a function of atomic positions. For this end, several minimization algorithms including steepest descent, conjugate gradient, Newton–Raphson, and simplex method can be employed.

Selection of an appropriate time step Δt is critical to the accuracy of MD simulations. Large jump in the potential energy can be expected for large time steps. This can result in a violation of energy conservation and the trajectories to be wrong. Very small time steps can prevent convergence and result in poor statistics due to the limited exploration of the phase space. In order to get the correct dynamics and enough statistics from the phase space, the time step should be a fraction of the atomic vibration.

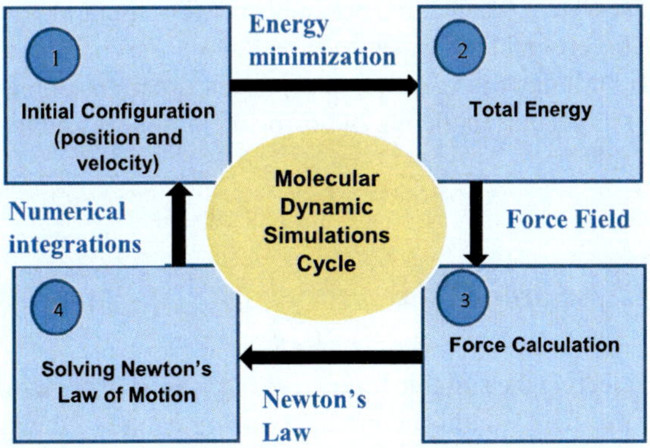

Fig. 3 Key steps in molecular dynamic simulations cycle

Table 1 Summary of commonly used ensembles utilized in molecular modeling of corrosion inhibitors [16]

Ensembles	Description
Microcanonical ensemble (NVE)	In the microcanonical, or NVE ensemble, the system is isolated from changes in moles (N), volume (V) and energy (E). It corresponds to an adiabatic process with no heat exchange. A microcanonical molecular dynamics trajectory may be an exchange of potential and kinetic energy, with total energy being conserved
Canonical Ensemble (NVT)	In the canonical ensemble, moles (N), volume (V) and temperature (T) are conserved. It is also sometimes called constant temperature molecular dynamics (CTMD). In NVT, the energy of endothermic and exothermic processes is exchanged with a thermostat
Isobaric-Isothermal Ensemble (NPT)	In the isothermal-isobaric ensemble, moles (N), pressure (P) and temperature (T) are conserved. In addition to a thermostat, a barostat is needed. It corresponds most closely to laboratory conditions with a flask open to ambient temperature and pressure

2.1 *Ensembles*

A collection of all possible systems which have different microscopic states but have an identical macroscopic or thermodynamic state is called an ensemble [16]. There exist different ensembles with different characteristics. The microcanonical, canonical and isobaric-isothermal ensembles are the most known and utilized in molecular modeling of corrosion inhibitors. For each of these ensembles, there is a set of fixed variables, the macroscopic thermodynamic observables that define the constraints on the system. The natural ensemble in MD simulations is the microcanonical ensemble. This ensemble represents the energy-conserving isolated system. Modification to the Hamiltonian is required to represent other ensembles. Commonly used ensembles in molecular modeling of corrosion inhibitors and their descriptions is presented in Table 1.

2.2 *Force Fields*

Selecting the suitable force field is the heart of the atomistic simulation. The selected force field should capture the bonding nature of the system to ensure the simulation accuracy. Further, the force field should be able to reproduce the equilibrium properties of the system. This includes the system density, thermal expansion and elastic moduli among other properties.

One of the most common ways to model the potential energy (U) of a system is defined on means of bonded and non-bonded terms. The bonded interactions included the bond stretching (U_b), angle bending (U_θ), and twisting a dihedral angle ($U_\varnothing$); whereas the non-boned interactions included the potential energy from electrostatic (U_{ele}) and van der Waals (U_{vdW}) interactions. The general expression of the potential energy (U) can be expressed as (Eq. 10) [17].

$$U = U_b + U_\theta + U_\varnothing + U_{ele} + U_{vdW} \tag{10}$$

Among several empirical force fields, Universal forcefield (UFF), consistent valence forcefield (CVFF), condensed-phase optimized molecular potentials for atomistic simulation studies (COMPASS), are commonly used in the modeling of corrosion inhibitors using MDS. More details on force fields utilized in molecular modeling of corrosion inhibitors are presented in Table 2.

For instance, the COMPASS force field potential energy expression is given as [18]:

$$\begin{aligned}
E_{total} = & \sum_b [k_2(b - b_o)^2 + k_3(b - b_o)^3 + k_4(b - b_o)^4] \\
& + \sum_\theta [k_2(\theta - \theta_o)^2 + k_3(\theta - \theta_o)^3 + k_4(\theta - \theta_o)^4] \\
& + \sum_\varphi [k_1(1 - \cos\varphi) + k_2(1 - \cos 2\varphi) + k_3(1 - \cos 3\varphi)] \\
& + \sum_\chi k_2\chi^2 + \sum_{b,b'} k(b - b_o)(b' - b'_o) \\
& + \sum_{b,\theta} k(b - b_o)(\theta - \theta_o) + \sum_{b,\theta} (b - b_o)[k_1 \cos\theta \\
& + k_2 \cos 2\varphi + k_3 \cos 3\varphi] + \sum_{\theta,\varphi} (\theta - \theta_o)[k_1 \cos\varphi \\
& + k_2 \cos 2\varphi + k_3 \cos 3\varphi] + \sum_{b,\theta} k(\theta' - \theta'_o)(\theta - \theta_o) \\
& + \sum_{\theta,\theta,\varphi} k(\theta - \theta_o)(\theta' - \theta'_o)\cos\varphi + \sum_{i,j} \frac{q_i q_j}{r_{ij}} \\
& + \sum_{i,j} \varepsilon_{ij}\left[2\left(\frac{r^o_{ij}}{r_{ij}}\right)^9 - 3\left(\frac{r^o_{ij}}{r_{ij}}\right)^6\right]
\end{aligned} \tag{11}$$

The first four terms account for bond stretch, angle bend, out-of-plane torsion and out-of-plane wag energies, terms five to ten are for cross-coupling and the last two terms show electrostatic interactions and van der Waals energies, respectively. b, θ, φ *and* χ *are* bond length, angle, torsion angle and out-of-plane wag or inversion angle, respectively. *qi* and *qj* are atomic charges and *rij* is the interatomic separation. Parameters of bo and θo are equilibrium values and k and k_1- k_3 are constants.

Table 2 Summary of force fields commonly used on molecular modeling of corrosion inhibitors [19]

Force field	Description
COMPASS[a]	This is a high quality ab initio force field that has been parameterized and validated using condensed phase properties. The forcefield enables accurate and simultaneous prediction of structural, conformational, vibrational, and thermophysical properties, that exist for organic molecules, inorganic small molecules, and polymers in isolation and in condensed phases, and under a wide range of conditions of temperature and pressure
CVFF[b]	CVFF was fit to small organic (amides, carboxylic acids, etc.) crystals and gas phase structures. It handles peptides, proteins, and a wide range of organic systems. As the default forcefield in Discover, it has been used extensively for many years. It is primarily intended for studies of structures and binding energies, although it predicts vibrational frequencies and conformational energies reasonably well
Universal	Universal forcefield (UFF) has full coverage of the periodic table. UFF is moderately accurate for predicting geometries and conformational energy differences of organic molecules, main-group inorganics, and metal complexes. It is recommended for organometallic systems and other systems for which other forcefields do not have parameters. It is mainly diagonal, harmonic force field where electrostatic and van der Waals interactions are described by atomic monopoles the Lennard–Jones potential, respectively
Dreiding	The Dreiding forcefield has good coverage for organic, biological and main-group inorganic molecules. It is only moderately accurate for geometries, conformational energies, intermolecular binding energies, and crystal packing. The Dreiding forcefield is a purely diagonal forcefield with harmonic valence terms and a cosine-Fourier expansion torsion term. The van der Waals interactions are described by the Lennard–Jones potential. Electrostatic interactions are described by atomic monopoles and a screened (distance-dependent) Coulombic term. Hydrogen bonding is described by an explicit Lennard–Jones 12–10 potential
PCFF[c]	It is primarily developed for application to polymers and organic materials. It is similar to the COMPASS force field; however, it is parametrized for a different range of functional groups

[a] COMPASS = Condensed-phase Optimized Molecular Potentials for Atomistic Simulation Studies
[b] CVFF = Consistent valence force field
[c] PCFF = Polymer Consistent Force Field

2.3 *Periodic Boundary Conditions*

Periodic boundary conditions (PBC) is one of the methods that can be utilized to simulate a system of infinite size while keeping a low number of actual particles. It is also a method that prevents the system from being dominated by the influence of particles constantly colliding with walls of the simulation box. This allows for the simulation of bulk properties, even in a smaller system with a limited number of particles. The corrosion inhibition systems are infinite (massive) systems, where thousands of atoms and molecules involved. The use of periodic boundary conditions

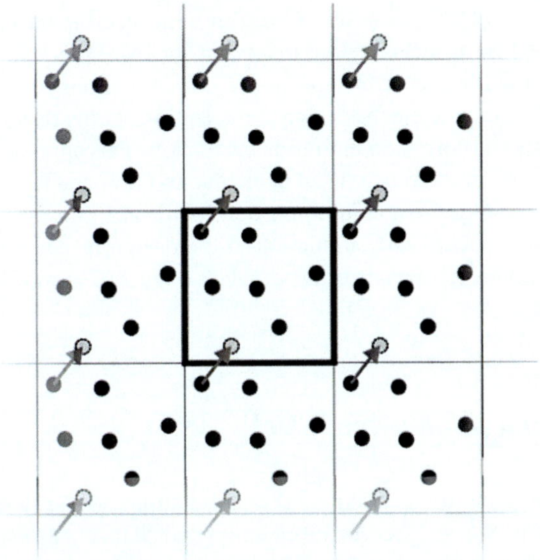

Fig. 4 Illustration of periodic boundary conditions. The central box is here considered to be the simulation box, surrounded by copies of itself. When a particle, here indicated by a black circle, leaves the simulation box and enters a neighboring box, another particle, gray, which is in principle the same as the previous black particle, enters from the other direction (Adapted from [16])

helps the simulations to track only small number of particles to reduce computational cost [16].

The general concept is quite simple and is illustrated in two dimensions in Fig. 4. When a particle, visualized as circles in Fig. 4, moves outside the simulation box it reenters from the opposite side. This can be considered as if the particle moves into another box, and that a new particle enters the simulation box from a third box. In effect, one obtains a seemingly infinite simulation system, free from wall effects, with constant number density, capable of reproducing bulk properties.

3 Application of MD Simulations to Corrosion Inhibitor Design

Molecular modeling provides the ability to simulate and analyze hypothetical processes at the atomic and molecular level. Such models have been used extensively in the realm of chemistry, as well as in solid-state physics and materials science [20]. Molecular dynamics simulation has been used extensively to model corrosion inhibitor-metal interface interactions and configurations which is very useful in corrosion inhibitor design. Important static and dynamic parameters such as adsorption (binding) energies of corrosion inhibitor film with metal surfaces, diffusion coefficient calculated from mean square displacement (MSD) equation and fractional volume of inhibitor film can be obtained from MD simulations [21–23]. These parameters can be used as a guide in the design and selection of effective corrosion inhibitor prior to synthesis and corrosion inhibition evaluation. This approach is cost effective and saves time and effort usually used in conduction trial and error experimentation during the process of corrosion inhibitor development. In this section, some examples of the application of MD simulation to the design of corrosion inhibitors for three important industrial metals iron/steel (Fe), copper (Cu) and aluminum (Al) are briefly discussed.

3.1 Corrosion Inhibition of Steel

Before discussing the application of MD simulations to the design of corrosion inhibitors for steel, we will present the electrochemical mechanism of steel corrosion in aqueous corrosive environments. The important electrochemical oxidation and reduction reactions involved in the steel corrosion process are presented in Eqs. (12–16) [24]:

Anodic reaction:

$$Fe \rightarrow Fe^{2+} + 2e^{-} \text{ (Iron dissolution)} \quad (12)$$

Cathodic reactions:

$$2H^{+} + 2e^{-} \rightarrow H_2 \text{ (Hydrogen reduction)} \quad (13)$$

$$O_2 + 2H_2O + 4e^{-} \rightarrow 4OH^{-} \text{ (Oxygen reduction in neutral and alkalis media)} \quad (14)$$

$$O_2 + 4H^{+} + 4e^{-} \rightarrow 2H_2O \text{ (Oxygen reduction in acidic media)} \quad (15)$$

Overall corrosion reaction in acidic media

$$Fe + 2H^{+} \rightarrow Fe^{2+} + H_2 \quad (16)$$

In order to reduce the rate of iron dissolution or hydrogen evolution, corrosion inhibitors are added to the aqueous solution. We present two examples of the use of molecular dynamics simulations for the design of steel corrosion inhibitors.

(i) ***Modeling of thiazole based corrosion inhibitor***

Molecular dynamics simulation was applied to model corrosion inhibitor for steel in an acidic medium based on thiazole molecules [25]. There corrosion inhibitor namely 1, 3-thiazole (TA), 2-amine-1,3-thiazole (2-ATA), and 4-amine-1,3-thiazole (4-ATA), were used for the simulation. Fe (110) surface was adopted for the simulation as it represents the most stable slab of Fe. COMPASS forcefield was used for the geometry optimization. The simulation was performed under 298 K, NVT ensemble, a time step of 1 fs and simulation time of 200 ps. The simulation was carried out in gas and aqueous phases. The adsorption (binding) energy of the molecules on the Fe (110) surface was as an important descriptor used to rank the performance of the three molecules as corrosion inhibitors. The binding energy of 4-ATA on the Fe (110) surface was more negative than the other two molecules investigated. This shows that 4-ATA could be selected as the best corrosion inhibitor for steel. Figure 5 shows the equilibrium adsorption configuration for the three corrosion inhibitors

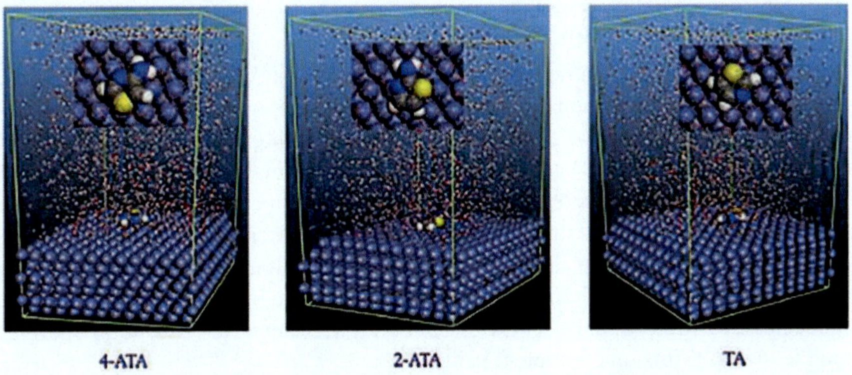

Fig. 5 Equilibrium adsorption configurations of the adsorption of the three corrosion inhibitors abbreviated as TA, 2-ATA and 4-ATA on Fe (110) surface in aqueous (water) phase. Reproduced with permission from Guo et al. [25]. Copyright 2017 Eslevier Inc

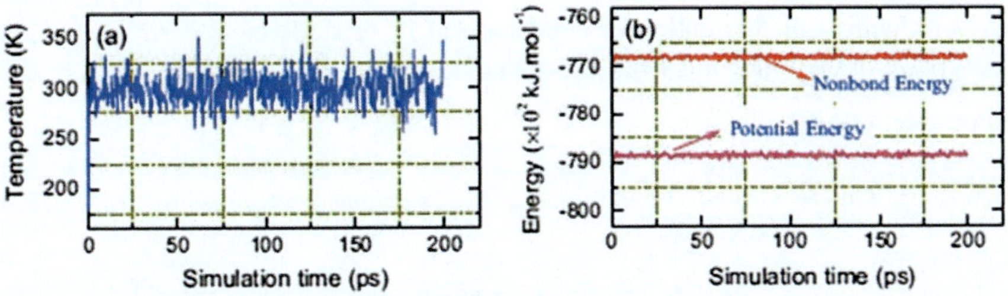

Fig. 6 **a** Temperature equilibrium curve and **b** Energy fluctuation curve during the simulation of 4-ATA on Fe (110) surface. Reproduced with permission from Guo et al. [25]. Copyright 2017 Eslevier Inc

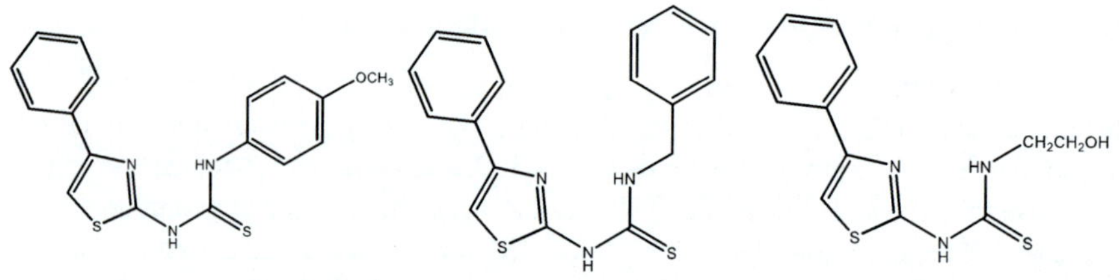

Fig. 7 The molecular structures of the three thiourea molecules investigated. Reproduced with permission from Guo et al. [26]. Copyright 2017 Eslevier Inc

after molecular dynamics simulation. Also, Fig. 6 shows the temperature equilibrium curve and the energy fluctuation curve during the MD simulations.

(ii) ***Modeling of thiourea based corrosion inhibitor***

Three thiourea derivatives with different substituent were theoretically evaluated using molecular dynamic simulation [26]. The three molecules were abbreviated Inh1, Inh2 and Inh3 as presented in Fig. 7. Fe (110) surface was selected as the adsorption substrate. COMPASS forcefield was used to optimize all the structures investigated. The MD simulation was performed under NVT ensemble with a time step of 1.0 fs and simulation time of 500 ps. The adsorption energy and the diffusion coefficient parameters calculated in acidic and in chloride solution using mean square displacement (MSD) were used to rank the performance of the three corrosion inhibitors. Figure 8a presents the equilibrium adsorption configuration of the three molecules as they interact with Fe (110) surface in the presence of water while Fig. 8b shows the MSD plots of Cl^- and H_3O^+ in the three inhibitor membrane at 303 K. The modeling results indicate that Inh1 was the best corrosion inhibitors among the three molecules investigated since it possesses the highest negative adsorption energies and the lowest diffusion coefficient value.

3.2 *Corrosion Inhibition of Copper*

The mechanism of dissolution of copper metal in aerated acidic corrosive media takes place in the following steps as presented below in Eqs. (17–22) [27]:

Anodic reactions:

$$Cu \leftrightarrow Cu^{+}e^{-} \text{ (Oxidation of metallic copper to cuprous ion)} \quad (17)$$

$$Cu^{+} + Cl^{-} \rightarrow CuCl \text{ (Precipitation of insoluble Copper chloride)} \quad (18)$$

The product CuCl in Eq. (18) does not provide enough protection to the copper surface and subsequently combine with more Cl^- ions to form $CuCl_2{}^-$ as depicted in Eq. (19):

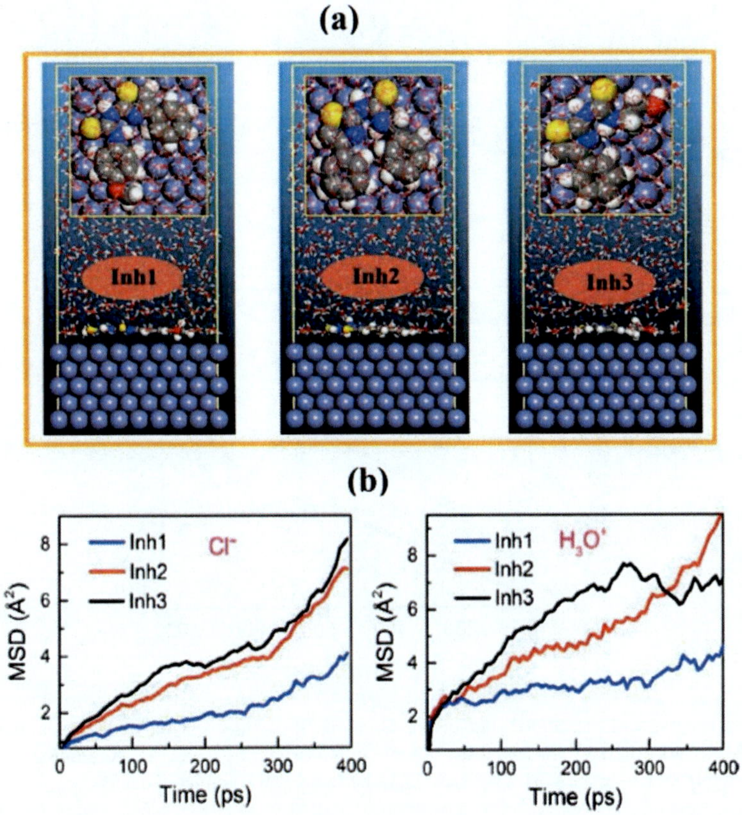

Fig. 8 **a** Equilibrium adsorption configurations of the adsorption of the three corrosion inhibitors abbreviated as Inh1, Inh2 and Inh3 on Fe (110) surface in aqueous (water) phase and **b** MSD plots of Cl^- and H_3O^+ in the three corrosion inhibitor membrane at 303 K. Reproduced with permission from Guo et al. [26]. Copyright 2017 Eslevier Inc

$$CuCl + Cl^- \rightarrow CuCl_2^- \quad (19)$$

This is followed by the oxidation of the complex $CuCl_2^-$ into cupric ions Cu^{2+} as shown in Eq. (20). Thus, the dissolution of copper occurs in this step.

$$CuCl_2^- \leftrightarrow Cu^{2+} + 2Cl^- + 2e^- \quad (20)$$

Cathodic reaction:

The cathodic reaction can be represented as follows:

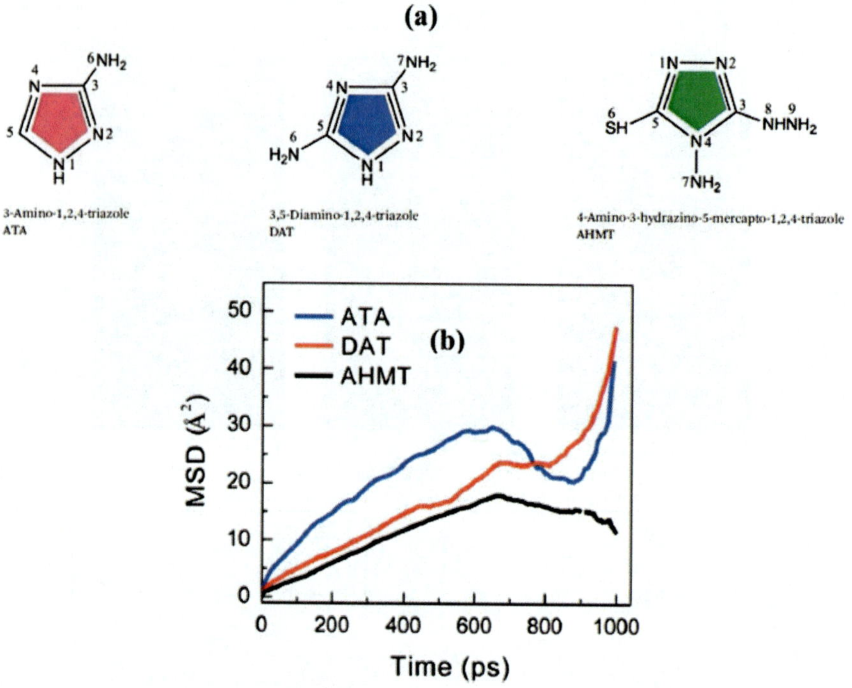

Fig. 9 **a** The molecular structures and the abbreviations of the three triazole molecules investigated and **b** MSD plots of H_3O^+ for the three triazole inhibitor membrane at 303 K. Reproduced with permission from Guo et al. [28]. Copyright 2014 by the Royal Chemical Society

$$O_2 + 4H^+ + 4e^- \rightarrow 2H_2O \text{ (Oxygen reduction in acidic media)} \quad (21)$$

The overall electrochemical corrosion reaction of copper in aerated acidic solution is given as:

$$Cu + 4H^+ + 4Cl^- + O_2 \rightarrow 2H_2O + 4Cl^- + 2Cu^{2+} \quad (22)$$

In order to reduce the corrosion rates of either the anodic or cathodic processes, corrosion inhibitors are added to the aqueous solution. We present an example of the use of molecular dynamics simulations for modeling of copper corrosion inhibitors.

(i) ***Modeling of triazole based corrosion inhibitors***

Three new triazole derivatives namely 3-Amino-1,2,4-triazole (ATA), 3–5-Diamino-1,2,4-triazole (DAT) and 4-Amino-3-hydrazino-5-mercapto-1,2,4-triazole (AHMT) were modeled using different theoretical techniques including molecular dynamic simulations [28]. Diffusion coefficients were computed and used in the ranking of the three corrosion inhibitors.

Figure 9a shows the molecular structures and the abbreviation of the investigated triazole molecules. Also, Fig. 9b shows the MSD plot of H_3O^+ in the three inhibitor membrane at 303 K during 1 ns MD simulations. From the MSD results obtained, AHMT was the best corrosion inhibitor since it has the lowest diffusion coefficient in the H_3O^+ media. The fractional free volume distribution of the three triazoles inhibitor on Cu (111) surface in the presence of H_3O^+ corrosive particles is shown in Fig. 10. Table 3 presents the corresponding values of the diffusion coefficient and the fractional free volume obtained after the molecular dynamic simulation. It is generally believed that a large free volume means there are abundant cavities among the corrosion inhibitor membranes. Thus, the migration probability of the corrosive particles (H_3O^+) is high. This induces high diffusion coefficient and by extension low inhibition efficiency. The result in Table 3 shows that AHMT has the smallest fractional free volume value among the three molecules investigated and thus can reduce the migration of the corrosive particle and prevent the corrosion of the copper.

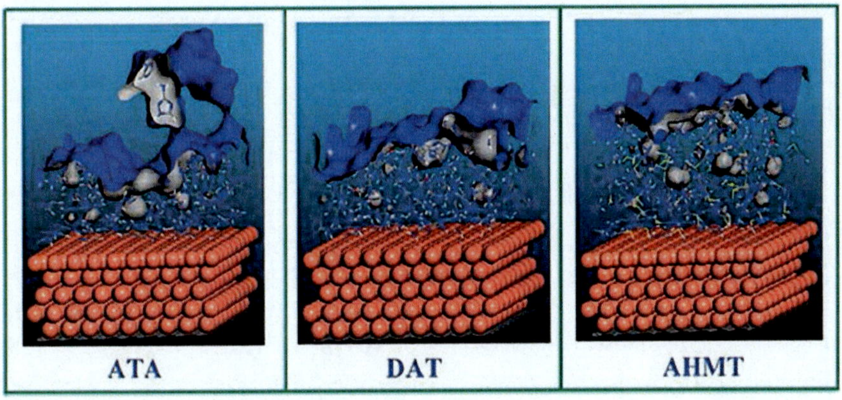

Fig. 10 The fractional free volume distribution of the three triazole inhibitors on Cu (111) surface simulated in the presence of H_3O^+ using molecular dynamics. With permission from Guo et al. [28] Reproduced with permission from Guo et al. [28]. Copyright 2014 by the Royal Chemical Society

Table 3 Calculated diffusion coefficient and fractional free volume values. Reproduced with permission from Guo et al. [28]. Copyright 2014 by the Royal Chemical Society

Corrosion inhibitor membrane	Diffusion coefficient ($\times 10^{-7}$ cm^2 s^{-1})	Fractional free volume
ATA	2.154	0.210
DAT	1.456	0.140
AHMT	1.275	0.069

3.3 Corrosion Inhibition of Aluminum

Aluminum (Al) shows good corrosion resistance when exposed to the atmosphere and many aqueous environments due to the formation of a protective oxide layer. However, when aluminum is exposed to acidic and alkaline conditions, the oxide layer can be destroyed, resulting in the corrosion of aluminum. The electrochemical mechanism for the corrosion of aluminum in chloride containing solution can be described according to Eqs. (23–27) [29]:

Anodic reaction:

Chloride attacks the oxide layer and exposes the metal. There are two main anodic reactions:

$$\mathrm{Al} \leftrightarrow \mathrm{Al}^{3+} + 3\mathrm{e}^{-} \tag{23}$$

$$\mathrm{Al}^{3+} + 3\mathrm{H_2O} \leftrightarrow \mathrm{Al\,(OH)_3} + 3\mathrm{H}^{+} \tag{24}$$

The reaction in Eq. 24, creates more acidic condition (pH = 3–4). The chloride ions available in solution facilitates the anodic dissolution of Al, resulting in the formation of aluminum chloride ($AlCl_3$). The $AlCl_3$ hydrolyzes to form the hydroxide and acid which shifts the pH to acidic values as shown in Eq. 25.

Cathodic reaction:

The following three possible reactions can occur at the cathodic sites:

$$\mathrm{AlCl_3} + 3\mathrm{H_2O} \leftrightarrow \mathrm{Al\,(OH)_3} + 3\mathrm{HCl} \tag{25}$$

$$3\mathrm{H}^{+} + 3\mathrm{e}^{-} \leftrightarrow 3/2\mathrm{H_2} \tag{26}$$

$$1/2\mathrm{O_2} + \mathrm{H_2O} + 2\mathrm{e}^{-} \leftrightarrow 2\mathrm{OH}^{-} \tag{27}$$

As seen in Eq. 27, the cathodic sites are expected to be more alkaline due to the formation of hydroxide. In order to reduce the corrosion rate of aluminum in chloride containing aqueous solution, several methods and techniques such as anodizing, coating and corrosion inhibitors are commonly used. Due to ease of application and cost, the use of corrosion inhibitors is the most practical approach of reducing the corrosion rate of Al in aqueous chloride solution.

We present an example of the use of molecular dynamics simulations for modeling of aluminum corrosion inhibitor. Generally, theoretical research for the modeling of corrosion inhibitors for aluminum corrosion is limited.

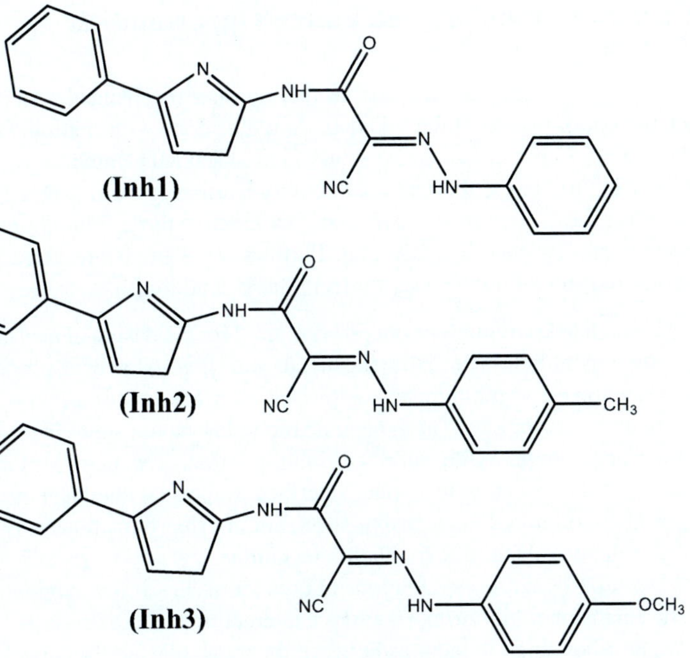

Fig. 11 Chemical structures of studied N-thiazolyl-2-cyanoacetamide derivatives. Reproduced with permission from Zhang et al. [30]. Copyright 2018 Eslevier Inc

(ii) ***Modeling of N-thiazolyl-2-cynoacetamide based corrosion inhibitors***

Cynoacetamides have proved as a good corrosion inhibitor for aluminum [30]. In order to understand the inhibition mechanism and promote the development of more derivatives of cynoacetamides as corrosion inhibitors for aluminum, molecular dynamics simulation was performed on three derivatives of N-thiazolyl-2-cynoacetamide. Figure 11 shows the molecular structures of the investigated cynoacetamides derivatives. The diffusion behavior of the corrosion particles in the corrosion inhibitors was determined. COMPASS force field was chosen in this study. MD simulation was conducted under NVT ensemble with a time step of 0.5 fs and a simulation time of 500 ps. Diffusion coefficients of three different corrosive particles namely H_2O, OH^- and O_2 in the three corrosion inhibitors were computed and used in the ranking of the three corrosion inhibitors. The results indicated that Inh3 was the best corrosion inhibitor for aluminum metal among the three molecules modeled. The theoretical results were in agreement with experimentally determined inhibition efficiencies of the three molecules.

4 Current Limitations and Future Opportunities

While MD simulations can predict the thermal and mechanical properties of the system, the interpretation of the MD data should be done with caution. For example, electronic effects are not treated explicitly in classical MD simulations. So the electronic contribution to the thermal conductivity is missing from MD. Also, for non-reactive force fields, electron transfer is not included in these simulations, and hence MD cannot predict chemical bonding. Further, force fields are fitted for specific conditions and should not be used far from these conditions.

(1) Although MD simulations is a powerful tool for the design of new and effective corrosion inhibitors for industrial metals and alloys. However, proper modeling of the corrosion inhibitor-metal surface interface is still a major problem in atomistic simulations. This is because modeling such systems involves complex interfaces, metal/oxide interface, inhibitor molecule/metal surface interface and metal or metal oxide/ water interface. Important chemical events like the metal dissolution, change in oxidation state and deprotonation are not described by molecular dynamics simulation technique.
(2) Molecular dynamics simulations allow identifying at low computational cost the corrosion inhibitor/metal surface interaction and are useful for calculating the non-bonding interactions between the molecule and the surface. However, information about covalent interaction between corrosion inhibitors and the metal surface cannot be obtained using force field based MD simulations.
(3) Molecular dynamics simulations can also be used to simulate the evolution of corrosion product morphologies, diffusion processes, effects of defects (such as grain boundaries) on the metal corrosion, and the rates at which oxidation or other processes occur. However, electronic level information such as formation and breaking of covalent bonds during metal corrosion and inhibition processes cannot be investigated using MD simulations.
(4) Reactive molecular dynamics simulation has recently emerged as a new computational method that can provide chemical bonding between corrosion inhibitor and the metal surface comparable to DFT based methods unlike the classical forcefield based MD simulation discussed in this chapter. This advances will provide a faster approach to model corrosion inhibition systems compared to the more computationally expensive DFT based methods.

5 Conclusion

Corrosion inhibitors are widely used for the mitigating of corrosion of industrial metals are alloys. Traditionally, the discovery of new corrosion inhibitors is carried out by costly trial and error approach. The recent advancement in computer hardware and software development has provided an opportunity for the rational design and discovery of corrosion inhibitors at a reduce cost. Molecular dynamics (MD) simulation can be utilized to analyze to compute important energetics and diffusion parameters of complex interfacial processes during corrosion inhibition modeling which are very difficult or sometimes impossible to be determined from experiments. This parameters can then be used in the screening of potential corrosion inhibitors and can serve as a guide for corrosion inhibitor design. In this chapter, a brief introduction and theory of MDS was presented. Some applications of MDS in the design of corrosion inhibitor for the most important industrial metals such as steel, copper and aluminum are highlighted. Strengths and limitations of MDS to corrosion inhibitor design are discussed.

References

1. NACE report, Impact.nace.org. (2016)
2. G.H. Koch, M.P.H. Brongers, N.G. Thompson, Y.P. Virmani, J.H. Payer, *Corrosion costs and preventive strategies in the United States*. United Department of Transportation, Federal highway Administration, Publication No. FHWA-RD-01–156 (2002)
3. A. Singh, K.R. Ansari, M.A. Quraishi, H. Lgaz, Effect of electron donating functional groups on corrosion inhibition of J55 steel in a sweet corrosive environment: experimental, density functional theory and molecular dynamic simulation. Molecules **12**(1), 1–19 (2019)
4. I.B. Onyeachu, D.S. Chauhan, K.R. Ansari, I.B. Obot, M.A. Quraishi, A.H. Alamri, Hexamethylene-1,6-bis(N-D-glucopyranosylamine) as a novel corrosion inhibitor for oil and gas industry: electrochemical and computational analysis. New J. Chem. **43**, 7282–7293 (2019)
5. I.B. Obot, I.B. Onyeachu, S.A. Umoren, Alternative corrosion inhibitor formulation for carbon steel in CO_2–saturated brine solution under high turbulent flow condition for use in oil and gas transportation pipelines. Corros. Sci. **159**, 108140 (2019)
6. S.A. Umoren, M.M. Solomon, I.B. Obot, R.K. Suleiman, A critical review on the recent studies on plant biomaterials as corrosion inhibitors for industrial metals. J. Ind. Eng. Chem. **76**, 91–115 (2019)
7. E. Barmatov, T. Hughes, M. Nagl, Efficiency of film-forming corrosion inhibitors in strong hydrochloric acid under laminar and turbulent flow conditions. Corros. Sci. **92**, 85–94 (2015)
8. C. Yang, Y. Liu, Y. Liu, L. Li, P. Fang, Z. Huang, M. Li, Y. Yang, Study of the synergistic inhibition effect of imidazolines and mercaptoethanol on the corrosion of carbon steel in Tarim oilfield, in *2nd International Conference on Energy and Power Engineering*, pp. 94–99 (2008.
9. J. Tang, Y. Hu, Z. Han, H. Wang, Y. Zhu, Y. Wang, Z. Nie, Y. Wang, Experimental and theoretical study on the synergistic inhibition effect of pyridine derivatives and sulfur–containing compounds on the corrosion of carbon steel in CO_2-saturated 3.5 wt.% NaCl solution. Molecules **23**(12), 3270 (2018)
10. Z. Belarbi, J.M. Dominguez Olivio, F. Farelas, M. Singer, D. Young, S. Nesic, Corrosion inhibitor (Decanethiol) for carbon steels exposed to aqueous CO_2, in *NACE 13184, NACE International Conference and Expo* (2019).
11. I.B. Obot, D.D. Macdonalda, Z.M. Gasem, Density functional theory (DFT) as a powerful tool for designing new organic corrosion inhibitors. Part 1: An overview. Corros. Sci. **99**, 1–30 (2015)
12. H. Ke, C.D. Taylor, Density functional theory: an essential partner in the integrated computational materials engineering approach. Corrosion **75**(7), 708–726 (2019)
13. I.B. Obot, K. Haruna, T.A. Saleh, Atomistic simulation: a unique and powerful computational tool for corrosion inhibition research. Arab. J. Sci. Eng. **44**, 1–32 (2019)
14. P. Ojeda, M.E. Garcia, A. Londoño, N.Y. Chen, Monte Carlo simulations of proteins in cages: Influence of confinement on the stability of intermediate states. Biophys. J . **96**, 1076–1082 (2009)

15. B. Rai, *Molecular Modeling for the Design of Novel Performance Chemicals and Materials* (Taylor and Francis, FL, USA, 2012)
16. B. Jensen, *Investigation into the Impact of Solid Surfaces in Aqueous Systems*, Ph.D. Thesis (University of Bergen, 2016)
17. H. Sun, P. Ren, J.R. Fried, The COMPASS force field: parameterization and validation for phosphazenes. Comput. Theor. Polym. Sci. **8**(1), 229–246 (1998)
18. H. Sun, COMPASS: an ab initio force-field optimized for condensed-phase applications—overview with details on alkane and benzene compounds. J. Phys. Chem. B **102**(38), 7338–7364 (1998)
19. http://www.chem.cmu.edu/courses/09-560/docs/msi/ffbsim/2_Forcefields.html#567119. Retrieved 16 June 2020
20. C.D. Taylor, P. Marcus, *Molecular Modeling of Corrosion Process Scientific Development and Engineering Applications* (John Wiley and Sons Inc., New Jersey, USA, 2015)
21. Z. Zhang, N. Tian, X. Li, L. Zhang, L. Wu, Y. Huang, Synergistic inhibition behavior between indigo carmine and cetyltrimethyl ammonium bromide on carbon steel corroded in a 0.5 M HCl solution. Appl. Surf. Sci. **357**, 845–855 (2015)
22. A. Guo, G. duan, K. He, B. Sun, C. Fan, S. Hu, Synergistic effect between 2-oleyl-1-oleylamidoethyl imidazole ammonium methyl sulfate and halide ion by molecular dynamics simulation. Comput. Theor. Chem. **1015**, 21–26 (2013)
23. Y. Yan, X. Wang, Y. Zhang, P. Wang, X. Cao, J. Zhang, Molecular dynamics simulation of corrosive species diffusion in imidazolines inhibitor films with different alkyl chain length. Corros. Sci. **73**, 123–129 (2013)
24. E. Ohaeri, U. Eduok, J. Szpunar, Hydrogen related degradation of pipeline steel: a review. Int. J. Hydrogen Energy **43**(31), 14584–14617 (2018)
25. L. Guo, X. Ren, Y. Zhou, S. Xu, Y. Gong, S. Zhang, Theoretical evaluation of the corrosion inhibition performance of 1,3-thiazole and its amino derivatives. Arab. J. Chem. **10**, 121–130 (2017)
26. L. Guo, S. Kaya, I.B. Obot, X. Zheng, Y. Qiang, Toward understanding the anticorrosive mechanism of some thiourea derivatives for carbon steel corrosion: a combined DFT and molecular dynamics investigation. J. Colloid Interface Sci. **506**, 478–485 (2017)
27. A. Fateh, M. Aliofkhazraei, A.R. Rezvanian, Review of corrosive environments for copper and its corrosion inhibitors. Arab. J. Chem. **13**, 481–544 (2020)
28. L. Guo, W. Dong, S. Zhang, Theoretical challenges in understanding the inhibition mechanism of copper corrosion in acid media in the presence of three triazole derivatives. RSC Adv. **4**, 41956–41967 (2014)
29. K. Xhanari, M. Finsgar, Organic corrosion inhibitors for aluminum and its alloys in chloride and alkaline solutions: A review. Arab. J. Chem. **12**, 4646–4663 (2019)
30. X.Y. Zhang, Q.X. Kang, Y. Wang, Theoretical study of N-thiazolyl-2-cyanoacetamide derivatives as corrosion inhibitor for aluminum in alkaline environments. Comput. Theor. Chem. **1131**, 25–32 (2018)

Hydrogen Embrittlement in Nickel-Base Superalloy 718

Hamza Khalid and B. Mansoor

Abstract Ni-base superalloy 718 is extensively employed in the oil, gas and aerospace industries since it exhibits an excellent combination of mechanical properties and corrosion resistance. However, upon exposure to hydrogenating environments, mechanical properties of alloy 718 can be severely degraded due to hydrogen embrittlement (HE). Hydrogen embrittlement—known since the 1870s—refers to markedly reduced ductility and fracture toughness in metals and alloys. Various theories have been put forward to explain the precise mechanism by which hydrogen deteriorates mechanical behavior of structural materials. Hydrogen enhanced localized plasticity (HELP) and hydrogen enhanced decohesion (HEDE) are two of the commonly suggested mechanisms. Evidence for HELP, widely reported with the support of strong experimental work, suggests that the presence of hydrogen in the metallic lattice enhances mobility of dislocations, leading to regions of enhanced plasticity. This in turn leads to regions of high localized strains resulting in early crack initiation. On the other hand, the hydrogen enhanced decohesion (HEDE) mechanism relies on the ability of hydrogen to reduce bond strength, especially at weaker interfaces, resulting in decohesion and early fracture. Despite being an active area of research for decades, a model that could predict how and when failure will occur due to hydrogen embrittlement remains elusive. Lately, it has been recognized that the various suggested mechanisms may be active simultaneously, with one or more of them dominating based on the weak sites in the material and the specific environmental conditions. In alloy 718, one of the weak sites is reported to be the interface of the δ-phase, which forms primarily along grain boundaries, and the matrix. In other works, nanovoid nucleation at intersection of dislocation slip bands (DSBs) are shown to result in crack initiation. Furthermore, grain boundaries can also act as weak sites causing intergranular hydrogen embrittlement by decohesion. In this

H. Khalid · B. Mansoor (✉)
Department of Materials Science and Engineering, Texas A&M University, College Station, TX, USA
e-mail: bilal.mansoor@qatar.tamu.edu

B. Mansoor
Department of Mechanical Engineering, Texas A&M University, College Station, TX, USA

Mechanical Engineering Program, Texas A&M University at Qatar, Doha, Qatar

I. Toor (ed.), *Recent Developments in Analytical Techniques for Corrosion Research*,
https://doi.org/10.1007/978-3-030-89101-5_13

review, we will first introduce the microstructure of alloy 718. Then, studies on the interplay between microstructural features and mechanical properties of alloy 718 in the presence of hydrogen are reviewed. Furthermore, the applicability, merits, and demerits of hydrogen embrittlement mechanisms for alloy 718 are discussed. Finally, some unanswered questions and avenues of future research are briefly discussed.

Keywords Hydrogen embrittlement · Super alloys · Alloy 718 · Fracture · Dislocation slip bands · Hydrogen enhanced localized plasticity · Hydrogen enhanced decohesion

1 Introduction

Ni alloy 718 is an alloy from a family of precipitation-hardened Ni-base superalloys that are extensively used in the oil, gas and aerospace industries where performance under demanding conditions is desired [1–3]. Alloy 718 is capable of maintaining its excellent mechanical properties and remarkable corrosion resistance under extreme temperature and pressure environments. A challenge faced in selecting precipitation-hardened Ni-base superalloys for use in aggressive environments like sour wells in the oil and gas production is their high sensitivity to hydrogen embrittlement. Alloy 718, just like other Ni-base superalloys and many other metals and alloys generally, drastically loses ductility and fracture toughness on exposure to hydrogenating environments and undergoes brittle fracture.

Hydrogen embrittlement has been known since at least the 1870s when it was found that iron and steels after interaction with acid temporarily lose ductility and toughness [4]. Upon slow heating of the metal, the ductility and toughness were regained. The evolution and ingress of hydrogen from the reaction between the acid and the metal was found to be the cause with subsequent egress of hydrogen from the metal on heating reversing the loss of ductility and fracture toughness. Hydrogen may enter a metal lattice in various scenarios but even in small quantities, the ductility and fracture toughness is markedly reduced that is characteristic of hydrogen embrittlement [5]. One, as discussed above, is corrosion reactions where hydrogen is evolved at the cathode. Another scenario is welding or electroplating of metals where there may be exposure to hydrogen due to a multitude of reasons. Hydrogen ingress into the material may also take place due to operating conditions where the environment has significant quantities of hydrogen, as is the case in deep gas wells and hydrogen tanks in aerospace applications.

The engineering stress-engineering strain plot obtained using a tensile test at a slow strain rate of 10^{-6} is shown in Fig. 1. It shows how ductility is reduced in a hydrogen charged sample compared to an uncharged sample.

In this particular case, the sample was charged at a charging current density of 7.7 mA/cm^2 for 168 h at 80 °C. The resultant loss of ductility is stark. Furthermore, the fracture of the metal changes from a classic ductile fracture with significant plastic deformation and micro void coalescence to one with very little plastic deformation;

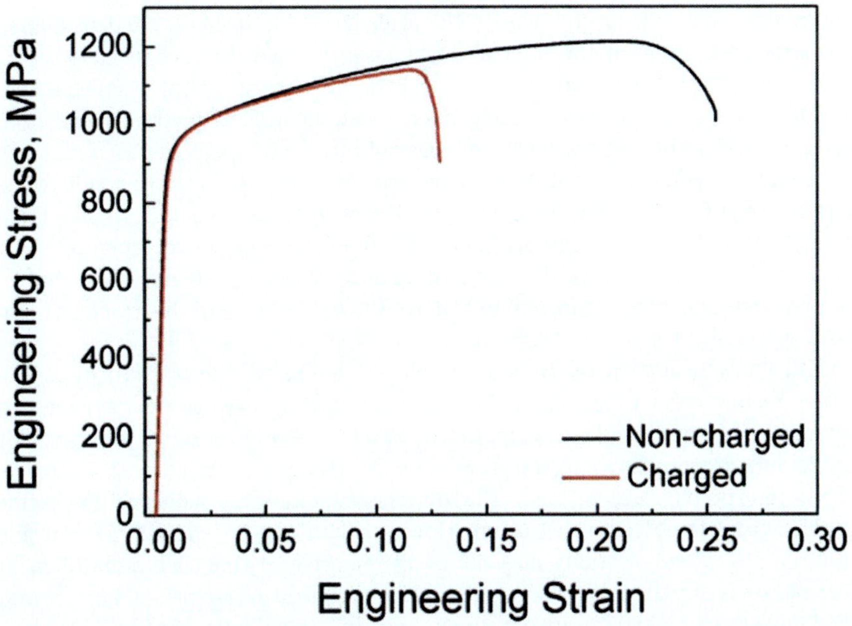

Fig. 1 An engineering stress-engineering strain curve before and after hydrogen charging. Hydrogen charging prior to testing was conducted in an electrochemical cell. After charging at 7.7 mA/cm^2 for 168 h at 80 °C, the ductility is significantly reduced, as total elongation decreases from 25% to about 12%. Adapted from [6]

thus, it is characterized as brittle and the phenomenon is called embrittlement. Understanding the role different microstructural features play in the fracture process and a precise and predictable mechanism by which embrittlement takes place in materials affected by hydrogen remains elusive since there is a significant variance in results across studies. Some of it is due to variation in alloy composition, microstructure, thermal and processing history, hydrogen concentration and charging conditions, in addition to inherent uncertainties associated with the interaction between hydrogen and the metallic atoms on small length and time scales. [6, 7]. Even though the fracture takes place at significantly lower levels of strain, upon careful observation, it has been observed that underneath the fracture surface, locally, dislocation cell structures representative of a material that may have undergone significantly more strain are present; see for example, work on pure nickel [8] and alloy 718 [9]. Such evidence of higher degree of localized plastic deformation in the vicinity of the fracture zone have also been obtained for iron, steels and palladium, among many other important metallic alloys [10]. Furthermore, Birnbaum et al. in a series of experiments in an environmental TEM provided direct evidence of enhanced dislocation mobility in pure nickel and iron as hydrogen was introduced into the material [11, 12]. Therefore, there is general consensus among the scientific community that fracture surface of materials exposed to hydrogen often show features of localized plastic deformation prior to fracture. The evidence of localized plastic deformation forms the cornerstone of one of the mechanisms called hydrogen enhanced localized plasticity (HELP), often invoked to explain hydrogen embrittlement. Another mechanism often invoked to describe early failure due to hydrogen is hydrogen enhanced decohesion (HEDE). This mechanism does not involve plasticity, rather it relies on the effect of hydrogen to reduce the cohesive strength of the atomic bonds of the material, resulting in cleavage. Often, decohesion may be along defects like grain boundaries and matrix-precipitate interfaces. In this mechanism, a combined effect of applied stress and a critical concentration of hydrogen results in early cleavage. Both modelling and experimental work have shown evidence of this mechanism in Ni alloys [13–18].

HELP has been shown to be active in alloy 718. Dislocation cell structure close to the fracture surface was reported by Li et al. which is taken as clear evidence of plasticity [9]. Additionally, it is reported by Zhang et al. in [6, 19] that transgranular cracks initiate by nanovoid nucleation at the intersection of dislocation slip bands (DSBs) and propagate along them. Thus, the enhanced mobility of dislocations seems to play a crucial role leading to fracture. However, there may be other HE mechanisms in alloy 718 where plasticity may not be of importance. One such embrittlement mechanism is initiation of cracks at the interface between the δ-phase and the matrix, likely due to its higher affinity for attracting and trapping hydrogen [20, 21]. Thus, decohesion of this interface under stress leads to failure [9, 22]. Another condition where plasticity may not play the primary role is at cryogenic temperatures [13]. For a closely related alloy 725, cracks are found to initiate preferentially at low sigma twin boundaries while propagating preferentially along random grain boundaries [7]. Some of these results have been explained using HEDE and are discussed in further detail later. However, evidence points in the direction that multiple HE mechanisms may be active and affect in a cooperative manner during failure of alloy 718 in the presence of hydrogen, just like many other metallic systems. It is likely that one HE mechanism may dominate, or multiple mechanisms might be acting synergistically in specific cases, based on factors including but not limited to hydrogen content, charging conditions, loading rate, temperature, and the microstructure.

In order to understand how hydrogen embrittlement affects alloy 718, it is important to first look at its microstructure to understand the role various microstructural features play in strengthening and HE mechanisms. Later, the tensile behavior of alloy 718 in hydrogen charged conditions is reviewed and theories explaining fracture are discussed.

2 Microstructure of Ni-Base Superalloys

Alloy 718 has been advanced empirically since its discovery in the 1940s. The primary matrix of alloys 718 consists of an FCC γ-phase solid solution of Ni, Cr, Fe and Mo. The composition range is given in Table 1. The FCC structure results in good tensile, creep and fatigue properties as compared to alloys with BCC structures.

Table 1 Nominal chemical compositions wt.% of alloy 718 (UNS N07718)

Element (wt.%)	Ni	Cr	Fe	Co	Mo	Nb	Ti	Al	C	Mn	Si	B	Cu
Alloy 718	50.0–55.0	17.0–21.0	Balance	1.0	2.80–3.30	4.75–5.50	0.65–1.15	0.20–0.80	0.08	0.35	0.35	0.006	0.30

Taken from ASM Handbook Vol 2

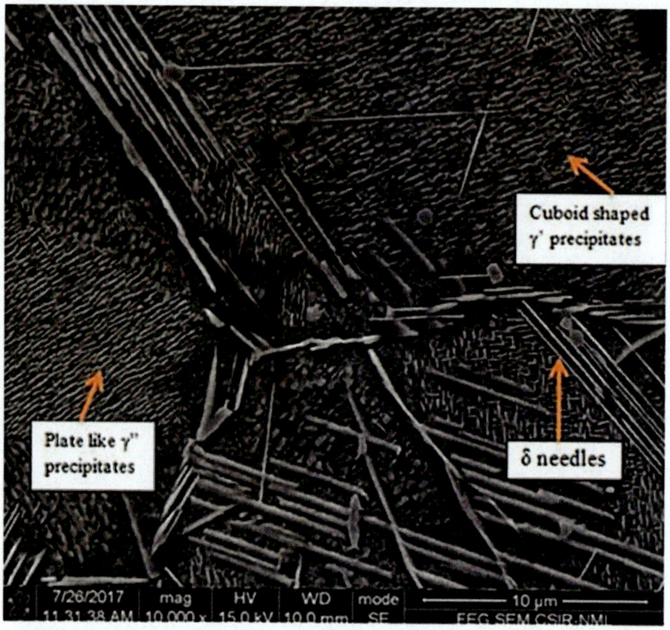

Fig. 2 SEM image of alloy 718 solutionized at 800 °C for 50 h. The three intermetallic phases γ', γ'' and δ can all be seen here. The γ' and γ'' phases are both coherent with the γ matrix, metastable and strengthen the alloy due to intense strain fields while the δ-phase is incoherent, stable and does not increase strength. Adapted from [25]

Alloy 718 offers a high tensile strength and fracture toughness, and good resistance to corrosion. However, it is susceptible to hydrogen embrittlement.

The primary strengthening mechanism of alloy 718 is precipitation hardening [23]. Metastable γ'—Ni_3 (Al, Ti) and γ''—Ni_3(Nb, Al, Ti) phases with LI_2 and $D0_{22}$ structures respectively are dispersed in the matrix (Figs. 2, 3 and 5) They are both coherent; the former is spherical, and the latter is disc shaped [23, 24]. The phases act to strengthen the alloy due to development of strain fields which act as barriers to dislocation motion. A stable δ-phase—Ni_3Nb, with a needle-like morphology and a $D0_a$ structure, is also present largely at the grain boundaries (Figs. 2 and 4); this phase is incoherent and does not provide the strengthening effect. Upon prolonged aging, the γ' and γ'' phases transform to δ-phase, reducing strength [25]. The δ-phase, however, does act to pin grain boundaries, thus limiting grain growth even at higher temperatures [26].

Carbides also precipitate during heat treatment between 760 and 980 °C (shown in Fig. 6 [27]). MC, M_6C and $M_{23}C_6$ carbides are all known to form at the grain boundaries [27]. They can be block shaped or thin films (Fig. 6) and increase the cohesive strength of the grain boundaries. However, they are brittle and offer little

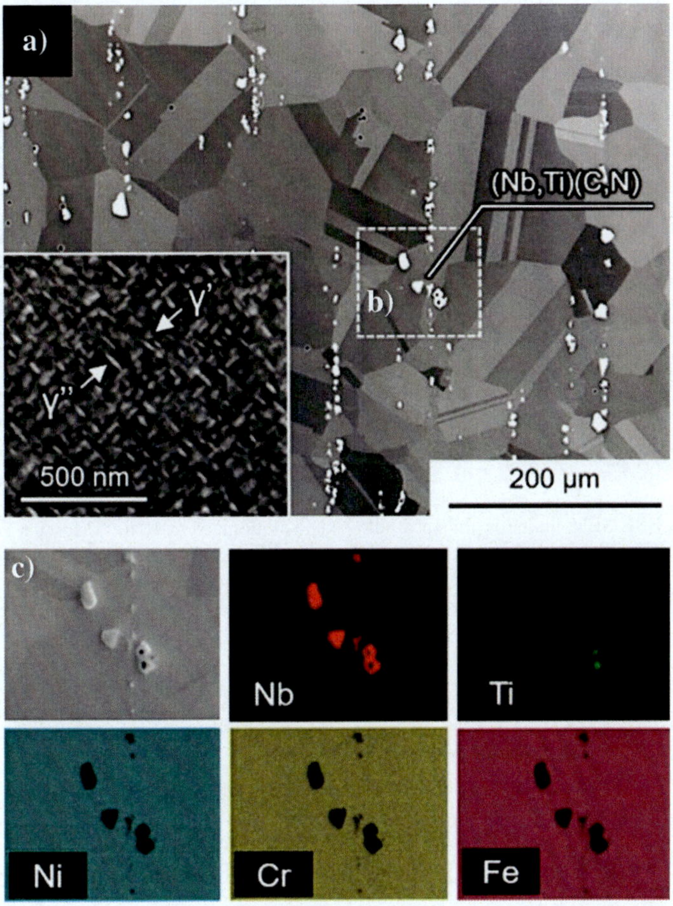

Fig. 3 **a** A backscattered electron SEM image showing alloy 718 microstructure. The alloy was solutionized at 1065 °C for 1 h and double aged at 760 °C and 650 °C for 10 h each. Carbides, nitrides, γ' and γ'' phase the δ-phase is not perceptible here since the solutionizing temperature is higher than its solvus temperature of 1035 °C. Annealing twins are also observed, while the grain size is rather large due to the dissolution of the δ-phase and the absence of its grain boundary pinning effect. **b** EDS results confirm the presence of Nb and Ti rich phases. Images from [26]

resistance to crack propagation [28, 29]. Furthermore, they are especially rich in Cr, thus reducing the amount of Cr_2O_3 formed at the surface making the alloy more susceptible to corrosion [30].

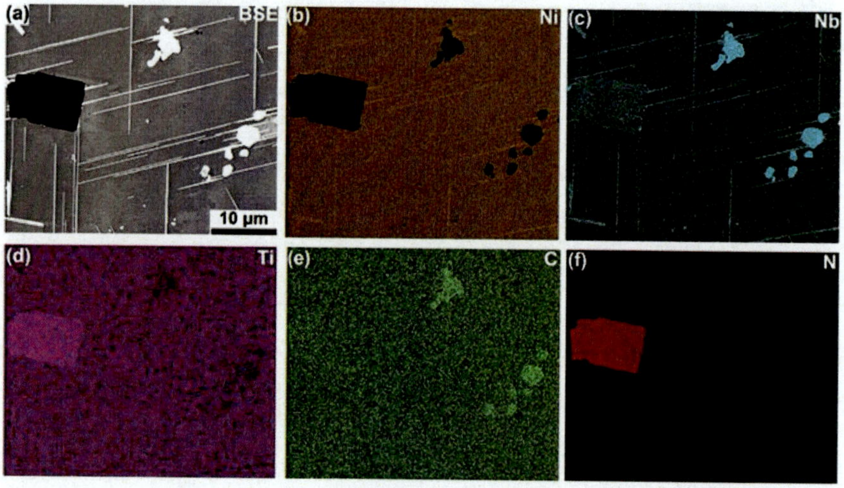

Fig. 4 **a** A BSE SEM image is shown of alloy 718 solutionized at 1040 °C for 1 h and aged at 950 °C for 120 h. This extensive aging time results in coarsening of the δ-phase to a width of 750 nm. **b–f** EDS elemental maps show that the needle-shaped δ-phase is rich in Ni and Nb; its chemical composition is indeed Ni_3Nb. Image taken from [20]

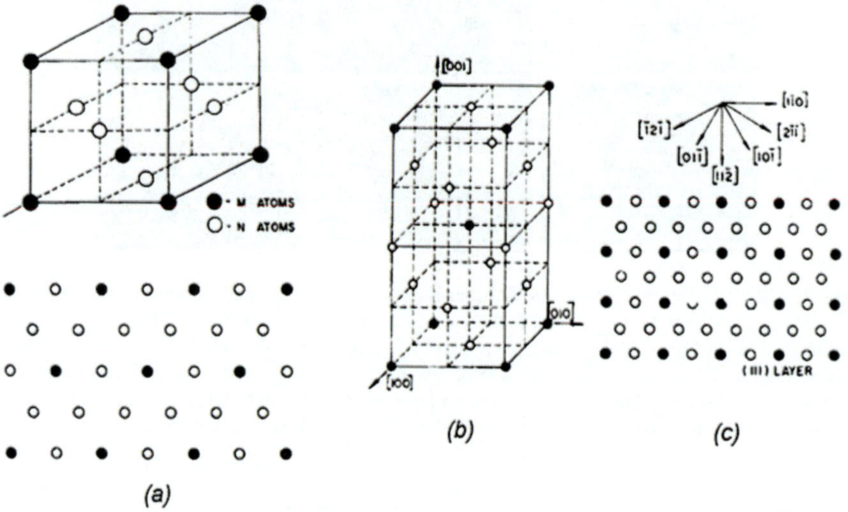

Fig. 5 **a** D022 structure of the γ' precipitate; **b**, **c** LI2 structure of coherent precipitate γ''. Adapted from [24]

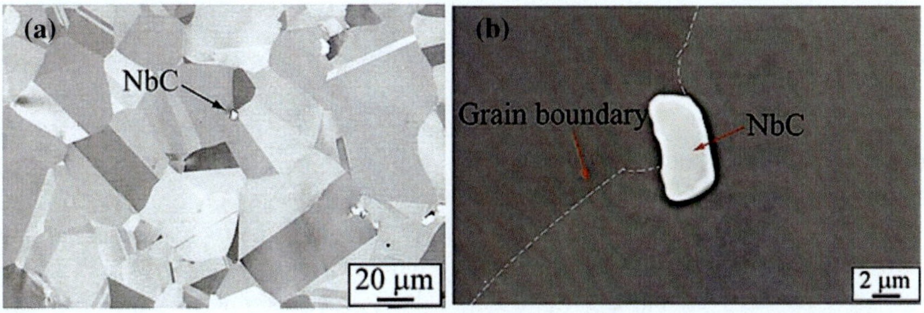

Fig. 6 NbC and other carbides precipitate at grain boundaries and provide a pinning effect by ensuring that the grains do not coarsen. **a** NbC precipitate at a triple junction; **b** a block shaped NbC precipitate. Taken from [27]

3 Microstructure-Property Relationship of H-charged Ni-Base Superalloys

In this section, we discuss the findings of mechanical tests performed on H-charged materials to understand the role played by various microstructural features in HE. Also, inferences to service conditions are provided where appropriate. Most work covered in this review relies on cathodic charging to introduce hydrogen into alloy 718 [6, 7, 31], while important findings such as those by Ogawa et al. where gaseous environment was used to charge the samples are also discussed [26]. Electrochemical cathodic charging is an easy and safe process that is commonly used to introduce hydrogen in metallic samples to investigate hydrogen embrittlement. It ensures a good, reliable concentration of hydrogen in the material and mimics cathodic protection. In an electrochemical cell the sample to be charged acts as the cathode and a Platinum wire can be used as the counter electrode [6, 7, 19]. Acidic or salt solutions are used as electrolytes [32]. When a potential difference is applied across the cell, the hydrogen ions present in the electrolyte move to the cathode, accept an electron, and evolve into H_2 gas. The H_2 gas continues to evolve at the cathode i.e., the metallic sample surface as long as the potential difference is applied across the cell. At low current densities, the absorption of hydrogen is only a function of the current density [33] while the depth to which a sample is H-charged can be calculated using the diffusivity of hydrogen. The other way of introducing hydrogen in a sample is by placing it into an environment of gaseous hydrogen. A problem with this method is that since the diffusivity of hydrogen in metals is low; to achieve a high enough concentration to observe hydrogen embrittlement, the temperature and/or pressure have to be high. This creates a hazard since hydrogen is a flammable gas.

The mechanical behavior of material H-charged using the two different techniques is broadly comparable. One interesting difference though is the apparent loss of tensile strength after cathodic charging, especially at higher charging currents [31]. In general, Ni-base alloys largely tend to retain their tensile strength, after exposure to hydrogen but some cases of slight increase and decrease in strength have also been reported. Anyhow, the effect of hydrogen on tensile strength remains small, especially compared to the striking loss in ductility and fracture toughness [26].

Increase in tensile strength is due to hydrogen's solid solution strengthening effect like other interstitial alloying elements such as carbon or nitrogen, where they act to resist the glide of dislocations [34]. The decrease in tensile strength may be due to pre-damage introduced by harsh electrochemical hydrogen charging [9, 22, 35]. Lu et al. investigated the condition of specimens before and after cathodic charging at a current density of 15 mA/cm^2 and found surface and subsurface cracks on the specimens after charging [35]. This pre-damage is linked to the relatively high hydrogen concentration gradient and resulting internal stresses and tends to increase the embrittlement and loss of strength. At greater current densities, hydrogen concentration gradient is larger which would in turn leads to higher degree of pre-damage. This conclusion is consistent with the findings of Li et al. who used current densities of 75 mA/cm^2, 220 mA/cm^2 and 590 mA/cm^2 and observed a decrease in tensile strength of 3.2%, 8.0% and 12.7% respectively (Fig. 7, plotted from data in [31]). Thus, it is suggested that to ensure a small degree of pre-damage and isolate the hydrogen effects only, lower values of current densities be used. To ensure a high and uniform concentration, longer charging times up to 168 h are recommended.

Multiple studies report the effect of precipitation phases on HE susceptibility of alloy 718. Liu et al. first reported the relationship between their volume fraction and the sensitivity to hydrogen embrittlement using percent reduction of area as a measure of ductility [36]. Using cathodic charging with current density of 50 mA/cm^2, they conclusively showed that presence of any of the γ', γ'' and δ phases decreases ductility, hence increasing the susceptibility of the alloy to hydrogen embrittlement. However,

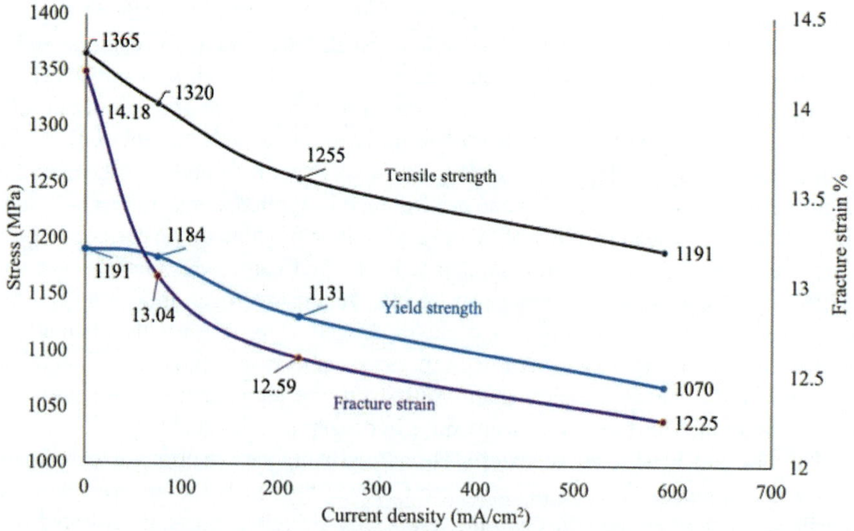

Fig. 7 Loss in strength and ductility are plotted as a function of charging current densities and shown to decrease. Charging duration was held constant at 2 h. Plotted using data from [31]

the effect of the δ-phase is the most prominent (Fig. 8, adapted from [36]). Similar results were reported by Stenerud et al. [29]. They used an as-received condition (AR, aged for 8 h each at 720 and 620 °C) and compared it to solution annealed (SA, 2 h at 1200°C) and solution annealed + aged (A1, 2 h at 1200 and 2 h at 800°C) conditions after hydrogen charging. Figure 9a shows that the solution annealed (SA) alloy has a much lower tensile strength yet higher ductility. Aging at 800 °C for two hours results in the precipitation of phases and thus increases tensile strength. However, it is clear that the aging time used for condition A1 is insufficient to reach the original tensile strength again. It is shown that the solution annealed (SA) specimen, without precipitation phases, was the least susceptible to hydrogen embrittlement while the specimens in the other two conditions were severely embrittled, as be seen in Fig. 9b. These findings are also consistent with the work of Kagay et al. who found a broadly consistent correlation of the presence of the δ-phase with susceptibility to hydrogen embrittlement over a series of heat treatments (Table 3 and Fig. 10, both taken from [37]). Based on these studies, it can be concluded that the presence of any of the precipitation phases increases the susceptibility of the material to HE, with the δ-phase having the most effect. It is of note that while the γ' and γ'' phases are coherent with the matrix, the δ-phase is an incoherent phase. The interface between the matrix and the δ-phase has been identified as a weak site; with the intersection of dislocation slip bands and the interface shown as a site for crack initiation and propagation [20]. A mechanism for this was suggested by Tarzimoghadam et al. who showed that the matrix has a higher affinity for absorbing hydrogen than the δ-phase [21]. As a result, hydrogen that evades the δ-phase gets trapped on interface between the matrix and the δ-phase, subsequently causing its decohesion (Table 2).

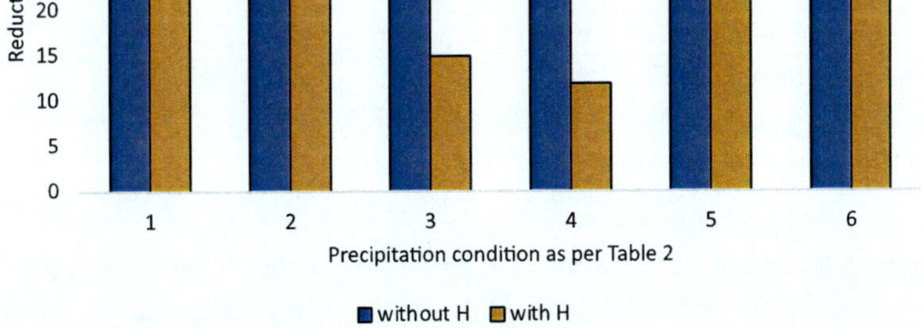

Fig. 8 It can be seen that hydrogen charging reduces reduction of area to failure for all precipitation conditions. Precipitation condition 6 shows that eliminating δ-phase results in the least loss of reduction in area after charging of all conditions with precipitate phases. Plotted using data in [36]

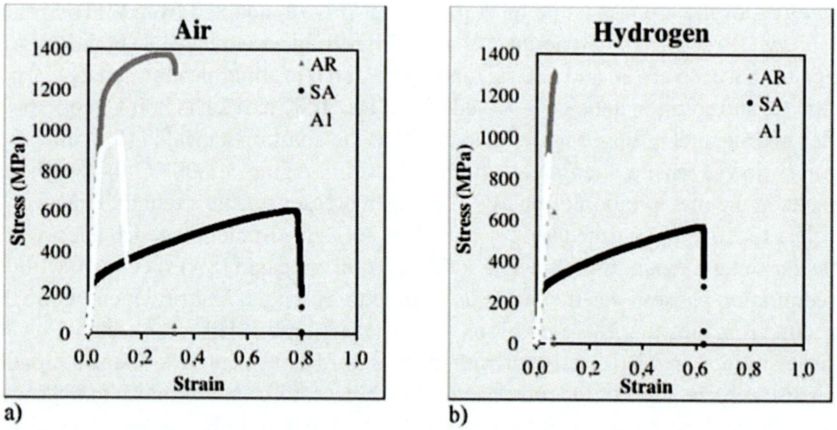

Fig. 9 **a** Stress–strain curves for the three microstructures in air. The SA specimen (black), without strengthening phases, has the least strength and the highest strain to failure. **b** Stress–strain curve after cathodic hydrogen charging. SA specimen's ductility is the least affected, showing the least susceptibility to HE; AR and A1 both are highly susceptible to HE. Plots taken from [29]. (AR = as-received, SA = solution annealed, A1 = solution annealed and aged. Please refer to text for heat treatment conditions used to obtain these microstructures.)

Table 2 The table shows the various heat treatments utilized to obtain different amounts of precipitate phases. Their susceptibility to hydrogen embrittlement is shown in Fig. 8

Precipitation conditions	Step 1		Step 2		Step 3		Expected precipitate phases
1	621 °C	8 h	13 °C/s up to 1030 °C	0 h	–	–	Only δ-phase
2	621 °C	8 h	14 °C/s up to 946 °C	0 h	–	–	δ + γ″ (partly dissolved) + γ′
3	621 °C	8 h	14 °C/s up to 931 °C	0 h	–	–	δ + γ″ (partly dissolved) + γ′
4	621 °C	8 h	–	0 h	–	–	δ + γ″ + γ′
5	621 °C	8 h	1040 °C	1 h	–	–	None[a]
6	1040 °C	1 h	720 °C	8 h	621 °C	8 h	Only γ″ + γ′ phases

[a] Solvus temperature of δ-phase is 1035 °C

One heat treatment in the work of Kagay et al. which is inconsistent with the correlation of the δ-phase and susceptibility to hydrogen embrittlement is their over-aged condition. A possible reason offered by authors is the presence of higher amount of γ′ and γ″ phases which may be playing a role in improving resistance to HE in the over-aged condition. However, this would be inconsistent with the findings of Liu et al. who show that the presence of any precipitates make the alloy more vulnerable

Table 3 Different metallurgical states and their susceptibilities to HE are compared. It is shown that the under-aged state, with the least amount of precipitate phase, is the least susceptible to HE since it retains 45.6% of its pre-charging strain to failure after charging. The high δ-phase condition 4 is the most susceptible. Stress–strain curves for this data, taken from [37], are plotted in Fig. 10

	Condition	Annealing temperature (°C)	Annealing time (h)	Aging temperature (°C)	Aging time (h)	Elongation (without H) (%)	Elongation (with H) (%)
1	Under-aged	1050	2.5	710	6	46.9	21.4
2	Peak-aged	1050	2.5	760	6	28.9	3.5
3	Over-aged	1050	2.5	802	8	31.2	5.8
4	High δ-phase	1050	2.5	950 and 760	4 and 6	29.2	2.3

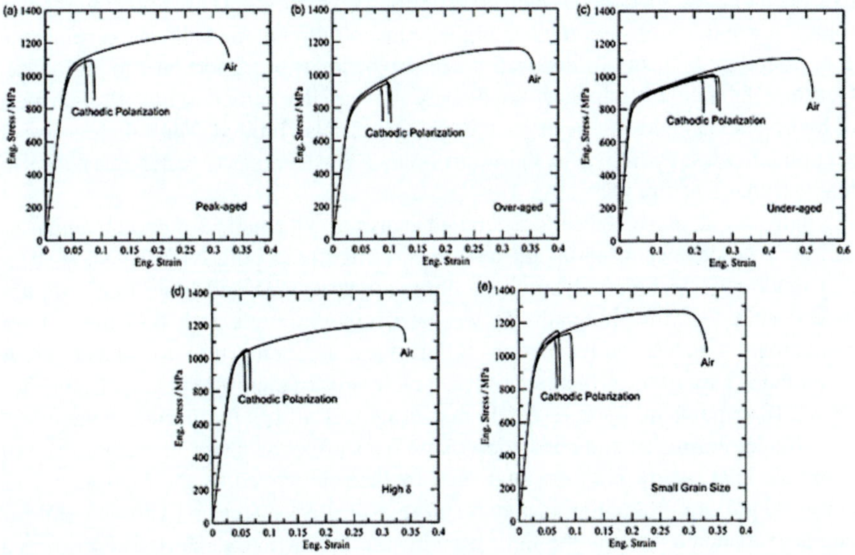

Fig. 10 The stress–strain curves show that after cathodic polarization, all metallurgical states lose ductility; **a** In peak-aged condition, there is severe drop in ductility; **b** In over-aged condition, the loss in ductility is less than peak-aged but is still significant; **c** With least amount of precipitation phases, under-aged condition is the least susceptible to HE; **d** This condition is the most susceptible due to the high amount of δ-phase; **e** Small grain size results in similar loss of ductility as the peak-aged condition in (**a**), thus leading the authors to conclude that grain size may have little effect on HE susceptibility. Stress–strain curves are adapted from [37]

to hydrogen embrittlement [36]. Anyhow, the evidence largely points in the direction that the presence of the δ-phase introduces vulnerable sites in the alloy, increasing its susceptibility to HE.

Although some studies report cracks initiating at the carbide-matrix interface and/or propagating through the carbide, they are not usually the first or the most prominent cracking sites [22, 31]. Other studies report no cracking at the carbides [7, 38, 39]. A reason suggested for the minimal, if any, role of carbides in HE is that they appear to act as deep, irreversible traps for hydrogen [20, 40]. As a result, hydrogen that diffuses into the carbides will not desorb and segregate to the interface easily, thus contributing little to early crack initiation or propagation.

Kagay et al. also investigated the effect of smaller average grain size of 57 μm compared with grain sizes of 120–145 μm in the other samples. They found that the finer grain sized material was still susceptible to hydrogen embrittlement; the susceptibility was not affected much by the different grain size ((a) vs. (e) in Fig. 10, taken from [37]). The role of grain size and grain boundaries was explored by Chen et al. and Oudriss et al. in a Ni–Fe alloy and pure Ni respectively [41, 42]. They find that reducing grain size improves resistance to HE. Similar results are obtained for steels [43, 44]. This is in spite of the fact that with a finer grain size, the diffusion of hydrogen into the specimen is enhanced as grain boundaries act as short circuit paths for hydrogen diffusion leading to a higher concentration in the bulk. However, Chen et al. found that density of dislocations in slip bands was proportional to grain size. Hence, with smaller grain size, the density was smaller and consequently transport of hydrogen via dislocations to grain boundaries is also limited. Thus, there is lesser accumulation of hydrogen in the grain boundaries leading to better resistance to intergranular fracture [41].

Oudriss et al. presented a more detailed analysis on the nature of grain boundaries and their propensity to act as diffusion short circuits in pure Ni [42]. They show that random grain boundaries with amorphous structures and large free volumes indeed act as fast diffusion paths. However, special low angle grain boundaries have geometrically necessary dislocations (GNDs) and vacancies, thus acting as traps. A systematic study of the effect of grain size on hydrogen embrittlement resistance for alloy 718 remains an open research area. Kagay et al. briefly discussed the effect of a smaller grain size and concluded that it does affect behavior in the presence of hydrogen [37]. Alloy 718 is complicated by the role played by the δ-phase whose ability to pin grain boundaries may result in a positive effect on hydrogen embrittlement resistance while at the same time its interface with the matrix is known as a weak site for crack initiation. In certain cases, these effects may be cancelling each other out. Further studies on grain size effect, nature of grain boundaries with various populations of strengthening or coarsening resistant phases could provide insights on the inter-relationship of these parameters.

Precipitation phases of alloy 718 play an important role in hydrogen embrittlement (Figs. 8, 9 and 10). It is clear that the presence of any precipitate deteriorates the resistance of HE. However, the metastable, coherent γ' and γ'' phases are strengthening phases and their elimination would drastically reduce the strength of the alloy. Furthermore, their role in HE is fairly limited, and their elimination only results in marginal improvement in HE resistance. On the other hand, the stable, incoherent δ-phase does not increase strength and its elimination starkly improves HE resistance. Thus, its elimination or control of morphology to ensure minimal presence is a recommended technique to engineer HE resistance. However, the ability of δ-phase to pin grain boundaries thus maintaining a fine grain size must not be overlooked.

Finer grain sized alloys usually possess higher strength due to the Hall–Petch effect. Elimination of this phase may increase HE resistance only at the expense of loss of strength. Thus, a compromise may be sought keeping the specific application in mind. Another engineering method to improve resistance to hydrogen embrittlement in alloy 718 is grain boundary engineering. Grain boundaries have already been identified as a weak site prone to intergranular cracking. However, it is also shown that coherent twin boundaries are resistant to crack propagation. Engineering grain boundaries to a higher fraction of special, low angle twin boundaries may improve the macroscopic ductility of the alloy.

4 Hydrogen Embrittlement Mechanisms in Alloy 718

Various mechanisms have been proposed to explain the hydrogen transport and interactions in metals that cause HE. Hydrogen has been observed to enhance dislocation mobility in a number of metals; it also stabilizes vacancies, decreases cohesive energy of grain boundaries, decreases stacking fault energy, increases slip planarity, enhances homogeneous dislocation nucleation and suppresses dislocation emission from crack tips [6, 14, 45–47]. One or more of these effects of hydrogen are responsible for, or contributing to, macroscopic embrittlement. Two of the proposed HE mechanisms i.e., Hydrogen Enhanced Localized Plasticity (HELP) and Hydrogen Enhanced Decohesion (HEDE) are discussed here [10, 47, 48]. A recent review article by Martin et al. in Ref. [10] presents extensive experimental evidence and simulation results on HELP's applicability in a wide range of structural materials. HEDE was introduced in 1926 and is briefly reviewed here [48, 49]. Indirect evidence supporting these two mechanisms has typically been obtained through uniaxial tensile or stress relaxation tests. Here, we discuss experimental studies deployed to uncover direct evidence of these two HE mechanisms –the need for the use of computational techniques is also discussed. We aim to explain fracture mechanisms in alloy 718; however, to present a complete picture, in cases where important breakthroughs are made on other materials, the findings are also briefly discussed and referenced.

First, we review fractographic studies of tensile tested alloy 718 in virgin and hydrogen-charged conditions. A close look at the fractographic surfaces reveals the characteristic effect of hydrogen on changing fracture behavior (see Fig. 11, from Ref. [6] and Fig. 12, taken from Ref. [38]). Since the diffusivity of hydrogen through alloy 718 is low, on the order of 10^{-15} m^2/s, the timescale of a few hours up to a week of cathodic hydrogen charging may not be enough to saturate mm-scale samples. Consequently, often it is possible to see hydrogen-rich regions close to the surface and relatively unaffected interior regions within the same sample; in Fig. 11, the hydrogen-affected brittle and unaffected ductile regions can be seen. In Fig. 11a, the free surface is on the left side of the image; the red and yellow dashed lines indicate the transition from brittle to mixed and mixed to ductile regions respectively. In Fig. 11b, a higher resolution image of the hydrogen-affected brittle region is shown where cracking along dislocation slip bands, a hydrogen-affected fracture

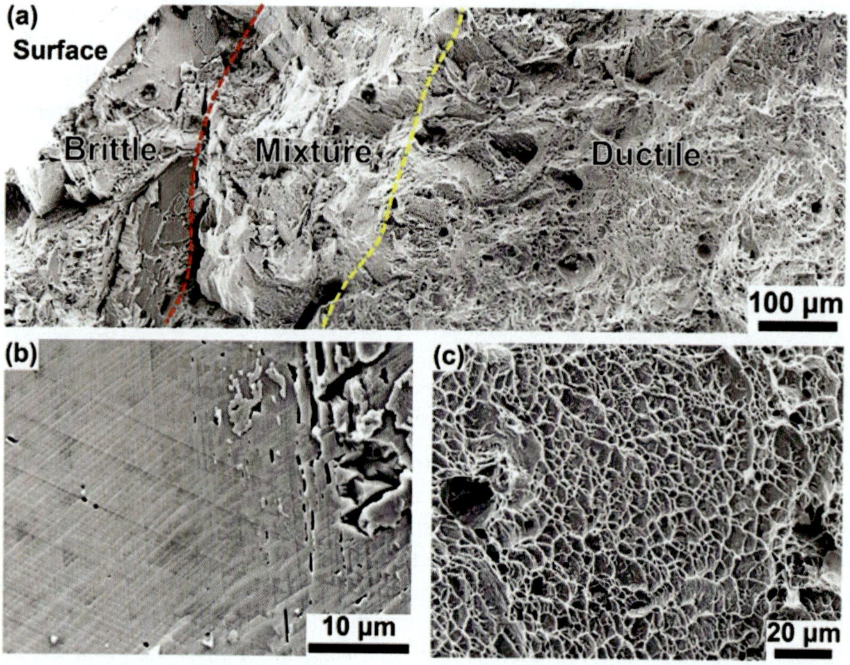

Fig. 11 A 718 sample charged with hydrogen at 7.7 mA/cm^2 for 168 h at 80 °C. The depth of penetration of H is estimated to be 140 μm from the surface. The diameter of the cylindrical sample is 3.8 mm. Images are sourced from [6]

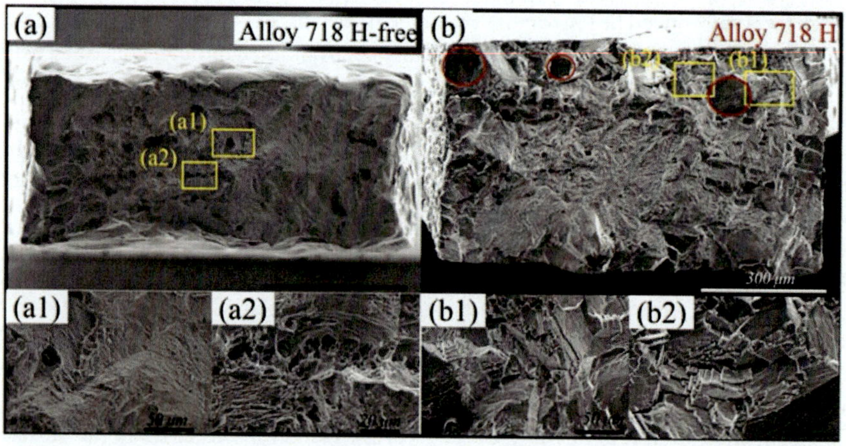

Fig. 12 Cross-section of 0.5 mm thick dog-bone samples strained to fracture in tensile tests. **a** 718 sample without hydrogen. Higher resolution (**a**1) and (**a**2) reveal hallmarks of ductile fracture. **b** 718 sample after hydrogen charging. Hydrogen charging was performed at 15 mA/cm^2 for 18 h at 75 °C. Higher resolution (**b**1) and (**b**2) indicate a ductile region in the center of the sample and brittle intergranular (circled in red) and transgranular fracture with flat features and steps. Fractographs adapted from [38]

mechanism, is apparent. In Fig. 11c, the higher resolution image of the unaffected ductile region shows dimples, characteristic of typical ductile fracture driven by microvoid coalescence. The depth of diffusion of hydrogen into the sample in this case was estimated to be 140 μm which correlates well with the extent of the brittle region.

Figure 12 shows fracture of 718, aged at 782 °C for 6.5 h, in hydrogen-free and hydrogen-affected conditions. Figure 12a has typical ductile features with dimples, pointing to fracture resulting from microvoid coalescence; this can be seen more clearly in higher resolution images in Fig. 12a1, a2. In Fig. 12b, the sample is charged with hydrogen prior to testing. Fracture towards the edges of the sample is largely brittle with steps and flat features; both intergranular (in red circles) and transgranular fracture can be seen. The flat features are distinctly visible in Fig. 12b2. Towards the center of the sample, a region of ductile fracture is seen which can again be attributed to the inability of hydrogen to fully saturate the sample. It is apparent that the macroscopic features of hydrogen-affected fracture are starkly different to those unaffected by hydrogen. Next, we look at the deformation features and origins of fracture at the micro and nano scales and discuss the suggested mechanisms i.e., HELP and HEDE that have been invoked to explain the embrittlement behavior in alloy 718. Much of the evidence discussed in the proceeding sections is based on experimental studies conducted on pure Ni, but wherever available alloy 718 results have been included to provide the reader a comprehensive picture.

4.1 Hydrogen Enhanced Localized Plasticity (HELP)

HELP was first proposed by Beachem et al. in 1972 to explain the decrease in local flow stress observed during torsion tests of the hydrogen-charged steel pipes [50]. HELP's basic premise is that hydrogen presence in sufficient concentrations uniquely affects the mechanical behavior by accelerating dislocation motion. It is suggested that even though macroscopically, fracture in hydrogen exposed materials appears brittle, the driving mechanism is enhanced localized plasticity assisted by hydrogen.

Hydrogen enhanced dislocation mobility has been observed directly through in-situ transmission electron microscopy (TEM) studies in a range of structural materials. Birnbaum et al. were the first, using in situ charging in an environmental cell of a TEM to successfully link the presence of hydrogen in the sample with enhanced mobility of dislocations [11, 12, 51–53]. In their technique, inside an environmental cell of a TEM, a pure Ni sample was pre-strained to a predetermined level, and then hydrogen was introduced. After introducing H_2 gas into the environmental chamber, images of higher dislocation density were also taken as direct evidence of enhanced dislocation mobility (Fig. 13) [12]. In addition, stress relaxation at a fixed strain was taken to be indirect evidence of strain softening due to enhanced dislocation mobility. Furthermore, the mobility of the dislocations was reduced again on the removal of hydrogen from the cell. In comparison, when inert gases like He and air were introduced in the environmental cell, evidence of enhanced dislocation mobility was not

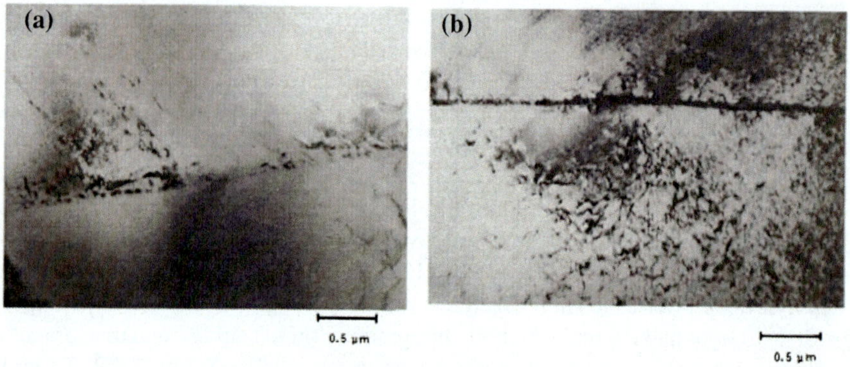

Fig. 13 **a** HVTEM image of a pure Ni sample; **b** same sample as in (**a**) held at the same stress and displacement as (**a**) but with addition of H_2 gas at 10^4 Pa in the environmental chamber. Higher dislocation density can be observed which is a consequence of higher dislocation mobility. TEM images taken from [12]

found. Similar observations were reported for iron and titanium, in addition to nickel, in the 1980s.

Wang et al. uncovered more direct evidence of enhanced plasticity in high pressure torsion of pure Ni where an advanced dislocation cell structure with finer cells were observed with hydrogen charging compared to the cell structure without hydrogen for the same level of strain. See Fig. 14 taken from Ref. [54]—the distance from the free surface of the disk sample subjected to high pressure torsion is shown for both conditions. In Fig. 14a, the hydrogen-free sample has larger (>1 μ) rectangular shaped dislocation cells with thick dislocation walls. On the other hand, in Fig. 14b, the dislocation cell structure is finer with comparatively thinner cell walls; this is indicative of a more advanced cell structure which has undergone a higher degree of plasticity.

Even though experiments leading to the discoveries in the preceding two paragraphs (results in Figs. 13 and 14) are for pure nickel and have not been reproduced for alloy 718, there is no evidence to suggest that they are not active in alloy 718. Both accelerated dislocation motion and the advanced dislocation structures have been shown for a large number of metals [10]. Furthermore, cracking along dislocation slip bands (DSBs) and dislocation cells near fracture zones, both shown for alloy 718, fit in the narrative suggested by HELP, which is in turn underpinned by accelerated dislocation motion. The specific findings for alloy 718 are discussed below.

Li et al. [9] imaged dislocation cells near the fracture zone of an alloy 718 tensile specimen charged with hydrogen (Fig. 15). This supports the notion that the macroscopically brittle fracture is not purely brittle, rather it does involve appreciable plasticity. The manifestation of HELP in hydrogen-assisted transgranular fracture in alloy 718 seems to be the initiation of cracks at intersection of dislocation slip bands (DSBs) and propagation along them [6, 19, 26]. HELP incorporates hydrogen's

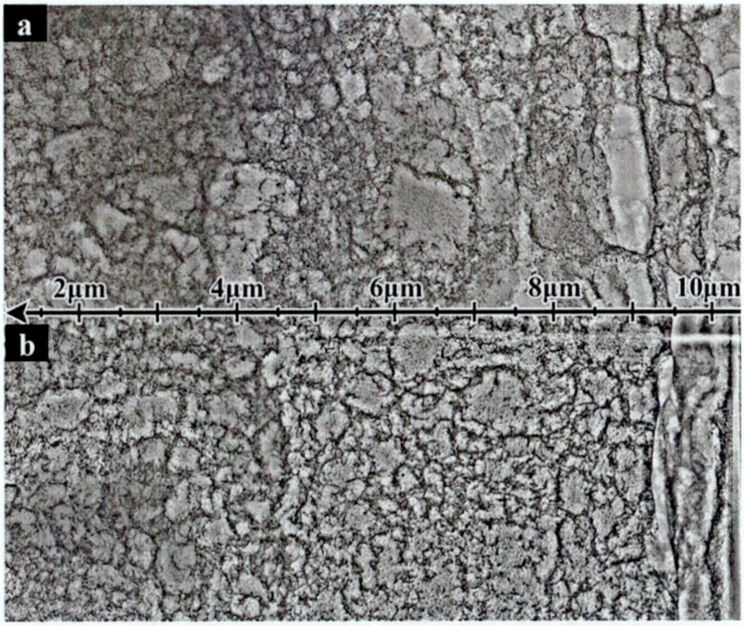

Fig. 14 SEM images taken after 0.23 shear strain on a pure Ni sample. **a** The dislocation cell structure in the hydrogen free sample; **b** dislocation cell structure in the thermally hydrogen charged sample. The cell structure in (**a**) is coarser with thicker walls while the cell structure in (**b**) is finer. Adapted from [54]

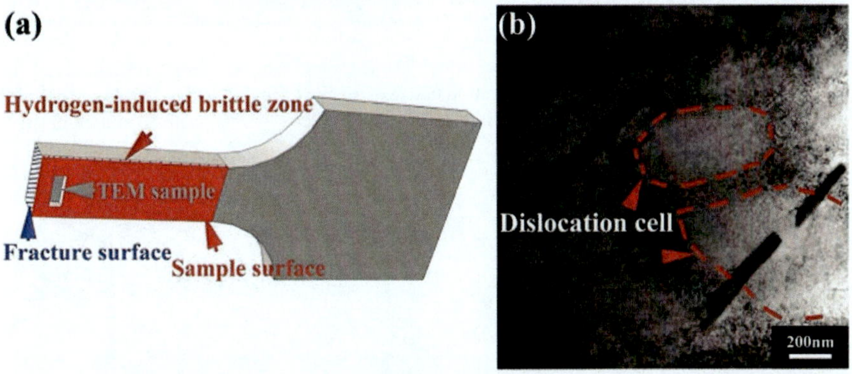

Fig. 15 **a** Schematic showing a broken alloy 718 tensile sample and precise location where TEM analysis was carried out; **b** dislocation cells can be observed which are hallmarks of plasticity close to the fracture surface. Figure taken from [9]

ability to reduce the stacking fault energy and thus increasing slip planarity. As a result, the dislocations form intense dislocation bands (DSBs) on active slip planes and nano-voids then nucleate at the intersection of DSBs to initiate cracks (see Fig. 16).

In Sect. 3, it was discussed that the presence of the incoherent δ-phase in alloy 718 results in a higher susceptibility to hydrogen embrittlement. Zhang et al. show that the interface between the δ-phase and the matrix is a weak site where cracks initiated at the intersection of the DSBs with the δ-phase (see Fig. 17 taken from [20]). This mechanism for crack initiation again relies on HELP; the accelerated motion of dislocations and high slip planarity results in formation of the dislocation slip bands not seen otherwise, and whose intersection with the δ-phase results in weak sites for crack initiation.

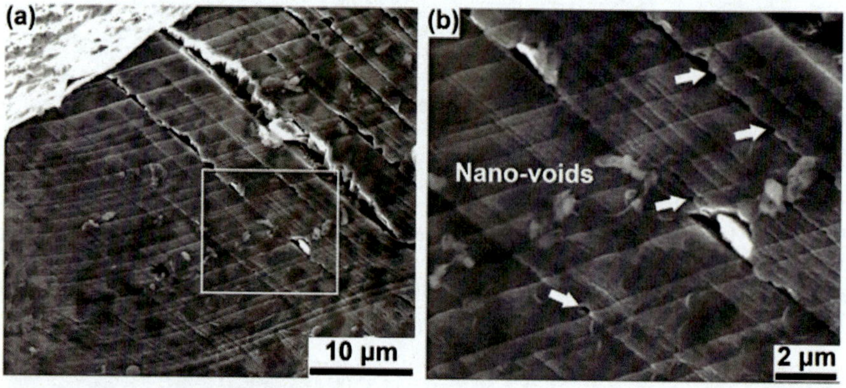

Fig. 16 SEM fractographs of hydrogen-charged 718 **a** Cracks can be seen to initiate along the dislocation slip bands; **b** at higher magnification image of the boxed region taken from (**a**), nanovoids formation at the intersection of DSBs is visible. Images adapted from the original source in [6]

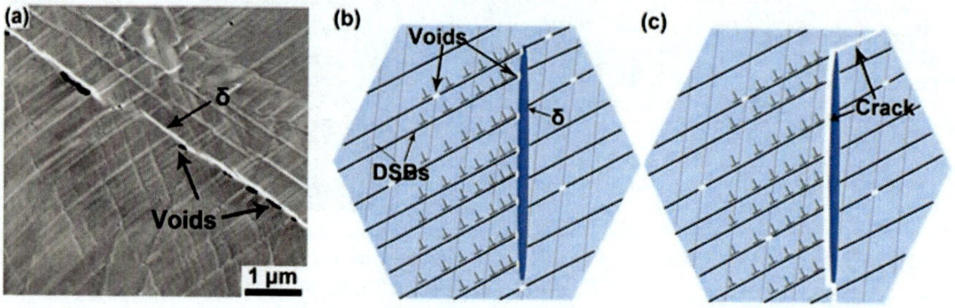

Fig. 17 **a** In the SEM image, voids are shown to nucleate at the intersection of dislocation slip bands (DSBs) and the δ-phase; schematics show **b** void nucleation and **c** crack formation. The image and schematics are sourced from [20]

Another important observation in hydrogen-affected alloy 718 is that fracture shows both intergranular and transgranular characteristics [26]. However, a systematic study on the precise role of grain boundaries vulnerable or resistant to hydrogen-affected intergranular fracture in alloy 718 is unavailable in open literature and remains an open question. Anyhow, studies available on hydrogen-affected Ni alloy 725—a material similar to the alloy 718—may provide some basis where fracture is reported to be largely intergranular [38, 55]. This may be due to interactions of CTBs with dislocations and their enhanced mobility—not seen otherwise. For example, Seita et al. reported a somewhat surprising result that intergranular fracture preferentially initiates at the usually stable and corrosion resistant coherent twin boundaries (CTBs) in Ni alloy 725 [7]. Seita et al. hypothesized that intergranular fracture in alloy 725 may be due to enhanced mobility of dislocations that then get trapped on coherent twin boundaries resulting in development of advanced dislocation structures within the twin boundary plane. Seita et al. also offered an alternative explanation that movement of dislocations to the twin boundaries increases the hydrogen concentration and the GB energy, making them more susceptible to decohesion. Thus, understanding of hydrogen-affected intergranular fracture beyond the role of coherent twin boundaries as preferential sites for crack initiation needs more work to offer an in-depth explanation. We have discussed a variety of results here for alloy 718 that all support the idea that the mechanism behind hydrogen-affected fracture is dislocation-mediated. We also discussed discoveries made using pure Ni that lend weight to the HELP mechanism and its applicability to alloy 718. In the next section, we discuss another hydrogen embrittlement mechanism which is not dependent on dislocations.

4.2 Hydrogen Enhanced Decohesion (HEDE)

In spite of a large body of evidence supporting HELP, sometimes hydrogen embrittlement also seems to occur in situations where dislocations might not be playing the central role. For example, reports of intergranular fracture and cleavage within grains in hydrogen-affected structural materials may not be dependent on the enhanced mobility of dislocations. In many such situations, hydrogen enhanced decohesion seems to explain the occurrences. In the particular case of hydrogen embrittled Ni-base alloys, while hydrogen enhanced localized plasticity seems plausible to explain the fracture behavior, there are still situations where it does not seem to be the dominant mechanism. One such example is the interface between the δ-phase and the matrix. It has been discussed previously that this interface is a weak zone for crack initiation and propagation. Tarzimoghadam et al. recreated the orthorhombic δ-phase in a Ni_3Nb alloy to study its role in hydrogen embrittlement [21]. Tarzimoghadam et al. (Fig. 18) failed to spot the formation of voids which is contrast to the findings of Zhang et al. (Fig. 17) who showed formation of nano-voids at the intersection of DSBs with this interface. Instead, Tarzimoghadam et al. reported a clean debonding of the δ-phase and the matrix without formation of nanovoids (Fig. 18) [20, 21]. They

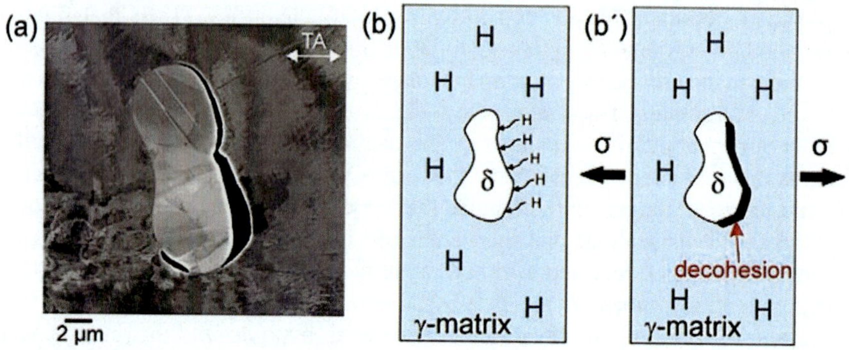

Fig. 18 **a** SEM image shows debonding of the δ-phase and the matrix; **b, c** the schematics show how diffusion of hydrogen to the interface leads to decohesion on application of normal stress. Figure is taken from [21]

suggest the reason is the high desorption energy of the δ-phase leading to hydrogen segregation at the interface and resultant decrease of cohesive energy. Possible explanations for the observed difference in the results of the two studies could include difference in material source, test schedules or the concentrations of hydrogen in the material. For instance, Zhang et al. charged their material at a current density of 7.7 mA/cm^2 for 168 h while Tarzimoghadam et al. charge at 10 mA/cm^2 for just 2 h. During tensile testing, Tarzimoghadam et al. deployed a slow strain rate of 1.6×10^{-5} while Zhang et al. used a comparatively rapid strain rate of 1×10^{-4}. A systematic study controlling such parameters could possibly reveal the interplay between these parameters and the dominant HE mechanism.

Another interesting approach was deployed by Harris et al. [13] to explore the importance of hydrogen enhanced dislocation mobility in pure nickel. They conducted tensile testing of polycrystalline Ni at 25 °C and −196 °C, charged it in an autoclave at 300 °C at a pressure of 96.5 MPa for 144 h, and showed that hydrogen embrittled the material at both temperatures (Fig. 19). Even though the loss of ductility was larger at 25 °C than at −196 °C, the mode of failure at −196 °C changed from transgranular to intergranular after the sample was charged with hydrogen (Fig. 20). At cryogenic temperatures, mobile hydrogen-deformation interactions can effectively be precluded, primarily due to the slower rates of diffusion of hydrogen and thus its inability to redistribute to defects. Harris et al. argued that since at −196 °C, true strain at crack initiation is significantly larger (1.75) without hydrogen than with hydrogen (0.20), and since mobile hydrogen-deformation interactions are effectively precluded at this temperature, the primary cause of crack initiation is hydrogen trapped in the grain boundaries causing their decohesion. It seems clear that at −196 °C, embrittlement is largely due to decohesion but at 25 °C, the much larger extent to which the material is embrittled indicates that both decohesion and enhanced plasticity might be active synergistically.

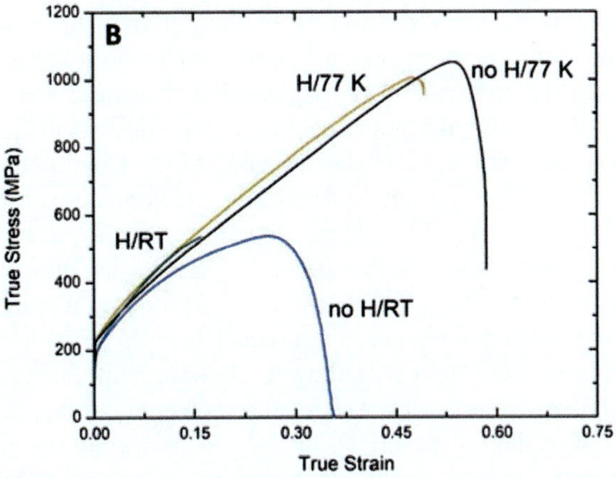

Fig. 19 The true stress-true strain curves showcase the loss of ductility in pure Ni after hydrogen charging; at −196 °C (77 K) it is much smaller than at 25 °C (RT). The stress–strain curve is taken from [13]

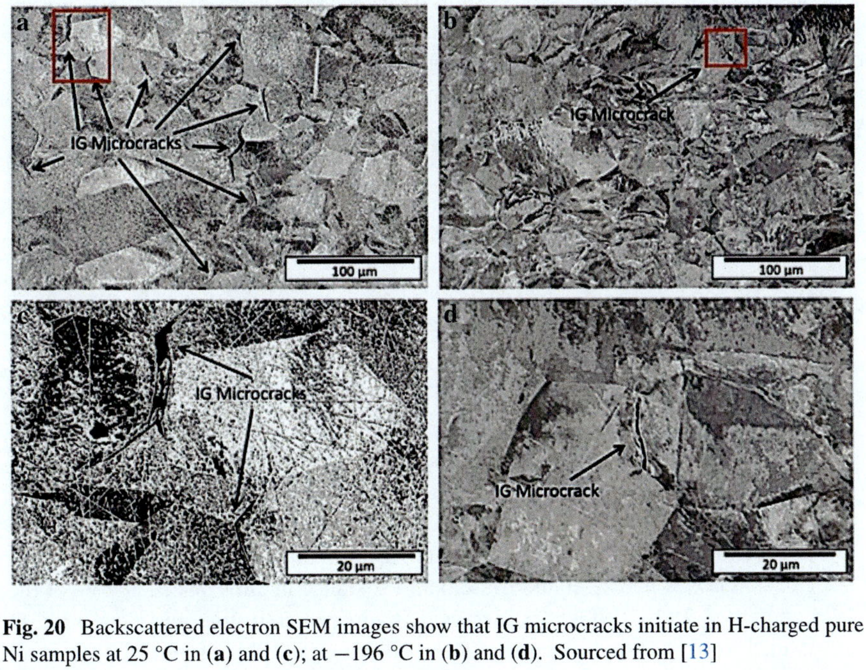

Fig. 20 Backscattered electron SEM images show that IG microcracks initiate in H-charged pure Ni samples at 25 °C in (**a**) and (**c**); at −196 °C in (**b**) and (**d**). Sourced from [13]

The question then becomes, what is the importance of mobile hydrogen-deformation interactions in intergranular fracture and how is it linked to debonding? Intergranular fracture, fracture along interfaces between phases and cleavage within grains may be due to debonding but conclusive experimental evidence is limited. Researchers have used simulations to show the effect of hydrogen on increasing interface energies and thus promoting decohesion. Computational techniques hold promise in providing evidence for the tendency of the alloy to cleave [17, 56]. First-principles investigations have yielded promising results by showing that in pure nickel, hydrogen reduced the cohesive energy of random grain boundaries by 40% while that of coherent twin boundaries was reduced negligibly [17, 57]. Debonding may also take place within grains leading to cleavage and thus leading to brittle fracture without formation of voids, as has been shown for α-iron [58]. A modelling study on alloy 718 to show these effects on the γ-matrix grain boundaries or phase boundaries between γ and the δ-phase could provide more evidence of decohesion.

Over the preceding paragraphs, it has been demonstrated that the presence of hydrogen in a metallic lattice has a large effect on its properties which results in a distinctly different fracture mode. Two mechanisms are reported here, one of which may dominate in some conditions while the other dominates in other conditions. However, it seems likely that both are active at the same time and their synergistic effect may be able to explain the macroscopic behavior adequately. A comprehensive review of HELP and HEDE and a unified HELP + HEDE mechanism has been presented for iron and steels [59]. A similar effort for Ni-base alloys and specifically Ni alloy 718 might be able to bring together the varying results and combine them into a single framework.

5 Conclusions

The hydrogen embrittlement of alloy 718 is analyzed based on studies available in open literature. Based on the analysis following general conclusions can be drawn:

1. Precipitation hardened Ni alloy 718 is highly susceptible to hydrogen embrittlement. Even though presence of γ', γ'' and δ phases all increase the susceptibility, the interface between the γ matrix and the δ-phase is especially vulnerable to cracking during monotonic loading. Elimination of the δ phase dramatically improves hydrogen embrittlement resistance in alloy 718. The role of the coherent γ' and γ'' phases in increasing hydrogen embrittlement susceptibility is unclear. The effect of grain size on HE susceptibility of alloy 718 has not been thoroughly and systematically investigated yet either.
2. The mechanisms behind hydrogen embrittlement of alloy 718 is an area of active research. Transgranular cracking initiates by nanovoid formation at the intersection of dislocation slip bands and propagates along them in alloy 718. Even

though transgranular cracking dominates the cracking mode in many conditions, intergranular cracking is also observed. The influence of the nature of grain boundaries on susceptibility to embrittlement is an open question.
3. Environmental and testing conditions, like temperature, strain rate and hydrogen concentration etc., are a likely contributor to the significant variations found in results. The authors' future efforts will focus on isolating these effects.

Acknowledgements The research was funded through NPRP Grant # 11S-1129-170045 from Qatar National Research Fund (A member of Qatar Foundation). The statements made herein are solely the responsibility of the authors. The authors thank the Mechanical Engineering Department at TAMU Qatar for the facilities and support provided.

References

1. J.J. Debarbadillo, S.K. Mannan, Alloy 718 for oilfield applications. Jom **64**(2), 265–270 (2012). https://doi.org/10.1007/s11837-012-0238-z
2. D.F. Paulonis, J.J. Schirra, Alloy 718 at Pratt & Whitney—Historical perspective and future challenges, in Proceedings of International Symposium Superalloys Variations Derivations, vol. 1 (2001), pp. 13–23. https://doi.org/10.7449/2001/superalloys_2001_13_23
3. J. Xu, H. John, G. Wiese, X. Liu, B.H. Incorporated, Oil-grade alloy 718 in oil field drilling application tension/compression (Weight) torsion Δp (OD/ID) bending, cyclic (tension/compression) weight on bit, in *7th International of Symposium Superalloy 718 Derivations* (2010), pp. 923–932
4. W.H. Johnson, On some remarkable changes produced in iron and steel by the action of hydrogen and acids. Nature **11**(281), 393 (1875). https://doi.org/10.1038/011393a0
5. S. Lynch, Hydrogen embrittlement phenomena and mechanisms. Corros. Rev. **30**(3–4), 105–123 (2012). https://doi.org/10.1515/corrrev-2012-0502
6. Z. Zhang, G. Obasi, R. Morana, M. Preuss, Hydrogen assisted crack initiation and propagation in a nickel-based superalloy. Acta Mater. **113**, 272–283 (2016). https://doi.org/10.1016/J.ACTAMAT.2016.05.003
7. M. Seita, J.P. Hanson, S. Gradečak, M.J. Demkowicz, The dual role of coherent twin boundaries in hydrogen embrittlement. Nat. Commun. **6** (2015). https://doi.org/10.1038/ncomms7164
8. M.L. Martin, B.P. Somerday, R.O. Ritchie, P. Sofronis, I.M. Robertson, Hydrogen-induced intergranular failure in nickel revisited. Acta Mater. **60**(6–7), 2739–2745 (2012). https://doi.org/10.1016/j.actamat.2012.01.040
9. X. Li, J. Zhang, E. Akiyama, Q. Fu, Q. Li, Hydrogen embrittlement behavior of Inconel 718 alloy at room temperature. J. Mater. Sci. Technol. **35**(4), 499–502 (2019). https://doi.org/10.1016/j.jmst.2018.10.002
10. M.L. Martin, M. Dadfarnia, A. Nagao, S. Wang, P. Sofronis, Enumeration of the hydrogen-enhanced localized plasticity mechanism for hydrogen embrittlement in structural materials. Acta Mater. **165**, 734–750 (2019). https://doi.org/10.1016/j.actamat.2018.12.014
11. T. Tabata, H.K. Birnbaum, Direct observations of the effect of hydrogen on the behavior of dislocations in iron. **17**(c), 240 (1983)
12. T. Matsumoto, J. Eastman, H.K. Birnbaum, Direct observations of enhanced dislocation mobility due to hydrogen. Perspect Hydrog. Met. **i**(c), 639–643 (1986). https://doi.org/10.1016/b978-0-08-034813-1.50090-5

13. Z.D. Harris, S.K. Lawrence, D.L. Medlin, G. Guetard, J.T. Burns, B.P. Somerday, Elucidating the contribution of mobile hydrogen-deformation interactions to hydrogen-induced intergranular cracking in polycrystalline nickel. Acta Mater. **158**, 180–192 (2018). https://doi.org/10.1016/j.actamat.2018.07.043
14. J. Song, W.A. Curtin, A nanoscale mechanism of hydrogen embrittlement in metals. Acta Mater. **59**(4), 1557–1569 (2011). https://doi.org/10.1016/j.actamat.2010.11.019
15. E.A. Clark, R. Yeske, H.K. Birnbaum, The effect of hydrogen on the surface energy of nickel. Metall. Trans. A **11**(11), 1903–1908 (1980). https://doi.org/10.1007/BF02655107
16. S. Jothi, S.V. Merzlikin, T.N. Croft, J. Andersson, S.G.R. Brown, An investigation of micro-mechanisms in hydrogen induced cracking in nickel-based superalloy 718. J. Alloys Compd. **664**, 664–681 (2016). https://doi.org/10.1016/j.jallcom.2016.01.033
17. A. Alvaro, I. Thue Jensen, N. Kheradmand, O.M. Løvvik, V. Olden, Hydrogen embrittlement in nickel, visited by first principles modeling, cohesive zone simulation and nanomechanical testing. Int. J. Hydrogen Energy **40**(47), 16892–16900 (2015) https://doi.org/10.1016/J.IJHYDENE.2015.06.069
18. Z. Tarzimoghadam, D. Ponge, J. Klöwer, D. Raabe, Hydrogen-assisted failure in Ni-based superalloy 718 studied under in situ hydrogen charging: The role of localized deformation in crack propagation. Acta Mater. **128**, 365–374 (2017). https://doi.org/10.1016/j.actamat.2017.02.059
19. Z. Zhang, G. Obasi, R. Morana, M. Preuss, In-situ observation of hydrogen induced crack initiation in a nickel-based superalloy. Scr. Mater. **140**, 40–44 (2017). https://doi.org/10.1016/J.SCRIPTAMAT.2017.07.006
20. Z. Zhang, K.L. Moore, G. McMahon, R. Morana, M. Preuss, On the role of precipitates in hydrogen trapping and hydrogen embrittlement of a nickel-based superalloy. Corros. Sci. **146**, 58–69 (2019). https://doi.org/10.1016/J.CORSCI.2018.10.019
21. Z. Tarzimoghadam et al., Multi-scale and spatially resolved hydrogen mapping in a Ni-Nb model alloy reveals the role of the δ phase in hydrogen embrittlement of alloy 718. Acta Mater. **109**, 69–81 (2016). https://doi.org/10.1016/j.actamat.2016.02.053
22. G.C. Obasi, Z. Zhang, D. Sampath, R. Morana, R. Akid, M. Preuss, Effect of microstructure and alloy chemistry on hydrogen embrittlement of precipitation-hardened Ni-based alloys. Metall. Mater. Trans. A Phys. Metall. Mater. Sci. **49**(4), 1167–1181 (2018). https://doi.org/10.1007/s11661-018-4483-9
23. M.C. Chaturvedi, Y.F. Han, Strengthening mechanisms in Inconel 718 superalloy. Met. Sci. **17**(3), 1–5 (1983). https://doi.org/10.1179/030634583790421032
24. M. Sundararaman, P. Mukhopadhyay, S. Banerjee, Some aspects of the precipitation of metastable intermetallic phases in INCONEL 718. Metall. Trans. A **23**(7), 2015–2028 (1992). https://doi.org/10.1007/BF02647549
25. D. Sindhura, M.V. Sravya, G.V.S. Murthy, Comprehensive microstructural evaluation of precipitation in Inconel 718. Metallogr. Microstruct. Anal. **8**(2), 233–240 (2019). https://doi.org/10.1007/s13632-018-00513-0
26. Y. Ogawa, O. Takakuwa, S. Okazaki, K. Okita, Y. Funakoshi, Pronounced transition of crack initiation and propagation modes in the hydrogen-related failure of a Ni-based superalloy 718 under internal and external hydrogen conditions. Corros. Sci., p. 108186 (2019). https://doi.org/10.1016/j.corsci.2019.108186
27. J.J. Ruan, N. Ueshima, K. Oikawa, Phase transformations and grain growth behaviors in superalloy 718. J. Alloys Compd. **737**, 83–91 (2018). https://doi.org/10.1016/j.jallcom.2017.11.327
28. S. Ranganath, C. Guo, S. Holt, Experimental investigations into the carbide cracking phenomenon on inconel 718 superalloy material, in *Proceedings of ASME International Manufacturing Science and Engineering Conference 2009, MSEC2009*, vol. 2 (2009), pp. 33–39. https://doi.org/10.1115/MSEC2009-84085
29. G. Stenerud, S. Wenner, J.S. Olsen, R. Johnsen, Effect of different microstructural features on the hydrogen embrittlement susceptibility of alloy 718. Int. J. Hydrogen Energy **43**(13), 6765–6776 (2018). https://doi.org/10.1016/j.ijhydene.2018.02.088

30. V. Shankar, K. Bhanu Sankara Rao, S.L. Mannan, Microstructure and mechanical properties of Inconel 625 superalloy. J. Nucl. Mater. **288**(2–3), 222–232 (2001). https://doi.org/10.1016/S0022-3115(00)00723-6
31. X. Li et al., Tensile mechanical properties and fracture behaviors of nickel-based superalloy 718 in the presence of hydrogen. Int. J. Hydrogen Energy **43**(43), 20118–20132 (2018). https://doi.org/10.1016/j.ijhydene.2018.08.179
32. N. Ehrlin, C. Bjerkén, M. Fisk, Cathodic hydrogen charging of Inconel 718. AIMS Mater. Sci. **3**(4), 1350–1364 (2016). https://doi.org/10.3934/matersci.2016.4.1350
33. G. Zheng, B.N. Popov, R.E. White, Hydrogen-atom direct-entry mechanism into metal membranes. J. Electrochem. Soc. **142**(1), 154–156 (1995). https://doi.org/10.1149/1.2043855
34. R.D. Louthan, G. Caskey, J. Donoval, Hydrogen embrittlement of metals. Phys. B Condens. Matter **289–290**, 443–446 (2000). https://doi.org/10.1016/S0921-4526(00)00431-2
35. X. Lu, D. Wang, D. Wan, Z.B. Zhang, N. Kheradmand, A. Barnoush, Effect of electrochemical charging on the hydrogen embrittlement susceptibility of alloy 718. Acta Mater. **179**(7491), 36–48 (2019). https://doi.org/10.1016/j.actamat.2019.08.020
36. L. Liu, K. Tanaka, A. Hirose, K.F. Kobayashi, Effects of precipitation phases on the hydrogen embrittlement sensitivity of inconel 718. Sci. Technol. Adv. Mater. **3**(4), 335–344 (2002). https://doi.org/10.1016/S1468-6996(02)00039-6
37. B. Kagay, K. Findley, S. Coryell, A.B. Nissan, Effects of alloy 718 microstructure on hydrogen embrittlement susceptibility for oil and gas environments. Mater. Sci. Technol. (United Kingdom) **32**(7), 697–707 (2016). https://doi.org/10.1080/02670836.2016.1139225
38. X. Lu, Y. Ma, D. Wang, On the hydrogen embrittlement behavior of nickel-based alloys: Alloys 718 and 725. Mater. Sci. Eng. A **792**(March) (2020). https://doi.org/10.1016/j.msea.2020.139785
39. F. Galliano, E. Andrieu, C. Blanc, J.-M. Cloue, D. Connetable, G. Odemer, Effect of trapping and temperature on the hydrogen embrittlement susceptibility of alloy 718. Mater. Sci. Eng. A **611**, 370–382 (2014). https://doi.org/10.1016/J.MSEA.2014.06.015
40. A. Turnbull, R.G. Ballinger, I.S. Hwang, M.M. Morra, M. Psaila-Dombrowski, R.M. Gates, Hydrogen transport in nickel-base alloys. Metall. Trans. A **23**(12), 3231–3244 (1992). https://doi.org/10.1007/BF02663432
41. S. Chen, M. Zhao, L. Rong, Effect of grain size on the hydrogen embrittlement sensitivity of a precipitation strengthened Fe-Ni based alloy. Mater. Sci. Eng. A **594**, 98–102 (2014). https://doi.org/10.1016/j.msea.2013.11.062
42. A. Oudriss et al., Grain size and grain-boundary effects on diffusion and trapping of hydrogen in pure nickel. Acta Mater. **60**(19), 6814–6828 (2012). https://doi.org/10.1016/j.actamat.2012.09.004
43. Y.H. Fan, B. Zhang, J.Q. Wang, E.H. Han, W. Ke, Effect of grain refinement on the hydrogen embrittlement of 304 austenitic stainless steel. J. Mater. Sci. Technol. **35**(10), 2213–2219 (2019). https://doi.org/10.1016/j.jmst.2019.03.043
44. N. Zan, H. Ding, X. Guo, Z. Tang, W. Bleck, Effects of grain size on hydrogen embrittlement in a Fe-22Mn-0.6C TWIP steel. Int. J. Hydrogen Energy **40**(33), 10687–10696 (2015). https://doi.org/10.1016/j.ijhydene.2015.06.112
45. A. Barnoush, H. Vehoff, In situ electrochemical nanoindentation: a technique for local examination of hydrogen embrittlement. Corros. Sci. **50**(1), 259–267 (2008). https://doi.org/10.1016/j.corsci.2007.05.026
46. D. Xie et al., Hydrogenated vacancies lock dislocations in aluminium. Nat. Commun. **7**, 1–7 (2016). https://doi.org/10.1038/ncomms13341
47. S. He, W. Ecker, R. Pippan, V.I. Razumovskiy, Hydrogen-enhanced decohesion mechanism of the special $\Sigma 5(0\ 1\ 2)[1\ 0\ 0]$grain boundary in Ni with Mo and C solutes. Comput. Mater. Sci. **167**(April), 100–110 (2019). https://doi.org/10.1016/j.commatsci.2019.05.029
48. X. Li, X. Ma, J. Zhang, E. Akiyama, Y. Wang, X. Song, Review of hydrogen embrittlement in metals: hydrogen diffusion, hydrogen characterization, hydrogen embrittlement mechanism and prevention. Acta Metall. Sin. (English Lett.) **33**(6), 759–773 (2020). https://doi.org/10.1007/s40195-020-01039-7

49. P.L. Bessemer, The effect of occluded hydrogen on the tensile strength of iron. J. Franklin Inst. **203**(2), 331–332 (1927). https://doi.org/10.1016/s0016-0032(27)92466-1
50. C.D. Beachem, A new model for hydrogen-assisted cracking (hydrogen 'embrittlement'). Metall. Trans. **3**(2), 441–455 (1972). https://doi.org/10.1007/BF02642048
51. I.M. Robertson, H.K. Birnbaum, An HVEM study of hydrogen effects on the deformation and fracture of nickel. Acta Metall. **34**(3), 353–366 (1986). https://doi.org/10.1016/0001-6160(86)90071-4
52. H.K. Birnbaum, On the mechanisms of Hydrogen related fracture in metals. Environ. Sensitive Fract. Met. Alloy. **XXX**, 105–113 (1987)
53. D.S. Shih, I.M. Robertson, H.K. Birnbaum, Hydrogen embrittlement of α titanium: In situ tem studies. Acta Metall. **36**(1), 111–124 (1988). https://doi.org/10.1016/0001-6160(88)90032-6
54. S. Wang, A. Nagao, K. Edalati, Z. Horita, I.M. Robertson, Influence of hydrogen on dislocation self-organization in Ni. Acta Mater. **135**, 96–102 (2017). https://doi.org/10.1016/j.actamat.2017.05.073
55. J.P. Hanson, M. Seita, E. Jones, S. Grade, M.J. Demkowicz, Hydrogen embrittlement behavior in precipitation hardened Ni-base alloys John, in *NACE International Corrosion. 2015 Conf. Expo*, no. 5853, pp. 1–8 (2015)
56. C.L. Fu, G.S. Painter, First principles investigation of hydrogen embrittlement in FeAl. J. Mater. Res. **6**(4), 719–723 (1991). https://doi.org/10.1557/JMR.1991.0719
57. A. Tehranchi, W.A. Curtin, Atomistic study of hydrogen embrittlement of grain boundaries in nickel: I. Fracture. J. Mech. Phys. Solids **101**, 150–165 (2017). https://doi.org/10.1016/j.jmps.2017.01.020
58. R. Matsumoto, S. Taketomi, S. Matsumoto, N. Miyazaki, Atomistic simulations of hydrogen embrittlement. Int. J. Hydrogen Energy **34**(23), 9576–9584 (2009). https://doi.org/10.1016/j.ijhydene.2009.09.052
59. M.B. Djukic, G.M. Bakic, V. Sijacki Zeravcic, A. Sedmak, B. Rajicic, The synergistic action and interplay of hydrogen embrittlement mechanisms in steels and iron: Localized plasticity and decohesion. Eng. Fract. Mech. **216**(June), 106528 (2019). https://doi.org/10.1016/j.engfracmech.2019.106528